Igor R. Shafarevich: Basic Algebraic Geometry 1

Igor R. Shafarevich

Basic Algebraic Geometry 1

Second, Revised and Expanded Edition

Springer-Verlag
Berlin Heidelberg New York
London Paris Tokyo
Hong Kong Barcelona
Budapest

Igor R. Shafarevich
Steklov Mathematical Institute
Ul. Vavilova 42, 117966 Moscow, Russia

Translator:
Miles Reid
Mathematics Institute, University of Warwick
Coventry CV4 7AL, England
e-mail: Miles@Maths.Warwick.Ac.UK.

With 21 Figures

The title of the original Russian edition:
Osnovy algebraicheskoj geometrii, tom 1
© Nauka, Moscow 1988

ISBN 3-540-54812-2 Springer-Verlag Berlin Heidelberg New York
ISBN 0-387-54812-2 Springer-Verlag New York Berlin Heidelberg

ISBN 3-540-08264-6 1. Auflage Springer-Verlag Berlin Heidelberg New York
ISBN 0-387-08264-6 1st edition Springer-Verlag New York Berlin Heidelberg

Library of Congress Cataloging-in-Publication Data
Shafarevich, I. R. (Igor' Rostislavovich), 1923– [Osnovy algebraicheskoĭ geometrii. English]
Basic algebraic geometry/Igor R. Shafarevich; [translator, Miles Reid]. – 2nd rev. and expanded ed. p. cm. "Springer study edition." Includes bibliographical references and indexes. ISBN 3-540-54812-2 (Berlin: acid-free: v. 1) – ISBN 0-387-54812-2 (New York: acid-free: v. 1). – ISBN 3-540-57554-5 (Berlin: acid-free: v. 2). – ISBN 0-387-57554-5 (New York: acid-free: v. 2). 1. Geometry, Algebraic. I. Title. QA564.S4513 1994 516.3'5. – dc20

Springer-Verlag Berlin Heidelberg New York
a member of Springer Science+Business Media GmbH

© Springer-Verlag Berlin Heidelberg 1977, 1994
Printed in Germany

This book was typeset by the translator using the $\mathcal{AMS}$-T$_{E}$X macro package and typefaces. together with Springer-Verlag's T$_{E}$X macro package CPMono01

SPIN: 12163717   46/3180 – 10 9 8   Printed on acid-free paper

Preface

The first edition of this book came out just as the apparatus of algebraic geometry was reaching a stage that permitted a lucid and concise account of the foundations of the subject. The author was no longer forced into the painful choice between sacrificing rigour of exposition or overloading the clear geometrical picture with cumbersome algebraic apparatus.

The 15 years that have elapsed since the first edition have seen the appearance of many beautiful books treating various branches of algebraic geometry. However, as far as I know, no other author has been attracted to the aim which this book set itself: to give an overall view of the many varied aspects of algebraic geometry, without going too far afield into the different theories. There is thus scope for a second edition. In preparing this, I have included some additional material, rather varied in nature, and have made some small cuts, but the general character of the book remains unchanged.

The three parts of the book now appear as two separate volumes. Book 1 corresponds to Part I, Chapters I–IV, of the first edition. Here quite a lot of material of a rather concrete geometric nature has been added: the first section, forming a bridge between coordinate geometry and the theory of algebraic curves in the plane, has been substantially expanded. More space has been given over to concrete algebraic varieties: Grassmannian varieties, plane cubic curves and the cubic surface. The main role that singularities played in the first edition was in giving rigorous definition to situations we wished to avoid. The present edition treats a number of questions related to degenerate fibres in families: degenerations of quadrics and of elliptic curves, the Bertini theorems. We discuss the notion of infinitely near points of algebraic curves on surfaces and normal surface singularities. Finally, some applications to number theory have been added: the zeta function of algebraic varieties over a finite field and the analogue of the Riemann hypothesis for elliptic curves.

Books 2 and 3 corresponds to Parts II and III, Chapters V–IX of the first edition. They treat the foundations of the theory of schemes, abstract algebraic varieties and algebraic manifolds over the complex number field. As in the Book 1 there are a number of additions to the text. Of these, the following are the two most important. The first is a discussion of the notion of moduli spaces, that is, algebraic varieties that classify algebraic or geometric objects of some type; as an example we work out the theory of the Hilbert polynomial and the Hilbert scheme. I am very grateful to V. I. Danilov for a series of recommendations on this subject. In particular the proof of Chap. VI, 4.3, Theorem 3 is due to him. The second addition is the definition and basic properties of a Kähler metric and a description (without proof) of Hodge's theorem.

For the most part, this material is taken from my old lectures and seminars, from notes provided by members of the audience. A number of improvements of proofs have been borrowed from the books of Mumford and Fulton. A whole series of misprints and inaccuracies in the first edition were pointed out by readers, and by readers of the English translation. Especially valuable was the advice of Andrei Tyurin and Viktor Kulikov; in particular, the proof of Chap. IV, 4.3, Theorem 3 was provided by Kulikov. I offer sincere thanks to all these.

Many substantial improvements are due to V. L. Popov, who edited the second edition, and I am very grateful to him for all the work and thought he has put into the book. I have the pleasure, not for the first time, of expressing my deep gratitude to the translator of this book, Miles Reid. His thoughtful work has made it possible to patch up many uneven places and inaccuracies, and to correct a few mathematical errors.

Prerequisites

The nature of the book requires the algebraic apparatus to be kept to a minimum. In addition to an undergraduate algebra course, we assume known basic material from field theory: finite and transcendental extensions (but not Galois theory), and from ring theory: ideals and quotient rings. In a number of isolated instances we refer to the literature on algebra; these references are chosen so that the reader can understand the relevant point, independently of the preceding parts of the book being referred to. Somewhat more specialised algebraic questions are collected together in the Algebraic Appendix at the end of Book 1. The preface to Books 2–3 contains recommended further reading in Algebraic Geometry.

Preface to the 1972 Edition

Algebraic geometry played a central role in 19th century math. The deepest results of Abel, Riemann, Weierstrass, and many of the most important works of Klein and Poincaré were part of this subject.

The turn of the 20th century saw a sharp change in attitude to algebraic geometry. In the 1910s Klein[1] writes as follows: "In my student days, under

---

[1] Klein, F.: Vorlesungen über die Entwicklung der Mathematik im 19. Jahrhundert, Grundlehren Math. Wiss. 24, Springer-Verlag, Berlin 1926. Jrb. 52, 22, p. 312

the influence of the Jacobi tradition, Abelian functions were considered as the unarguable pinnacle of math. Every one of us felt the natural ambition to make some independent progress in this field. And now? The younger generation scarcely knows what Abelian functions are." (From the modern viewpoint, the theory of Abelian functions is an analytic aspect of the theory of Abelian varieties, that is, projective algebraic group varieties; compare the historical sketch.)

Algebraic geometry had become set in a way of thinking too far removed from the set-theoretic and axiomatic spirit that determined the development of math at the time. It was to take several decades, during which the theories of topological, differentiable and complex manifolds, of general fields, and of ideals in sufficiently general rings were developed, before it became possible to construct algebraic geometry on the basis of the principles of set-theoretic math.

Towards the middle of the 20th century algebraic geometry had to a large extent been through such a reconstruction. Because of this, it could again claim the place it had once occupied in math. The domain of application of its ideas had grown tremendously, both in the direction of algebraic varieties over arbitrary fields and of more general complex manifolds. Many of the best achievements of algebraic geometry could be cleared of the accusation of incomprehensibility or lack of rigour.

The foundation for this reconstruction was algebra. In its first versions, the use of precise algebraic apparatus often led to a loss of the brilliant geometric style characteristic of the preceding period. However, the 1950s and 60s have brought substantial simplifications to the foundation of algebraic geometry, which have allowed us to come significantly closer to the ideal combination of logical transparency and geometric intuition.

The purpose of this book is to treat the foundations of algebraic geometry across a fairly wide front, giving an overall account of the subject, and preparing the ground for a study of the more specialised literature. No prior knowledge of algebraic geometry is assumed on the part of the reader, neither general theorems, nor concrete examples. Therefore along with development of the general theory, a lot of space is devoted to applications and particular cases, intended to motivate new ideas or new ways of formulating questions.

It seems to me that, in the spirit of the biogenetic law, the student who repeats in miniature the evolution of algebraic geometry will grasp the logic of the subject more clearly. Thus, for example, the first section is concerned with very simple properties of algebraic plane curves. Similarly, Part I of the book considers only algebraic varieties in an ambient projective space, and the reader only meets schemes and the general notion of a variety in Part II.

Part III treats algebraic varieties over the complex number field, and their relation to complex analytic manifolds. This section assumes some acquaintance with basic topology and the theory of analytic functions.

I am extremely grateful to everyone whose advice helped me with this book. It is based on lecture notes from several courses I gave in Moscow University. Many participants in the lectures or readers of the notes have provided me with useful remarks. I am especially indebted to the editor B. G. Moishezon for a large number of discussions which were very useful to me. A series of proofs contained in the book are based on his advice.

Translator's Note

Shafarevich's book is the fruit of lecture courses at Moscow State University in the 1960s and early 1970s. The style of Russian mathematical writing of the period is very much in evidence. The book does not aim to cover a huge volume of material in the maximal generality and rigour, but gives instead a well-considered choice of topics, with a human-oriented discussion of the motivation and the ideas, and some sample results (including a good number of hard theorems with complete proofs). In view of the difficulty of keeping up with developments in algebraic geometry during the 1960s, and the extraordinary difficulties faced by Soviet mathematicians of that period, the book is a tremendous achievement.

The student who wants to get through the technical material of algebraic geometry quickly and at full strength should perhaps turn to Hartshorne's book [35]; however, my experience is that some graduate students (by no means all) can work hard for a year or two on Chaps. II–III of Hartshorne, and still know more-or-less nothing at the end of it. For many students, it's just not feasible both to do the research for a Ph. D. thesis and to master all the technical foundations of algebraic geometry at the same time. In any case, even if you have mastered everything in scheme theory, your research may well take you into number theory or differential geometry or representation theory or math physics, and you'll have just as many new technical things to learn there. For all such students, and for the many specialists in other branches of math who need a liberal education in algebraic geometry, Shafarevich's book is a must.

The previous English translation by the late Prof. Kurt Hirsch has been used with great profit by many students over the last two decades. In preparing the new translation of the revised edition, in addition to correcting a few typographical errors and putting the references into English alphabetical order, I have attempted to put Shafarevich's text into the language used by the present generation of English-speaking algebraic geometers. I have in a few cases corrected the Russian text, or even made some fairly arbitrary changes when the original was already perfectly all right, in most case with the author's explicit or implicit approval. The footnotes are all mine: they are mainly pedantic in nature, either concerned with minor points of terminology, or giving references for proofs not found in the main text; my references do not necessarily follow Shafarevich's ground-rule of being a few pages accessible to the general reader, without obliging him or her to read a whole book, and so may not be very useful to the beginning graduate student. It's actually quite demoralising to realise just how difficult or obscure the literature can be on some of these points, at the same time as many of the easier points are covered in any number of textbooks. For example: (1) the "principle of conservation of number" (algebraic equivalence implies numerical equivalence); (2) the Néron–Severi theorem (stated as Chap. 3, 4.4, Theorem D); (3) a punctured neighbourhood of a singular point of a normal variety over $\mathbb{C}$ is connected; (4) Chevalley's theorem that every algebraic group is an extension of an Abelian variety by an affine (linear) group. A practical solution for the reader is to take the statements on trust for the time being.

The two volumes have a common index and list of references, but only the second volume has the references for the historical sketch.

Table of Contents Volume 1

BOOK 1. Varieties in Projective Space

Table of Contents Volume 2

BOOK 2. Schemes and Varieties

BOOK 3. Complex Algebraic Varieties and Complex Manifolds

Chapter VII. The Topology of Algebraic Varieties 117

BOOK 1

*Varieties
in Projective Space*

Chapter I. Basic Notions

1. Algebraic Curves in the Plane

Chapter I discusses a number of the basic ideas of algebraic geometry; this first section treats some examples to prepare the ground for these ideas.

1.1. Plane Curves

An *algebraic plane curve* is a curve consisting of the points of the plane whose coordinates x, y satisfy an equation

$$f(x, y) = 0, \tag{1}$$

where f is a nonconstant polynomial. Here we fix a field k and assume that the coordinates x, y of points and the coefficients of f are elements of k. We write $\mathbb{A}^2$ for the *affine plane*, the set of points (a, b) with $a, b \in k$; because the affine plane $\mathbb{A}^2$ is not the only ambient space in which algebraic curves will be considered – we will be meeting others presently – an algebraic curve as just defined is called an *affine* plane curve.

The degree of the equation (1), that is, the degree of the polynomial $f(x, y)$, is also called the *degree* of the curve. A curve of degree 2 is called a *conic*, and a curve of degree 3 a *cubic*.

It is well known that the polynomial ring $k[X, Y]$ is a unique factorisation domain (UFD), that is, any polynomial f has a unique factorisation $f = f_1^{k_1} \cdots f_r^{k_r}$ (up to constant multiples) as a product of irreducible factors f_i, where the irreducible f_i are nonproportional, that is, $f_i \neq \alpha f_j$ with $\alpha \in k$ if $i \neq j$. Then the algebraic curve X given by $f = 0$ is the union of the curves X_i given by $f_i = 0$. A curve is *irreducible* if its equation is an irreducible polynomial. The decomposition $X = X_1 \cup \cdots \cup X_r$ just obtained is called a decomposition of X into irreducible components.

In certain cases, the notions just introduced turn out not to be well defined, or to differ wildly from our intuition. This is due to the specific nature of the field k in which the coordinates of points of the curve are taken. For example if $k = \mathbb{R}$ then following the above terminology we should call the point $(0, 0)$ a "curve", since it is defined by the equation $x^2 + y^2 = 0$. Moreover, this "curve" should have "degree" 2, but also any other even number, since the same point $(0,0)$ is also defined by the equation $x^{2n} + y^{2n} = 0$. The curve is irreducible if we take its equation to be $x^2 + y^2 = 0$, but reducible if we take it to be $x^6 + y^6 = 0$.

Problems of this kind do not arise if k is an algebraically closed field. This is based on the following simple fact.

Lemma. Let k be an arbitrary field, $f \in k[x,y]$ an irreducible polynomial, and $g \in k[x,y]$ an arbitrary polynomial. If g is not divisible by f then the system of equations $f(x,y) = g(x,y) = 0$ has only a finite number of solutions.

Proof. Suppose that x appears in f with positive degree. We view f and g as elements of $k(y)[x]$, that is, as polynomials in one variable x, whose coefficients are rational functions of y. It is easy to check that f remains irreducible in this ring: if f splits as a product of factors, then after multiplying each factor by the common denominator $a(y) \in k[y]$ of its coefficients, we obtain a relation that contradicts the irreducibility of f in $k[x,y]$. For the same reason, g is not divisible by f in the new ring $k(y)[x]$. Hence there exist two polynomials $\widetilde{u}, \widetilde{v} \in k(y)[x]$ such that $f\widetilde{u} + g\widetilde{v} = 1$. Multiplying this equality through by the common denominator $a \in k[y]$ of all the coefficients of $\widetilde{u}$ and $\widetilde{v}$ gives $fu + gv = a$, where $u = a\widetilde{u}$, $v = a\widetilde{v} \in k[x,y]$, and $0 \neq a \in k[y]$. It follows that if $f(\alpha, \beta) = g(\alpha, \beta) = 0$ then $a(\beta) = 0$, that is, there are only finitely many possible values for the second coordinate β. For each such value, the first coordinate α is a root of $f(x, \beta) = 0$. The polynomial $f(x, \beta)$ is not identically 0, since otherwise $f(x,y)$ would be divisible by $y - \beta$, and hence there are also only a finite number of possibilities for α. The lemma is proved.

An algebraically closed field k is infinite; and if f is not a constant, the curve with equation $f(x,y) = 0$ has infinitely many points. Because of this, it follows from the lemma that an irreducible polynomial $f(x,y)$ is uniquely determined, up to a constant multiple, by the curve $f(x,y) = 0$. The same holds for an arbitrary polynomial, under the assumption that its factorisation into irreducible components has no multiple factors. We can always choose the equation of a curve to be a polynomial satisfying this condition. The notion of the degree of a curve, and of an irreducible curve, is then well defined.

Another reason why algebraic geometry only makes sense on passing to an algebraically closed field arises when we consider the number of points of intersection of curves. This phenomenon is already familiar from algebra: the theorem that the number of roots of a polynomial equals its degree is only valid if we consider roots in an algebraically closed field. A generalisation of this theorem is the so-called Bézout theorem: the number of points of intersection of two distinct irreducible algebraic curves equals the product of their degrees. The lemma shows that, in any case, this number is finite. The theorem on the number of roots of a polynomial is a particular case, for the curves $y - f(x) = 0$ and $y = 0$.

Bézout's theorem holds only after certain amendments. The first of these is the requirement that we consider points with coordinates in an algebraically closed field. Thus Figure 1 shows three cases for the relative position of two curves of degree 2 (ellipses) in the real plane. Here Bézout's theorem holds in case (c), but not in cases (a) and (b).

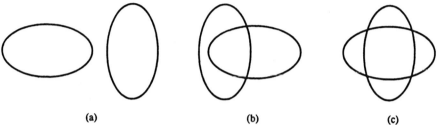

        (a)   (b)   (c)

Figure 1. Intersections of Conics

We assume throughout what follows that k is algebraically closed; in the contrary case, we always say so. This does not mean that algebraic geometry does not apply to studying questions concerned with algebraically nonclosed fields k_0. However, applications of this kind most frequently involve passing to an algebraically closed field k containing k_0. In the case of $\mathbb{R}$, we pass to the complex number field $\mathbb{C}$. This often allows us to guess or to prove purely real relations. Here is the most elementary example of this nature. If P is a point outside a circle C then there are two tangent lines to C through P. The line joining their points of contact is called the *polar line* of P with respect to C (Figure 2, (a)). All these constructions can be expressed in terms of algebraic relations between the coordinates of P and the equation of C. Hence they are also applicable to the case that P lies inside C. Of course, the points of tangency of the lines now have complex coordinates, and can't be seen in the picture. But since the original data was real, the set of points obtained (that is, the two points of tangency) should be invariant on replacing all the numbers by their complex conjugates; that is, the two points of tangency are complex conjugates. Hence the line L joining them is real. This line is also called the polar line of P with respect to C. It is also easy to give a purely real definition of it: it is the locus of points outside the circle whose polar line passes through P (Figure 2, (b)).

Here are some other situations in which questions arise involving algebraic geometry over an algebraically nonclosed field, and whose study usually requires passing to an algebraically closed field.

(1) $k = \mathbb{Q}$. The study of points of an algebraic curve $f(x,y) = 0$, where $f \in \mathbb{Q}[x,y]$, and the coordinates of the points are in $\mathbb{Q}$. This is one of the fun-

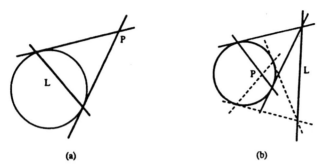

(a)   (b)

Figure 2. The Polar Line of a Point with Respect to a Conic

damental problems of number theory, the theory of indeterminate equations. For example, Fermat's last theorem requires us to describe points $(x, y) \in \mathbb{Q}^2$ of the curve $x^n + y^n = 1$.

(2) Finite fields. Let $k = \mathbb{F}_p$ be the field of residues modulo p. Studying the points with coordinates in k on the algebraic curve given by $f(x, y) = 0$ is another problem of number theory, on the solutions of the congruence $f(x, y) \equiv 0 \bmod p$.

(3) $k = \mathbb{C}(z)$. Consider the algebraic surface in $\mathbf{A}^3$ given by $F(x, y, z) = 0$, with $F(x, y, z) \in \mathbb{C}[x, y, z]$. By putting z into the coefficients and thinking of F as a polynomial in x, y, we can consider our surface as a curve over the field $\mathbb{C}(z)$ of rational functions in z. This is an extremely fertile method in the study of algebraic surfaces.

1.2. Rational Curves

As is well known, the curve given by

$$y^2 = x^2 + x^3 \tag{1}$$

has the property that the coordinates of its points can be expressed as rational functions of one parameter. To deduce these expressions, note that the line through the origin $y = tx$ intersects the curve (1) outside the origin in a single point. Indeed, substituting $y = tx$ in (1), we get $x^2(t^2 - x - 1) = 0$; the double root $x = 0$ corresponds to the origin $0 = (0, 0)$. In addition to this, we have another root $x = t^2 - 1$; the equation of the line gives $y = t(t^2 - 1)$. We thus get the required parametrisation

$$x = t^2 - 1, \quad y = t(t^2 - 1), \tag{2}$$

and its geometric meaning is evident: t is the slope of the line through 0 and (x, y); and (x, y) are the coordinates of the point of intersection of the line $y = tx$ with the curve (1) outside 0. We can see this parametrisation even more intuitively by drawing another line, not passing through 0 (for example, the line $x = 1$) and projecting the curve from 0, by sending a point P of the curve to the point Q of intersection of the line $0P$ with this line (see Figure 3). Here the parameter t plays the role of coordinate on the given line. Either from this geometric description, or from (2), we see that t is uniquely determined by the point (x, y) (for $x \neq 0$).

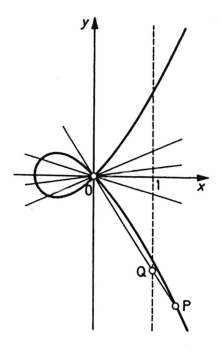

Figure 3. Projection of a Cubic

We now give a general definition of algebraic plane curves for which a representation in these terms is possible. We say that an irreducible algebraic curve X defined by $f(x, y) = 0$ is *rational* if there exist two rational functions $\varphi(t)$ and $\psi(t)$, at least one nonconstant, such that

$$f(\varphi(t), \psi(t)) \equiv 0, \tag{3}$$

as an identity in t. Obviously if $t = t_0$ is a value of the parameter, and is not one of the finitely many values at which the denominator of φ or ψ vanishes, then $(\varphi(t_0), \psi(t_0))$ is a point of X. We will show subsequently that for a suitable choice of the parametrisation φ, ψ, the map $t_0 \mapsto (\varphi(t_0), \psi(t_0))$ is a one-to-one correspondence between the values of t and the points of the curve, provided that we exclude certain finite sets from both the set of values of t and the points of the curve. Then conversely, the parameter t can be expressed as a rational function $t = \chi(x, y)$ of the coordinates x and y.

If the coefficients of the rational functions φ and ψ belong to some subfield k_0 of k and $t_0 \in k_0$ then the coordinates of the point $(\varphi(t_0), \psi(t_0))$ also belong to k_0. This observation points to one possible application of the notion of rational curve. Suppose that $f(x, y)$ has rational coefficients. If we know that the curve given by 1.1, (1) is rational, and that the coefficients of φ and ψ are in $\mathbb{Q}$, then the parametrisation $x = \varphi(t), y = \psi(t)$ gives us all the rational points of this curve, except possibly a finite number, as t runs through all rational values. For example, all the rational solutions of the indeterminate equation (1) can be obtained from (2) as t runs through all rational values.

Another application of rational curves relates to integral calculus. We can view the equation of the curve 1.1, (1) as determining y as an algebraic function of x. Then any rational function $g(x, y)$ is a (usually complicated) function of x. The rationality of the curve 1.1, (1) implies the following important fact: for any rational function $g(x, y)$, the indefinite integral

$$\int g(x, y)\mathrm{d}x \tag{4}$$

can be expressed in elementary functions. Indeed, since the curve is rational, it can be parametrised as $x = \varphi(t), y = \psi(t)$ where φ, ψ are rational functions. Substituting these expressions in the integral (4), we reduce it to the form $\int g(\varphi(t), \psi(t))\varphi'(t)\mathrm{d}t$, which is an integral of a rational function. It is known that an integral of this form can be expressed in elementary functions. Substituting the expression $t = \chi(x, y)$ for the parameter in terms of the coordinates, we get an expression for the integral (4) as an elementary function of the coordinates.

We now give some examples of rational curves. Curves of degree 1, that is, lines, are obviously rational. Let us prove that an irreducible conic X is rational. Choose a point (x_0, y_0) on X. Consider the line through (x_0, y_0) with slope t. Its equation is

$$y - y_0 = t(x - x_0). \tag{5}$$

We find the points of intersection of X with this line; to do this, solve (5) for y and substitute this in the equation of X. We get the equation for x

$$f(x, y_0 + t(x - x_0)) = 0, \tag{6}$$

which has degree 2, as one sees easily. We know one root of this quadratic equation, namely $x = x_0$, since by assumption (x_0, y_0) is on the curve. Divide the equation (6) by the coefficient of x^2, and write A for the coefficient of x in the resulting equation; the other root is then determined by $x + x_0 = -A$. Since t appears in the coefficients of equation (6), A is a rational function of t. Substituting the expression $x = -x_0 - A$ in (5), we get an expression for y also as a rational function of t. These expressions for x and y satisfy the equation of the curve, as can be seen from their derivation, and thus prove that the curve is rational.

The above parametrisation has an obvious geometric interpretation. A point (x, y) of X is sent to the slope of the line joining it to (x_0, y_0); and the parameter t is sent to the point of intersection of the curve with the line through (x_0, y_0) with slope t. This point is uniquely determined precisely because we are dealing with an irreducible curve of degree 2. In the same way as the parametrisation of the curve (1), this parametrisation can be interpreted as the projection of X from the point (x_0, y_0) to some line not passing through this point (Figure 4).

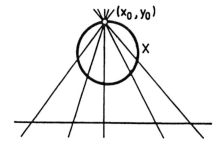

Figure 4. Projection of a Conic

Note that in constructing the parametrisation we have used a point (x_0, y_0) of X. If the coefficients of the polynomial $f(x, y)$ and the coordinates of (x_0, y_0) are contained in some subfield k_0 of k, then so do the coefficients of the functions giving the parametrisation. Thus we can, for example, find the general form for the solution in rational numbers of an indeterminate equation of degree 2 if we know just one solution.

The question of whether there exists one solution is rather delicate. For the rational number field $\mathbb{Q}$ it is solved by Legendre's theorem (see for example Borevich and Shafarevich [13], Chap. 1, 7.2.).

We consider another application of the parametrisation we have found. The second degree equation $y^2 = ax^2 + bx + c$ defines a rational curve, as we have just seen. It follows from this that for any rational function $g(x, y)$, the integral $\int g\left(x, \sqrt{ax^2 + bx + c}\right) dx$ can be expressed in elementary functions. The parametrisation we have given provides an explicit form of the substitutions that reduce this integral to an integral of a rational function. It is easy to see that this leads to the well-known *Euler substitutions*.

The examples considered above lead us to the following general question: how can we determine whether an arbitrary algebraic plane curve is rational? This question relates to quite delicate ideas of algebraic geometry, as we will see later.

1.3. Relation with Field Theory

We now show how the question at the end of 1.2 can be formulated as a problem of field theory. To do this, we assign to every irreducible plane curve a certain field, by analogy with the way we assign to an irreducible polynomial in one variable the smallest field extension in which it has a root.

Let X be the irreducible curve given by 1.1, (1). Consider rational functions $u(x,y) = p(x,y)/q(x,y)$, where p and q are polynomials with coefficients in k such that the denominator $q(x,y)$ is not divisible by $f(x,y)$. We say that such a function $u(x,y)$ is a *rational function* defined on X; and two rational functions $p(x,y)/q(x,y)$ and $p_1(x,y)/q_1(x,y)$ defined on X are *equal* on X if the polynomial $p(x,y)q_1(x,y) - q(x,y)p_1(x,y)$ is divisible by $f(x,y)$. It is easy to check that rational functions on X, up to equality on X, form a field. This field is called the *function field* or *field of rational functions* of X, and denoted by $k(X)$.

A rational function $u(x,y) = p(x,y)/q(x,y)$ is defined at all points of X where $q(x,y) \neq 0$. Since by assumption q is not divisible by f, by 1.1, Lemma, there are only finitely many points of X at which $u(x,y)$ is not defined. Hence we can also consider elements of $k(X)$ as functions on X, but defined everywhere except at a finite set. It can happen that a rational function u has two different expressions $u = p/q$ and $u = p_1/q_1$, and that for some point $(\alpha, \beta) \in X$ we have $q(\alpha, \beta) = 0$ but $q_1(\alpha, \beta) \neq 0$. For example, the function $u = (1-y)/x$ on the circle $x^2 + y^2 = 1$ at the point $(0,1)$ has an alternative expression $u = x/(1+y)$ whose denominator does not vanish at $(0,1)$. If u has an expression $u = p/q$ with $q(P) \neq 0$ then we say that u is *regular* at P.

Every element of $k(X)$ can obviously be written as a rational function of x and y; now x, y are algebraically dependent, since they are related by $f(x,y) = 0$. It is easy to check from this that $k(X)$ has transcendence degree 1 over k.

If X is a line, given say by $y = 0$, then every rational function $\varphi(x,y)$ on X is a rational function $\varphi(x,0)$ of x only, and hence the function field of X equals the field of rational functions in one variable, $k(X) = k(x)$.

Now assume that the curve X is rational, say parametrised by $x = \varphi(t)$, $y = \psi(t)$. Consider the substitution $u(x,y) \mapsto u(\varphi(t), \psi(t))$ that takes any rational function $u = p(x,y)/q(x,y)$ on X into the rational function in t obtained by substituting $\varphi(t)$ for x and $\psi(t)$ for y. We check first that this substitution makes sense, that is, that the denominator $q(\varphi(t), \psi(t))$ is not identically 0 as a function of t. Assume that $q(\varphi(t), \psi(t)) = 0$, and compare this equality with 1.2, (3). Recalling that the field k is algebraically closed, and therefore infinite, by making t take different values in k, we see that $f(x,y) = 0$ and $q(x,y) = 0$ have infinitely many common solutions. But by 1.1, Lemma, this is only possible if f and q have a common factor.

Thus our substitution sends any rational function $u(x,y)$ defined on X into a well-defined element of $k(t)$. Moreover, since φ and ψ satisfy the relation

1.2, (3), the substitution takes rational functions u, u_1 that are equal on X to the same rational function in t. Thus every element of $k(X)$ goes to a well-defined element of $k(t)$. This map is obviously an isomorphism of $k(X)$ with some subfield of $k(t)$. It takes an element of k to itself.

At this point we make use of a theorem on rational functions. This is the result known as Lüroth's theorem, that asserts that a subfield of the field $k(t)$ of rational functions containing k is of the form $k(g(t))$, where $g(t)$ is some rational function; that is, the subfield consists of all the rational functions of $g(t)$. If $g(t)$ is not constant, then sending $f(u) \mapsto f(g(t))$ obviously gives an isomorphism of the field of rational functions $k(u)$ with $k(g(t))$. Thus Lüroth's theorem can be given the following statement: a subfield of the field of rational functions $k(t)$ that contains k and is not equal to k is itself isomorphic to the field of rational functions. Lüroth's theorem can be proved from simple properties of field extensions (see van der Waerden [73], 10.2 (§73)). Applying it to our situation, we see that if X is a rational curve then $k(X)$ is isomorphic to the field of rational functions $k(t)$. Suppose, conversely, that for some curve X given by 1.1, (1), the field $k(X)$ is isomorphic to the field of rational functions $k(t)$. Suppose that under this isomorphism x corresponds to $\varphi(t)$ and y to $\psi(t)$. The polynomial relation $f(x, y) = 0 \in k(X)$ is respected by the field isomorphism, and gives $f(\varphi(t), \psi(t)) = 0$; therefore X is rational.

It is easy to see that any field $K \supset k$ having transcendence degree 1 over k and generated by two elements x and y is isomorphic to a field $k(X)$, where X is some irreducible algebraic plane curve. Indeed, x and y must be connected by a polynomial relation, since K has transcendence degree 1 over k. If this dependence relation is $f(x, y) = 0$, with f an irreducible polynomial, then we can obviously take X to be the algebraic curve defined by this equation. It follows from this that the question on rational curves posed at the end of 1.2 is equivalent to the following question of field theory: when is a field $K \supset k$ with transcendence degree 1 over k and generated by two elements x and y isomorphic to the field of rational functions of one variable $k(t)$? The requirement that K is generated over k by two elements is not very natural from the algebraic point of view. It would be more natural to consider field extensions generated by an arbitrary finite number of elements. However, we will prove later that doing this does not give a more general notion (compare 3.3, Theorem 5 and Appendix, §5, Proposition 1).

In conclusion, we note that the preceding arguments allow us to solve the problem of obtaining a generically one-to-one parametrisation of a rational curve. Let X be a rational curve. By Lüroth's theorem, the field $k(X)$ is isomorphic to the field of rational functions $k(t)$. Suppose that this isomorphism takes x to $\varphi(t)$ and y to $\psi(t)$. This gives the parametrisation $x = \varphi(t)$, $y = \psi(t)$ of X.

Proposition. *The parametrisation* $x = \varphi(t)$, $y = \psi(t)$ *has the following properties:*

(i) Except possibly for a finite number of points, any $(x_0, y_0) \in X$ has a representation $(x_0, y_0) = (\varphi(t_0), \psi(t_0))$ for some t_0.

(ii) Except possibly for a finite number of points, this representation is unique.

Proof. Suppose that the function that maps to t under the isomorphism $k(X) \to k(t)$ is $\chi(x, y)$. Then the inverse isomorphism $k(t) \to k(X)$ is given by the formula $u(t) \mapsto u(\chi(x, y))$. Writing out the fact that the correspondences are inverse to one another gives

$$x = \varphi(\chi(x, y)), \quad y = \psi(\chi(x, y)), \tag{1}$$
$$t = \chi(\varphi(t), \psi(t)). \tag{2}$$

Now (1) implies (i). Indeed, if $\chi(x, y) = p(x, y)/q(x, y)$ and $q(x_0, y_0) \neq 0$, we can take $t_0 = \chi(x_0, y_0)$; there are only finitely many points $(x_0, y_0) \in X$ at which $q(x_0, y_0) = 0$, since $q(x, y)$ and $f(x, y)$ are coprime. Suppose that (x_0, y_0) is such that $\chi(x_0, y_0)$ is distinct from the roots of the denominators of $\varphi(t)$ and $\psi(t)$; there are only finitely many points (x_0, y_0) for which this fails, for similar reasons. Then formula (1) gives the required representation of (x_0, y_0). In the same way, it follows from (2) that the value of the parameter t, if it exists, is uniquely determined by the point (x_0, y_0), except possibly for the finite number of points at which $q(x_0, y_0) = 0$. The proposition is proved.

Note that we have proved (i) and (ii) not for any parametrisation of a rational curve, but for a specially constructed one. For an arbitrary parametrisation, (ii) can be false: for example, the curve 1.2, (1) has, in addition to the parametrisation given by 1.2, (2), another parametrisation $x = t^4 - 1$, $y = t^2(t^4 - 1)$, obtained from 1.2, (2) on replacing t by t^2. Obviously here the values t and $-t$ of the parameter correspond to the same point of the curve.

1.4. Rational Maps

A rational parametrisation is a particular case of a more general notion. Let X and Y be two irreducible algebraic plane curves, and $u, v \in k(X)$. The map $\varphi(P) = (u(P), v(P))$ is defined at all points P of X where both u and v are defined; it is called a *rational map* from X to Y if $\varphi(P) \in Y$ for every $P \in X$ at which φ is defined. If Y has the equation $g = 0$ then $g(u, v) \in k(X)$ must vanish at all but finitely many points of X, and therefore we must have $g(u, v) = 0 \in k(X)$.

For example, the projection from a point P considered in 1.2 is a rational map of X to the line. A rational parametrisation of a rational curve X is a rational map of the line to X.

A rational map $\varphi \colon X \to Y$ is *birational*, or is a *birational equivalence* of X to Y, if φ has a rational inverse, that is, if there exists a rational map $\psi \colon Y \to X$ such that $\varphi \circ \psi$ and $\psi \circ \varphi$ are the identity (at the points where they are defined). In this case, we say that X and Y are *birational*, or *birationally equivalent*.

A birational map is not constant, that is, at least one of the functions defining it is not an element of k. Indeed, a constant map is defined everywhere, and sends X to a single point $Q \in Y$. Taking any point $Q' \neq Q$ at which the inverse ψ of φ is defined contradicts the definition.

It follows that for any point $Q \in Y$ the inverse image $\varphi^{-1}(Q)$ of Q (the set of points $P \in X$ such that $\varphi(P) = Q$) is finite; this follows at once from 1.1, Lemma. Let S be the finite set of points of X at which a birational map $\varphi \colon X \to Y$ is not defined, $U = X \setminus S$ its complement, and T and V the same for $\psi \colon Y \to X$. It follows from what we said above that the complement in X of $\varphi^{-1}(V) \cap U$ and in Y of $\psi^{-1}(U) \cap V$ are finite, and φ establishes a one-to-one correspondence between $\varphi^{-1}(V) \cap U$ and $\psi^{-1}(U) \cap V$.

Birational equivalence is a fundamental equivalence relation in algebraic geometry, and we usually classify algebraic curves up to birational equivalence. We have seen that the rational curves are exactly the curves birational to the line.

Suppose that the equation $f(x, y)$ of an irreducible curve of degree n is a polynomial all of whose terms are monomials in x and y of degree $n - 1$ and n only. Then the projection from the origin defines a birational map of our curve and the line: this can be proved by a direct generalisation of the arguments for the curve 1.2, (1).

Now suppose that the equation f has terms of degrees $n - 2$, $n - 1$ and n, that is, $f = u_{n-2} + u_{n-1} + u_n$, where u_i is homogeneous of degree i. Again we set $y = tx$ and cancel the factor of x^{n-2} from the equation, thus reducing it to the form $a(t)x^2 + b(t)x + c(t) = 0$, where $a(t) = u_n(1, t)$, $b(t) = u_{n-1}(1, t)$ and $c(t) = u_{n-2}(1, t)$. Setting $s = 2ax + b$ to complete the square (assuming that the ground field has characteristic $\neq 2$), we see that our curve is birational to the curve given by $s^2 = p(t)$, where $p = b^2 - 4ac$. A curve of this type is called a *hyperelliptic curve*. If $p(t)$ has even degree $2m$ then rewriting it in the form $p(t) = q(t)(t - \alpha)$ and dividing both sides of the equation through by $(t - \alpha)^{2m}$ shows that the curve is birational to the curve given by

$$\eta^2 = h(\xi), \quad \text{where } \xi = \frac{1}{t - \alpha}, \, \eta = \frac{s}{(t - \alpha)^m} \text{ and } h(\xi) = \frac{q(t)}{(t - \alpha)^{2m-1}},$$

in which h is a polynomial of degree $\leq 2m - 1$ in ξ.

These ideas apply in particular to any cubic curve, if we take the origin to be any point of the curve. We see that, if $\operatorname{char} k \neq 2$, an irreducible cubic curve is birational to a curve given by $y^2 = f(x)$ where f is a polynomial of degree ≤ 3. If $f(x)$ has degree ≤ 2 then the cubic is rational. If it has degree 3 then we can assume that its leading coefficient is 1. Then the equation takes the form

$$y^2 = x^3 + ax^2 + bx + c.$$

This is called the *Weierstrass normal form* of the equation of a cubic. If char $k \neq 3$ then after making a translation $x \mapsto x - a/3$ we can reduce the equation to the form

$$y^2 = x^3 + px + q. \tag{1}$$

Let X and Y be two irreducible algebraic plane curves that are birational, and suppose that the maps between them are given by

$$(u, v) = (\varphi(x, y), \psi(x, y)) \quad \text{and} \quad (x, y) = (\xi(u, v), \eta(u, v)).$$

As in our study of rational curves, we can establish a relation between the function fields $k(X)$ and $k(Y)$ of these two curves. For this, we send a rational function $w(x, y) \in k(X)$ to $w(\xi(u, v), \eta(u, v))$, viewed as a rational function on Y. It is easy to check that this defines a map $k(X) \to k(Y)$ that is an isomorphism between these two fields. Conversely, if $k(X)$ and $k(Y)$ are isomorphic, then under this isomorphism $x, y \in k(X)$ correspond to functions $\xi(u, v), \eta(u, v) \in k(Y)$, and $u, v \in k(Y)$ to functions $\varphi(x, y), \psi(x, y) \in k(X)$, and it is again trivial to check that the pairs of functions φ, ψ and ξ, η define birational maps between the curves X and Y. Thus two curves are birational if and only if their rational function fields are isomorphic.

We see that the problem of classifying algebraic curves up to birational equivalence is a geometric aspect of the natural algebraic problem of classifying finitely generated extension fields of k of transcendence degree 1 up to isomorphism. In this problem, it is also natural not to restrict to fields of transcendence degree 1, but to consider fields of any finite transcendence degree. We will see later that this wider formulation of the problem also has a geometric interpretation. However, for this we have to leave the framework of the theory of algebraic curves, and consider algebraic varieties of any dimension.

1.5. Singular and Nonsingular Points

We borrow a definition from coordinate geometry: a point P is a *singular point* or *singularity* of the curve defined by $f(x, y) = 0$ if $f'_x(P) = f'_y(P) = f(P) = 0$, where f'_x denotes the partial derivative $\partial f / \partial x$. If we translate P to the origin, we can say that $(0, 0)$ is singular if f does not have constant or linear terms. A point is *nonsingular* if it is not singular, that is, if $f'_x(P)$ or $f'_y(P) \neq 0$. A curve all of whose points are nonsingular is *nonsingular* or *smooth*. It is well known that an irreducible conic is nonsingular; the simplest example of a singular curve is the curve of 1.2, (1).

For an irreducible curve, either f'_x vanishes at only finitely many points of the curve, or f'_x is divisible by f. However, since f'_x has smaller degree than f, the latter is only possible if $f'_x = 0$. The same holds for f'_y. But $f'_x = f'_y = 0$ implies, if char $k = 0$, that $f \in k$, and, if char $k = p > 0$, that f involves x and y only as pth powers; in this last case, taking pth roots of the coefficients of f and using the well-known characteristic p identity $(\alpha + \beta)^p = \alpha^p + \beta^p$, we deduce that

$$f = \sum a_{ij} x^{pi} y^{pj} = \left(\sum b_{ij} x^i y^j \right)^p \quad \text{where } b_{ij}^p = a_{ij},$$

which contradicts the irreducibility of the curve. This shows that an irreducible curve has only a finite number of singular points.

If $P = (0,0)$ and the leading terms in the equation of the curve have degree r, then r is called the *multiplicity* of P, and we say that P is an r-*tuple point*, or *point of multiplicity* r. Thus a nonsingular point has multiplicity 1. If $P = (0,0)$ has multiplicity 2 and the terms of degree 2 in the equation of the curve are $ax^2 + bxy + cy^2$ then there are two possibilities: (a) $ax^2 + bxy + cy^2$ factorises into two distinct linear factors; or (b) $ax^2 + bxy + cy^2$ is a perfect square. In case (a) the singularity is called a *node* (see Figure 3), and in case (b) a *cusp* (Figure 5).

Figure 5. A Cusp

It follows from the definition that a curve of degree n cannot have a singularity of multiplicity $> n$. If a singular point has multiplicity n then the equation of the curve is a homogeneous polynomial in x and y of degree n, and therefore factorises as a product of linear factors, so that the curve is reducible. In 1.4 we proved that if an irreducible curve of degree n has a point of multiplicity $n - 1$ it is rational, and if it has a point of multiplicity $n - 2$ then it is hyperelliptic. The cubic curve written in Weierstrass normal form 1.4, (1) is nonsingular if and only if the cubic polynomial on the right-hand side has no multiple roots, that is, $4p^3 + 27q^2 \neq 0$. In this case it is called an *elliptic curve*.

If $k = \mathbb{R}$ and P is a nonsingular point of the curve with equation $f(x,y) = 0$, and $f'_y(P) \neq 0$, say, then by the implicit function theorem we can write y as a function of x in some neighbourhood of P. Substituting this expression for y, this represents any rational function on the curve as a function of x near P.

When k is a general field, x can still be used to describe all the rational functions on the curve, admittedly to a more modest extent. For simplicity, set $P = (0,0)$. Then $f = \alpha x + \beta y + g$, where g contains only terms of degree ≥ 2 and $\beta \neq 0$. We distinguish the terms in f that involve x only, writing $f = x\varphi(x) + y\beta + yh$, with $h(0,0) = 0$. Thus on the curve $f = 0$ we have $y(\beta + h) = -x\varphi(x)$, or, in other words, $y = xv$, where $v = -\varphi(x)/(\beta + h)$ is a regular function at P (because $\beta + h(P) \neq 0$).

Let u be any rational function on our curve that is regular at P and has $u(P) = 0$. Then $u = p/q$, where $p, q \in k[x, y]$ with $p(P) = 0$ and $q(P) \neq 0$. Substituting our expression for y in this gives $p(x, y) = p(x, xv) = xr$ (because p has no constant term), where r is a regular function on the curve, and hence $u = xr/q = xu_1$. If $u_1(P) = 0$ then we can repeat the argument, getting $u = x^2 u_2$, and so on. We now prove that, provided u is not identically 0 on the curve, this process must stop after a finite number of steps.

For this, return to the expression $u = p/q$, in which, by assumption, p is not divisible by f. Hence there exist $\xi, \eta \in k[x, y]$ and a polynomial $a \in k[x]$ with $a \neq 0$ such that $f\xi + p\eta = a$ (we have already used this argument in the proof of 1.1, Lemma). Suppose $a = x^k a_0$ with $a_0(0) \neq 0$. Then $p\eta = a$ on the curve, and a representation $p = x^l w$ with $l > k$ would give a contradiction: $x^k(x^{l-k}w - a_0) = 0$ on the curve, that is, $x^{l-k}w - a_0 = 0$. If $w = c/d$ with $c, d \in k[x, y]$ and $d(P) \neq 0$ then $x^{l-k}c - a_0 d = 0$ on the curve, that is, $x^{l-k}c - a_0 d$ is divisible by f. But this is impossible, since x^{l-k} vanishes at P and $a_0 d$ does not. Since any rational function is a ratio of regular functions, we have proved the following theorem.

Theorem 1. *At any nonsingular point P of an irreducible algebraic curve, there exists a regular function t that vanishes at P and such that every rational function u that is not identically 0 on the curve can be written in the form*

$$u = t^k v, \tag{1}$$

with v regular at P and $v(P) \neq 0$. The function u is regular at P if and only if $k \geq 0$ in (1). $\square$

A function t with this property is called a *local parameter* on the curve at P. Obviously two different local parameters are related by $t' = tv$, where v is regular at P and $v(P) \neq 0$. We saw in the proof of the theorem that if $f'_y(P) \neq 0$ then x can be taken as a local parameter.

The number k in (1) is called the *multiplicity of the zero* of u at P. It is independent of the choice of the local parameter.

Let X and Y be algebraic curves with equations $f = 0$ and $g = 0$, and suppose that X is irreducible and not contained in Y, and that $P \in X \cap Y$ is a nonsingular point of X. Then g defines a function on X that is not identically zero; the multiplicity of the zero of g at P is called the *intersection multiplicity*[2] of X and Y at P. The notion of intersection multiplicity is one of the amendments needed in a correct statement of Bézout's theorem: for the theorem that the number of roots of a polynomial is equal to its degree is false unless we count roots with their multiplicities. Here we analyse intersection multiplicities in the case that X is a line.

---

[2] This is discussed at length later in the book; see Chap. IV, 1.1 for the general definition of intersection multiplicity, which is symmetric in X and Y, and for the fact that it coincides with the simple notion used here.

Let $P = (\alpha, \beta) \in X$, and suppose that the equation of X is written in the form $f(x,y) = a(x - \alpha) + b(y - \beta) + g$, where the polynomial g expanded in powers of $x - \alpha$ and $y - \beta$ has only terms of degree ≥ 2. We write the equation of a line L through P in the form

$$x = \alpha + \lambda t, \quad y = \beta + \mu t. \tag{2}$$

t is a local parameter on L at P. The restriction of f to L is of the form

$$f(\alpha + \lambda t, \beta + \mu t) = (a\lambda + b\mu)t + t^2 \varphi(t).$$

From this we see that if P is singular, that is, if $a = b = 0$, then every line through P has intersection multiplicity > 1 with X at P. On the other hand, if the curve is nonsingular, then there is only one such line, namely that for which $a\lambda + b\mu = 0$, with equation $a(x - \alpha) + b(y - \beta) = 0$. Obviously $a = f'_x(P)$, $b = f'_y(P)$, and hence this equation can we expressed

$$f'_x(P)(x - \alpha) + f'_y(P)(y - \beta) = 0. \tag{3}$$

The line given by this equation is called the *tangent line* to X at the nonsingular point P.

We now determine when a line has intersection multiplicity ≥ 3 with a curve at a nonsingular point $P = (\alpha, \beta)$. For this, we write the equation in the form

$$\begin{aligned} f(x,y) = a(x - \alpha) + b(y - \beta) \\ + c(x - \alpha)^2 + d(x - \alpha)(y - \beta) + e(y - \beta)^2 + h, \end{aligned} \tag{4}$$

where h is a polynomial which has only terms of degree ≥ 3 when expanded in power of $x - \alpha$ and $y - \beta$. Restricting f to the line L given by (2), we get that $f = (a\lambda + b\mu)t + (c\lambda^2 + d\lambda\mu + e\mu^2)t^2 + t^3 \psi(t)$. Therefore the intersection multiplicity will be ≥ 3 if the two conditions $a\lambda + b\mu = c\lambda^2 + d\lambda\mu + e\mu^2 = 0$ hold. The first of these, as we have seen, means that L is the tangent line to X at P, and the second that moreover $cu^2 + duv + ev^2$ is divisible by $au + bv$ as a homogeneous polynomial in u, v. Together they show that $q = au + bv + cu^2 + duv + ev^2$ is reducible: it is divisible by $au + bv$. Conversely, if q is reducible, then $q = rs$, and r and s must have degree 1, and one of them, say r, must vanish when $u = v = 0$. But then r is proportional to $au + bv$ and $cu^2 + duv + ev^2$ is divisible by it. Thus the reducibility of the conic $q = au + bv + cu^2 + duv + ev^2$ is a necessary and sufficient condition for there to exist a line L through P with intersection multiplicity ≥ 3 at P. Such a point is called an *inflexion point* or *flex* of X.

We know from coordinate geometry the condition for a conic to be reducible. We assume that k has characteristic $\neq 2$; then recalling that $a = f'_x(P)$, $b = f'_y(P)$, $c = (1/2)f''_x(P)$, $d = f''_{xy}(P)$ and $e = (1/2)f''_{yy}(P)$, we can write this condition in the form

$$\begin{vmatrix} f''_{xx} & f''_{xy} & f'_x \\ f''_{xy} & f''_{yy} & f'_y \\ f'_x & f'_y & 0 \end{vmatrix}(P) = 0. \tag{5}$$

1.6. The Projective Plane

We return to Bézout's theorem stated in 1.1. Even if we consider points with coordinates in an algebraically closed field and take account of multiplicities of intersections, this fails in very simple cases, and still needs one further amendment. This can already be seen in the example of two lines, which have no points of intersection if they are parallel. However, on the projective plane, parallel lines do intersect, in a point of the line at infinity.

In the same way, any two circles in the plane, although they are curves of degree 2, have at most 2 points of intersection, and never 4 as predicted by Bézout's theorem. This follows from the fact that the quadratic term in the equation of all circles is always the same, namely $x^2 + y^2$, so that subtracting the equation of one circle from that of the other gives a linear equation, and therefore the intersection of two circles is the same thing as the intersection of a circle and a line. Moreover, if the circles are not tangent, their multiplicity of intersection is 1 at each point of intersection.

To understand what lies behind this failure of Bézout's theorem, write the equation of the circle $(x - a)^2 + (y - b)^2 = r^2$ in homogeneous coordinates by setting $x = \xi/\zeta$ and $y = \eta/\zeta$. We get the equation $(\xi - a\zeta)^2 + (\eta - b\zeta)^2 = r^2\zeta^2$, from which we see that the circle intersects the line at infinity $\zeta = 0$ in the points $\xi^2 + \eta^2 = 0$, that is, in the two circular points at infinity $(1, \pm i, 0)$. Thus all circles have the two points $(1, \pm i, 0)$ at infinity in common. Taken together with the two finite points of intersection, we thus get 4 points of intersection, in agreement with Bézout's theorem. This type of phenomenon motivates passing from the affine to the projective plane.

Recall that a point of the projective plane $\mathbb{P}^2$ is determined by 3 elements (ξ, η, ζ) of the field k, not all simultaneously zero. Two triples (ξ, η, ζ) and (ξ', η', ζ') determine the same point if there exists $\lambda \in k$ with $\lambda \neq 0$ such that $\xi = \lambda\xi'$, $\eta = \lambda\eta'$ and $\zeta = \lambda\zeta'$. Any triple (ξ, η, ζ) defining a point P is called a set of *homogeneous coordinates* of P, and we write $P = (\xi : \eta : \zeta)$.

There is an inclusion $\mathbb{A}^2 \subset \mathbb{P}^2$ which sends $(x, y) \in \mathbb{A}^2$ to $(x : y : 1)$. We get in this way all points with $\zeta \neq 0$: a point $(\xi : \eta : \zeta) \in \mathbb{P}^2$ with $\zeta \neq 0$ corresponds to the point $(\xi/\zeta, \eta/\zeta) \in \mathbb{A}^2$. The points of the complementary set $\zeta = 0$ are called *points at infinity*. This notion is related to the choice of the coordinate ζ. In fact, $\mathbb{P}^2$ contains 3 sets that are copies of the affine plane in this way: $\mathbb{A}_1^2$ (given by $\xi \neq 0$), $\mathbb{A}_2^2$ (given by $\eta \neq 0$), and $\mathbb{A}_3^2$ (given by $\zeta \neq 0$). These intersect, of course: if a point $P \in \mathbb{A}_3^2$ has coordinates $x = \xi/\zeta$, $y = \eta/\zeta$ and $\eta \neq 0$ then in $\mathbb{A}_2^2$ the same point has coordinates $x' = \xi/\eta$, $y' = \zeta/\eta$, so that $x' = x/y$, $y' = 1/y$; if $\xi \neq 0$ then in $\mathbb{A}_1^2$ it has coordinates $x'' = \eta/\xi$, $y'' = \zeta/\xi$, so that $x'' = y/x$, $y'' = 1/x$. Every point $P \in \mathbb{P}^2$ is contained in at least one of the pieces A_1^2, A_2^2 or A_3^2, and can be written down in the affine coordinates of that piece.

An algebraic curve in $\mathbb{P}^2$, or a *projective algebraic plane curve* is defined in homogeneous coordinates by an equation $F(\xi, \eta, \zeta) = 0$, where F is a homogeneous polynomial. Then whether $F(\xi, \eta, \zeta) = 0$ holds or not is independent of the choice of the homogeneous coordinates of a point; that is, it is preserved on passing from ξ, η, ζ to $\xi' = \lambda\xi$, $\eta' = \lambda\eta$, $\zeta' = \lambda\zeta$ with $\lambda \neq 0$. A homogeneous polynomial is also called a *form*. An affine algebraic curve of degree n with equation $f(x, y) = 0$ defines a homogeneous polynomial $F(\xi, \eta, \zeta) = \zeta^n f(\xi/\zeta, \eta/\zeta)$, and hence a projective curve with equation $F(\xi, \eta, \zeta) = 0$. It is easy to see that intersecting this curve with the affine plane A_3^2 gives us the original affine curve, to which it therefore only adds points at infinity with $\zeta = 0$. If the equation of the projective curve is $F(\xi, \eta, \zeta) = 0$, then that of the corresponding affine curve is $f(x, y) = 0$, where $f(x, y) = F(x, y, 1)$. Since every point $P \in \mathbb{P}^2$ is contained in one of the affine sets A_1^2, A_2^2 or A_3^2, we can use this correspondence to write out the properties of curves, defined above for affine curves, in terms of homogeneous coordinates. We do this now for the notions of tangent line, singular point and inflexion point of an algebraic curve. We always assume that $P \in A_3^2$.

In affine coordinates, the equation of the tangent is

$$\frac{\partial f}{\partial x}(P)(x - \alpha) + \frac{\partial f}{\partial y}(P)(y - \beta) = 0.$$

By assumption $f(x, y) = F(x, y, 1)$, where $F(\xi, \eta, \zeta) = 0$ is the homogeneous equation of our curve. Hence writing F_x' etc. for the partial derivatives, we get $\partial f/\partial x = F_x'(x, y, 1)$ and $\partial f/\partial y = F_y'(x, y, 1)$, and by the well-known theorem of Euler on homogeneous functions, we have

$$F_\xi' \xi + F_\eta' \eta + F_\zeta' \zeta = nF.$$

Since $P = (\alpha : \beta : 1)$ is a point of the curve, $F_\xi'(P)\alpha + F_\eta'(P)\beta + F_\zeta'(P) = 0$, so that the equation of the tangent is $F_\xi'(P)x + F_\eta'(P)y + F_\zeta'(P) = 0$, or in homogeneous coordinates

$$F_\xi'(P)\xi + F_\eta'(P)\eta + F_\zeta'(P)\zeta = 0.$$

The conditions in affine coordinates for a singular point are $f_x' = f_y' = f = 0$. Hence in homogeneous coordinates $F_\xi' = F_\eta' = F = 0$, and by Euler's theorem, since $\zeta = 1$, also $F_\zeta' = 0$. If the characteristic of the field k is 0 then it is enough to require the conditions $F_\xi'(P) = F_\eta'(P) = F_\zeta'(P) = 0$, since then also $F(P) = 0$.

The condition defining an inflexion point is given by the relation 1.5, (5). Here again $f(x, y) = F(x, y, 1)$, so that $f_x' = F_x'$, $f_y' = F_y'$, $f_{xx}'' = F_{xx}''$, $f_{xy}'' = F_{xy}''$, $f_{yy}'' = F_{yy}''$. From now on, in the homogeneous polynomial F we write ξ for x and η for y. We substitute these expressions in the determinant of 1.5, (5), and use Euler's theorem

$$F''_{\xi\xi}\xi + F''_{\xi\eta}\eta + F''_{\xi\zeta}\zeta = (n-1)F'_\xi$$

$$F''_{\xi\eta}\xi + F''_{\eta\eta}\eta + F''_{\zeta\eta}\zeta = (n-1)F'_\eta$$

$$F'_\xi\xi + F'_\eta\eta + F'_\zeta\zeta = nF.$$

Multiply the last column of our determinant by $(n-1)$, and subtract from it ξ times the first column and η times the second. Using the above identities and recalling that $F(P) = 0$, we get the determinant

$$\begin{vmatrix} F''_{\xi\xi} & F''_{\xi\eta} & F''_{\xi\zeta} \\ F''_{\xi\eta} & F''_{\eta\eta} & F''_{\zeta\eta} \\ F'_\xi & F'_\eta & F'_\zeta \end{vmatrix}(P).$$

Now perform the same operation on the rows of the determinant. The condition for P to be an inflexion point then takes the form

$$\begin{vmatrix} F''_{\xi\xi} & F''_{\xi\eta} & F''_{\xi\zeta} \\ F''_{\eta\xi} & F''_{\eta\eta} & F''_{\eta\zeta} \\ F''_{\zeta\xi} & F''_{\zeta\eta} & F''_{\zeta\zeta} \end{vmatrix}(P) = 0. \tag{1}$$

The determinant on the left-hand side of equation (1) is called the *Hessian form* of F, and denoted by $H(F)$.

We now proceed to considering rational functions. Making the substitution $x = \xi/\zeta$, $y = \eta/\zeta$ and clearing denominators, we can rewrite a rational function $f = p(x,y)/q(x,y)$ on $\mathbb{A}_3^2$ in the form $P(\xi,\eta,\zeta)/Q(\xi,\eta,\zeta)$, where P and Q are homogeneous polynomials of the same degree. Hence its value at a point $(\xi : \eta : \zeta)$ does not change on multiplying the homogeneous coordinates through by a common multiple, and hence f can be viewed as a partially defined function on $\mathbb{P}^2$.

Given a rational map $\varphi \colon \mathbb{A}_3^2 \to \mathbb{A}_3^2$ defined by $(x,y) \mapsto (u(x,y), v(x,y))$, we first rewrite it, as just explained, in the form

$$\frac{U(\xi,\eta,\zeta)}{R(\xi,\eta,\zeta)}, \quad \frac{V(\xi,\eta,\zeta)}{S(\xi,\eta,\zeta)},$$

where U, V, R, S are homogeneous polynomials, with $\deg U = \deg R$ and $\deg V = \deg S$. Next we put the two components over a common denominator, that is, in the form $(A/C, B/C)$, with $\deg A = \deg B = \deg C$. Finally, introducing homogeneous coordinates $\xi'/\zeta' = A/C$, $\eta'/\zeta' = B/C$, we write the map in the form

$$(\xi : \eta : \zeta) \mapsto \big(A(\xi : \eta : \zeta) : B(\xi : \eta : \zeta) : C(\xi : \eta : \zeta)\big),$$

where A, B, C are homogeneous polynomials of the same degree. Now φ is naturally a rational map $\mathbb{P}^2 \to \mathbb{P}^2$. The map is regular at a point P if one of A, B, C does not vanish at P. Studying properties related to points P in the affine set $\mathbf{A}_3^2$, say, we can divide each of A, B, C by ζ^n, where n is their common degree, and write the map in the form $(x, y) \mapsto (u(x, y), v(x, y), w(x, y))$, where u, v and w are polynomials. This map is regular at P if the 3 polynomials do not vanish simultaneously at P.

As a first illustration we prove the following important result.

Theorem 2. *A rational map from a projective plane curve C to $\mathbf{P}^2$ is regular at every nonsingular point of C (see 1.5 for the definition).*

Proof. Suppose that the nonsingular point P is in the affine piece $\mathbf{A}_3^2$ with coordinates denoted by x, y. We write the map as above in the form $(x, y) \mapsto (u_0 : u_1 : u_2)$ where u_0, u_1, u_2 are polynomials, and apply 1.5, Theorem 1 to these. Restricting the u_i to C, we can write them in the form $u_i = t^{k_i} v_i$, where t is a local parameter, $v_i(P) \neq 0$ and $k_i \geq 0$ for $i = 0, 1, 2$. Suppose that k_0, say, is the smallest of the numbers k_0, k_1, k_2. Then the same map can be rewritten in the form $(x, y) \mapsto (v_0 : t^{k_1 - k_0} v_1 : t^{k_2 - k_0} v_2)$, with $k_1 - k_0 \geq 0$, $k_2 - k_0 \geq 0$, and $v_0(P) \neq 0$. It follows that it is regular at P. The theorem is proved.

Corollary. *A birational map between nonsingular projective plane curves is regular at every point, and is a one-to-one correspondence.* □

As an example, consider a birational map of the projective line to itself. Just as with any rational map, this can be written as a rational function $x \mapsto p(x)/q(x)$, with $p(x)$, $q(x) \in k[x]$ (here we assume that x is a coordinate on our line, for example the line given by $y = 0$). The points that map to a given point α are those for which $p(x)/q(x) = \alpha$, that is, $p(x) - \alpha q(x) = 0$. Hence from the fact that the map is birational, it follows that p and q are linear, that is, the map is of the form $x \mapsto (ax + b)/(cx + d)$ with $ad - bc \neq 0$. As a consequence, we get that *a birational map of the line to itself has at most two fixed points*, the roots of the equation $x(cx + d) = ax + b$.

Now consider the elliptic curve given by 1.4, (1), and assume that $4p^3 + 27q^2 \neq 0$. All its finite points are nonsingular. Passing to homogeneous coordinates, we can write its equation in the form $\eta^2 \zeta = \xi^3 + p\xi\zeta^2 + q\zeta^3$. Hence it has a unique point on the line at infinity $\zeta = 0$, namely the point $o = (0 : 1 : 0)$. Dividing through by η^3 we write the equation of the curve in the form $v = u^3 + puv^2 + qv^3$, in coordinates u, v, where $u = \xi/\eta$ and $v = \zeta/\eta$. The point $o = (0, 0)$ in these coordinates is also nonsingular. Hence our curve is nonsingular. The map $(x, y) \mapsto (x, -y)$ is obviously a birational map of the curve to itself. Its fixed points in the finite part of the plane are the points with $y = 0$, $x^3 + px + q = 0$, that is, there are 3 such points. The point o is also a fixed point, since $u = x/y$, $v = 1/y$, and in coordinates u, v, the map is written $(u, v) \mapsto (-u, -v)$. We have constructed on an elliptic curve an automorphism having 4 fixed points. It follows from this that *an elliptic curve is not birational to a line, that is, is not rational.* This shows that the problem of birational classification of curves is not trivial: not all curves are birational to one another.

Passing to projective curves is the final amendment required in the statement of Bézout's theorem. One version of this is as follows:

Theorem. *Let X and Y be projective curves, with X nonsingular and not contained in Y. Then the sum of the multiplicities of intersection of X and Y at all points of $X \cap Y$ equals the product of the degrees of X and Y.*

We will prove this theorem and a series of generalisations in a later section (Chap. III, 2.2, and Chap. IV, 2.1). Here we verify the two simplest cases, when X is a line or a conic.

Let X be a line. By 1.1, Lemma, X and Y have a finite number of points of intersection. We choose a convenient coordinate system, so that the line $\zeta = 0$ does not pass through the points of intersection, and is not equal to X, and $\eta = 0$ is the line X. Then the points of intersection of X and Y are contained in the affine plane with coordinates $x = \xi/\zeta$, $y = \eta/\zeta$, and the equation of X is $y = 0$. Let $f(x, y) = 0$ be the equation of the curve Y and $f = f_0 + f_1(x, y) + \cdots + f_n(x, y)$ its expression as a sum of homogeneous polynomials. The point $(1 : 0 : 0)$ is not contained in Y by the choice of the coordinate system, and hence $f_n(1, 0) \neq 0$, that is, f contains the term ax^n with $a \neq 0$. Hence $f(x, 0)$, the restriction of f to X, has degree n. The function $x - \alpha$ is a local parameter of X at the point $x = \alpha$, and the multiplicity of intersection of X and Y at this point equals the multiplicity of the root $x = \alpha$ of the polynomial $f(x, 0)$. Therefore the sum of these multiplicities equals n.

Let X be a conic. Take any point $P \in X$ with $P \notin Y$, and choose coordinates so that $\zeta = 0$ is the tangent line to X at P, and $\xi = 0$ some other line through P. An easy calculation in coordinates shows that X is a parabola in the affine plane with coordinates $x = \xi/\zeta$, $y = \eta/\zeta$ (since it touches the line at infinity), with equation $y = px^2 + qx + r$ and $p \neq 0$. As before, $f = f_0 + \cdots + f_n(x, y)$, and now $f_n(0, 1) \neq 0$, that is, $f(x, y)$ contains the term ay^n with $a \neq 0$. The conic X has no other points of intersection with the line $\zeta = 0$ except P, and hence all the points of intersection of X and Y are contained in the finite part of the plane. At any point with $x = \alpha$ the function $x - \alpha$ is a local parameter on X, and the multiplicity of intersection of X and Y at this point is equal to the multiplicity of the root $x = \alpha$ of the polynomial $f(x, px^2 + qx + r)$. Since $f(x, y)$ contains the term ay^n with $a \neq 0$, the degree of $f(x, px^2 + qx + r)$ is $2n$, so that the sum of multiplicities of all the points of intersection equals $2n$.

This proves the theorem in the case X is a line or conic.

Already this simple particular case of Bézout's theorem has beautiful geometric applications. One of these is the proof of Pascal's theorem, which asserts that for a hexagon inscribed in a conic, the 3 points of intersection of pairs of opposite sides are collinear. Let l_1 and m_1, l_2 and m_2, l_3 and m_3

be linear forms that are the equations of the opposite sides of a hexagon (see Figure 6). Consider the cubic with the equation $f_\lambda = l_1 l_2 l_3 + \lambda m_1 m_2 m_3$ where λ is an arbitrary parameter. This has six points of intersection with the conic, the vertexes of the hexagon. Moreover, we can choose the value of λ so that $f_\lambda(P) = 0$ for any given point $P \in X$, distinct from these 6 points of intersection. We get a cubic f_λ having 7 points of intersection with a conic X, and by Bézout's theorem this must decompose as the conic X plus a line L. This line L must contain the points of intersection $l_1 \cap m_1$, $l_2 \cap m_2$ and $l_3 \cap m_3$. (This proof is due to Plücker.)

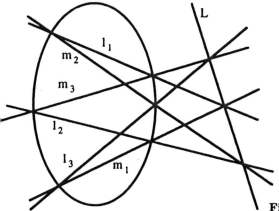

Figure 6. Pascal's Theorem

Exercises to §1

1. Find a characterisation in real terms of the line through the points of intersection of two circles in the case that both these points are complex. Prove that it is the locus of points having the same power with respect to both circles. (The power of a point with respect to a circle is the square of the distance between it and the points of tangency of the tangent lines to the circle.)

2. Which rational functions $p(x)/q(x)$ are regular at the point at infinity of $\mathbb{P}^1$? What order of zero do they have there?

3. Prove that an irreducible cubic curve has at most one singular point, and that the multiplicity of a singular point is 2. If the singularity is a node then the cubic is projectively equivalent to the curve in 1.2, (1); and if a cusp then to the curve $y^2 = x^3$.

4. What is the maximum multiplicity of intersection of two nonsingular conics at a common point?

5. Prove that if the ground field has characteristic p then every line through the origin is a tangent line to the curve $y = x^{p+1}$. Prove that over a field of characteristic 0, there are at most a finite number of lines through a given point tangent to a given irreducible curve.

6. Prove that the sum of multiplicities of two singular points of an irreducible curve of degree n is at most n, and the sum of multiplicities of any 5 points is at most $2n$.

7. Prove that for any two distinct points of an irreducible curve there exists a rational function that is regular at both, and takes the value 0 at one and 1 at the other.

8. Prove that for any nonsingular points $P_1, \ldots, P_r$ of an irreducible curve and numbers $m_1, \ldots, m_r \geq 0$ there exists a rational function that is regular at all these points, and has a zero of multiplicity m_i at P_i.

9. For what values of m is the cubic $x_0^3 + x_1^3 + x_2^3 + m x_0 x_1 x_2 = 0$ in $\mathbb{P}^2$ nonsingular? Find its inflexion points.

10. Find all the automorphisms of the curve of 1.2, (1).

11. Prove that on the projective line and on a conic of $\mathbb{P}^2$, a rational function that is regular at every point is a constant.

12. Give an interpretation of Pascal's theorem in the case that pairs of vertexes of the hexagon coincide, and the lines joining them become tangents.

2. Closed Subsets of Affine Space

Throughout what follows, we work with a fixed algebraically closed field k, which we call the *ground field*.

2.1. Definition of Closed Subsets

At different stages of the development of algebraic geometry, there have been changing views on the basic object of study, that is, on the question of what is the "natural definition" of an algebraic variety; the objects considered to be most basic have been projective or quasiprojective varieties, abstract algebraic varieties, schemes or algebraic spaces.

In this book, we consider algebraic geometry in a gradually increasing degree of generality. The most general notion considered in the first chapters, embracing all the algebraic varieties studied here, is that of quasiprojective variety. In the final chapters this role will be taken by schemes. At present we define a class of algebraic varieties that will play a foundational role in all the subsequent definitions. Since the word variety will be reserved for the more general notions, we use a different word here.

We write $\mathbf{A}^n$ for the n-dimensional affine space over the field k. Thus its points are of the form $\alpha = (\alpha_1, \ldots, \alpha_n)$ with $\alpha_i \in k$.

Definition. A *closed subset* of $\mathbf{A}^n$ is a subset $X \subset \mathbf{A}^n$ consisting of all common zeros of a finite number of polynomials with coefficients in k. We will sometimes say simply *closed set* for brevity.

From now on we will write $F(T)$ to denote a polynomial in n variables, allowing T to stand for the set of variables $T_1, \ldots, T_n$. If a closed set X consists of all common zeros of polynomials $F_1(T), \ldots, F_m(T)$, then we refer to $F_1(T) = \cdots = F_m(T) = 0$ as the *equations* of the set X.

A set X defined by an infinite system of equations $F_\alpha(T) = 0$ is also closed. Indeed, the ideal $\mathfrak{A}$ of the polynomial ring in $T_1, \ldots, T_n$ generated by all the polynomials $F_\alpha(T)$ is finitely generated (the Hilbert Basis Theorem, see Atiyah and Macdonald [7], Theorem 7.5), that is, $\mathfrak{A} = (G_1, \ldots, G_m)$. One checks easily that X is defined by the system of equations $G_1 = \cdots = G_m = 0$.

It follows from this that the intersection of any number of closed sets is closed. Indeed, if X_α are closed sets, then to get a system of equations defining $X = \bigcap X_\alpha$, we need only take the union of the systems defining all the X_α.

The union of a finite number of closed sets is again closed. It is obviously enough to check this for two sets. If $X = X_1 \cup X_2$, where X_1 is defined by the system of equations $F_i(T) = 0$ for $i = 1, \ldots, m$ and X_2 by $G_j(T) = 0$ for $j = 1, \ldots, l$ then it is easy to check that X is defined by the system $F_i(T)G_j(T) = 0$ for $i = 1, \ldots, m$ and $j = 1, \ldots, l$.

Let $X \subset \mathbf{A}^n$ be a closed subset of affine space. We say that a set $U \subset X$ is *open* if its complement $X \setminus U$ is closed. Any open set $U \ni x$ is called a *neighbourhood* of x. The intersection of all the closed subsets of X containing a given subset $M \subset X$ is closed. It is called the *closure* of M and denoted by $\overline{M}$. A subset is *dense* in X if $\overline{M} = X$. This means that M is not contained in any closed subset $Y \subsetneq X$.

Example 1. The whole affine space $\mathbf{A}^n$ is closed, since it is defined by the empty set of equations, or by $0 = 0$.

Example 2. The subset $X \subset \mathbf{A}^1$ consisting of all points except 0 is not closed: every polynomial $F(T)$ that vanishes at all $T \neq 0$ must be identically 0.

Example 3. Let us determine all the closed subsets $X \subset \mathbf{A}^1$. Such a set is given by a system of equations $F_1(T) = \cdots = F_m(T) = 0$ in one variable T. If all the F_i are identically 0 then $X = \mathbf{A}^1$. If the F_i don't have any common factor, then they don't have any common roots, and X does not contain any points. If the highest common factors of all the F_i is $D(T)$ then

$D(T) = (T - \alpha_1) \cdots (T - \alpha_n)$ and X consists of the finitely many points $T = \alpha_1, \ldots, T = \alpha_n$.

Example 4. Let us determine all the closed subsets $X \subset \mathbf{A}^2$. A closed subset is given by a system of equations

$$F_1(T) = \cdots = F_m(T) = 0, \tag{1}$$

where now $T = (T_1, T_2)$. If all the F_i are identically 0 then $X = \mathbf{A}^2$. Suppose this is not the case. If the polynomials $F_1, \ldots, F_m$ do not have a common factor then, as follows from 1.1, Lemma, the system (1) has only a finite set of solutions (possibly empty). Finally, suppose that the highest common factor of all the $F_i(T)$ is $D(T)$. Then $F_i(T) = D(T)G_i(T)$, where now the polynomials $G_i(T)$ do not have a common factor. Obviously then $X = X_1 \cup X_2$ where X_1 is given by $G_1(T) = \cdots = G_m(T) = 0$ and X_2 is given by the single equations $D(T) = 0$. As we have seen, X_1 is a finite set. The closed sets defined in $\mathbf{A}^2$ by one equation are the algebraic plane curves. Thus a closed set $X \subset \mathbf{A}^2$ either consists of a finite set of points (possibly empty), or the union of an algebraic plane curve and a finite set of points, or the whole of $\mathbf{A}^2$.

Example 5. If $\alpha \in \mathbf{A}^r$ is the point with coordinates $(\alpha_1, \ldots, \alpha_r)$ and $\beta \in \mathbf{A}^s$ the point with coordinates $(\beta_1, \ldots, \beta_s)$, we take α, β into the point $(\alpha, \beta) \in \mathbf{A}^{r+s}$ with coordinates $(\alpha_1, \ldots, \alpha_r, \beta_1, \ldots, \beta_s)$. Thus we identify $\mathbf{A}^{r+s}$ as the set of pairs (α, β) with $\alpha \in \mathbf{A}^r$ and $\beta \in \mathbf{A}^s$. Let $X \subset \mathbf{A}^r$ and $Y \subset \mathbf{A}^s$ be closed sets. The set of pairs $(x, y) \in \mathbf{A}^{r+s}$ with $x \in X$ and $y \in Y$ is called the *product* of X and Y, and denoted by $X \times Y$. This is again a closed set. Indeed, if X is given by $F_i(T) = 0$ and Y by $G_j(U) = 0$ then $X \times Y \subset \mathbf{A}^{r+s}$ is defined by $F_i(T) = G_j(U) = 0$.

Example 6. A set $X \subset \mathbf{A}^n$ defined by one equation $F(T_1, \ldots, T_n) = 0$ is called a *hypersurface*.

2.2. Regular Functions on a Closed Subset

Let X be a closed set in the affine space $\mathbf{A}^n$ over the ground field k.

Definition. A function f defined on X with values in k is *regular* if there exists a polynomial $F(T)$ with coefficients in k such that $f(x) = F(x)$ for all $x \in X$.

If f is a given function, the polynomial F is in general not uniquely determined. We can add to F any polynomial entering in the system of equations of X without altering f. The set of all regular functions on a given closed set X forms a ring and an algebra over k; the operations of addition, multiplication and scalar multiplication by elements of k, are defined as in analysis, by performing the operations on the value of the functions at each point $x \in X$. The ring obtained in this way is denoted by $k[X]$ and is called the *coordinate ring* of X.

We write $k[T]$ for the polynomial ring with coefficients in k in variables $T_1, \ldots, T_n$. We can obviously associate with each polynomial $F \in k[T]$ a function $f \in k[X]$, by viewing F as a function on the set of points of X; in this way we get a homomorphism from $k[T]$ to $k[X]$. The kernel of this homomorphism consists of all polynomials $F \in k[T]$ that take the value 0 at every point $x \in X$. This is an ideal of $k[T]$, just as the kernel of any ring homomorphism; it is called the *ideal of the closed set X*, and denoted by $\mathfrak{A}_X$. Obviously

$$k[X] = k[T]/\mathfrak{A}_X.$$

Thus $k[X]$ is determined by the ideal $\mathfrak{A}_X \subset k[T]$.

Example 1. If X is a point then $k[X] = k$.

Example 2. If $X = \mathbf{A}^n$ then $\mathfrak{A}_X = 0$ and $k[X] = k[T]$.

Example 3. Let $X \subset \mathbf{A}^2$ be given by the equation $T_1 T_2 = 1$. Then $k[X] = k[T_1, T_1^{-1}]$, and it consists of all the rational functions in T_1 of the form $G(T_1)/T_1^n$ with $G(T_1)$ a polynomial and $n \geq 0$.

Example 4. We prove that if X and Y are any closed sets then $k[X \times Y] = k[X] \otimes_k k[Y]$. Define a homomorphism $\varphi \colon k[X] \otimes_k k[Y] \to k[X \times Y]$ by the condition

$$\varphi\left(\sum_i f_i \otimes g_i\right)(x, y) = \sum_i f_i(x) g_i(y).$$

The right-hand side is obviously a regular functions on $X \times Y$, and it is clear that φ is onto, since, in the notation of 2.1, Example 5, the functions α_i and β_j are contained in the image of φ, and these generate $k[X \times Y]$. To prove that φ is one-to-one, it is enough to check that if $\{f_i\}$ are linearly independent in $k[X]$ and $\{g_j\}$ in $k[Y]$ then $\{f_i \otimes g_j\}$ are linearly independent in $k[X \times Y]$. Now an equality

$$\sum_{i,j} c_{ij} f_i(x) g_j(y) = 0$$

implies the relation $\sum_j c_{ij} g_j(y) = 0$ for any fixed y, and in turn that $c_{ij} = 0$.

Since $k[X]$ is a homomorphic image of the polynomial ring $k[T]$, it satisfies the Hilbert basis theorem: any ideal of $k[X]$ is finitely generated. It also satisfies the following analogue of the Nullstellensatz (Appendix, §6, Proposition 1): if a function $f \in k[X]$ is zero at every point $x \in X$ at which functions $g_1, \ldots, g_m$ vanish then $f^r \in (g_1, \ldots, g_m)$ for some $r > 0$. Indeed, suppose that f is given by a polynomial $F(T)$, the g_i by polynomials $G_i(T)$, and let $F_j = 0$ for $j = 1, \ldots, l$ be the equations of X. Then $F(T)$ vanishes at all points $\alpha \in \mathbf{A}^n$ at which all the polynomials $G_1, \ldots, G_m, F_1, \ldots, F_l$ vanish; for since $F_j(\alpha) = 0$ it follows that $\alpha \in X$, and then by assumption $F(\alpha) = 0$. Applying the Nullstellensatz in the polynomial ring we deduce that $F^r \in (G_1, \ldots, G_m, F_1, \ldots, F_l)$ for some $r > 0$, and hence $f^r \in (g_1, \ldots, g_m)$ in $k[X]$.

How is the ideal $\mathfrak{A}_X$ of a closed set X related to a system $F_1 = \cdots = F_m = 0$ of defining equations of X? Clearly $F_i \in \mathfrak{A}_X$ by definition of $\mathfrak{A}_X$, and hence $(F_1, \ldots, F_m) \subset \mathfrak{A}_X$; however, it's not always true that $(F_1, \ldots, F_m) = \mathfrak{A}_X$. For example, if $X \subset \mathbf{A}^1$ is defined by the equation T^2 then it consists just of the point $T = 0$, so that $\mathfrak{A}_X$ consists of all polynomials with no constant term. That is, $\mathfrak{A}_X = (T)$, whereas $(F_1, \ldots, F_m) = (T^2)$. We can however always define the same closed set X by a system of equation $G_1 = \cdots = G_l = 0$ in such a way that $\mathfrak{A}_X = (G_1, \ldots, G_l)$. For this it is enough to recall that any ideal of $k[T]$ is finitely generated. Let $G_1, \ldots, G_l$ be a basis of the ideal $\mathfrak{A}_X$, that is, $\mathfrak{A}_X = (G_1, \ldots, G_l)$. Then obviously the equations $G_1 = \cdots = G_l = 0$ define the same set X and have the required property. It is sometimes even convenient to consider a closed set as defined by the infinite system of equations $F = 0$ for all polynomials $F \in \mathfrak{A}_X$. Indeed, if $(F_1, \ldots, F_m) = \mathfrak{A}_X$ then these equations are all consequences of $F_1 = \cdots = F_m = 0$.

Relations between closed subsets are often reflected in their ideals. For example, if X and Y are closed sets in the affine space $\mathbf{A}^n$ then $X \supset Y$ if and only if $\mathfrak{A}_X \subset \mathfrak{A}_Y$. It follows from this that with any closed subset Y contained in X we can associate the ideal $\mathfrak{a}_Y$ of $k[X]$, consisting of the images under the homomorphism $k[T] \to k[X]$ of polynomials $F \in \mathfrak{A}_Y$. Conversely, any ideal $\mathfrak{a}$ of $k[X]$ defines an ideal $\mathfrak{A}$ in $k[T]$, consisting of all inverse images under $k[T] \to k[X]$ of elements of $\mathfrak{a}$. Clearly $\mathfrak{A} \supset \mathfrak{A}_X$. The equations $F = 0$ for all $F \in \mathfrak{A}$ define the closed set $Y \subset X$.

It follows from the Nullstellensatz that Y is the empty set if and only if $\mathfrak{a}_Y = k[X]$. The ideal $\mathfrak{a}_Y \subset k[X]$ can alternatively be described as the set of all functions $f \in k[X]$ that vanish at all points of the subset Y.

In particular, each point $x \in X$ is a closed subset, and hence defines an ideal $\mathfrak{m}_x \subset k[X]$. By definition this ideal is the kernel of the homomorphism $k[X] \to k$ that takes a function $f \in k[X]$ to its value $f(x)$ at x. Since $k[X]/\mathfrak{m}_x = k$ is a field, the ideal $\mathfrak{m}_x$ is maximal. Conversely, every maximal ideal $\mathfrak{m} \subset k[X]$ corresponds in this way to some point $x \in X$. Indeed, it defines a closed subset $Y \subset X$; for any point $y \in Y$ we have $\mathfrak{m}_y \supset \mathfrak{m}$, and then $\mathfrak{m}_y = \mathfrak{m}$ since $\mathfrak{m}$ is maximal. For $u \in k[X]$ the set of points $x \in X$ at which $u(x) = 0$ is closed; it is denoted by $V(u)$, and called a *hypersurface* in X.

2.3. Regular Maps

Let $X \subset \mathbf{A}^n$ and $Y \subset \mathbf{A}^m$ be closed subsets.

Definition. A map $f \colon X \to Y$ is *regular* if there exist m regular functions $f_1, \ldots, f_m$ on X such that $f(x) = (f_1(x), \ldots, f_m(x))$ for all $x \in X$.

Thus any regular map $f \colon X \to \mathbf{A}^m$ is given by m functions $f_1, \ldots, f_m \in k[X]$; in order to know that this maps into the closed subset $Y \subset \mathbf{A}^m$, it is obviously enough to check that $f_1, \ldots, f_m$ as elements of $k[X]$ satisfy the equations of Y, that is

$$G(f_1, \ldots, f_m) = 0 \in k[X] \qquad \text{for all } G \in \mathfrak{A}_Y.$$

Example 1. A regular function on X is exactly the same thing as a regular map $X \to \mathbf{A}^1$.

Example 2. A linear map $\mathbf{A}^n \to \mathbf{A}^m$ is a regular map.

Example 3. The projection map $(x, y) \mapsto x$ defines a regular map of the curve defined by $xy = 1$ to $\mathbf{A}^1$.

Example 4. The preceding example can be generalised as follow: let $X \subset \mathbf{A}^n$ be a closed subset and F a regular function on X. Consider the subset $X' \subset X \times \mathbf{A}^1$ defined by the equation $T_{n+1} F(T_1, \ldots, T_n) = 1$. The projection $\varphi(x_1, \ldots, x_{n+1}) = (x_1, \ldots, x_n)$ defines a regular map $\varphi \colon X' \to X$.

Example 5. The map $f(t) = (t^2, t^3)$ is a regular map of the line $\mathbf{A}^1$ to the curve given by $y^2 = x^3$.

Example 6. The zeta function of a variety over $\mathbf{F}_p$. We give an example that is very important for number theory. Suppose that the coefficients of the equations $F_i(T)$ of a closed subset $X \subset \mathbf{A}^n$ belong to the field $\mathbf{F}_p$ with p elements, where p is a prime number.

As we said in 1.1, the points of X with coordinates in $\mathbf{F}_p$ correspond to solutions of the system of congruences $F_i(T) \equiv 0 \bmod p$. Consider the map $\varphi \colon \mathbf{A}^n \to \mathbf{A}^n$ defined by

$$\varphi(\alpha_1, \ldots, \alpha_n) = (\alpha_1^p, \ldots, \alpha_n^p).$$

This is obviously a regular map. The important thing is that it takes $X \subset \mathbf{A}^n$ to itself. Indeed, if $\alpha \in X$, that is, $F_i(\alpha) = 0$, then since $F_i(T) \in \mathbf{F}_p[T]$, it follows from properties of fields of characteristic p that $F_i(\alpha_1^p, \ldots, \alpha_n^p) = \bigl(F_i(\alpha_1, \ldots, \alpha_n)\bigr)^p = 0$. The map $\varphi \colon X \to X$ obtained in this way is called the *Frobenius map*. Its significance is that the points of X with coordinates in $\mathbb{F}_p$ are characterised among all points of X as the fixed points of φ. Indeed, the solutions of the equation $\alpha_i^p = \alpha_i$ are exactly all the elements of $\mathbb{F}_p$.

In exactly the same way, the elements $\alpha \in \mathbb{F}_{p^r}$ of the field with p^r elements are characterised as the solutions of $\alpha^{p^r} = \alpha$, and hence the points $x \in X$ with coordinates in $\mathbb{F}_{p^r}$ are the fixed points of the map φ^r. For each r, write ν_r for the number of points $x \in X$ with coordinates in $\mathbb{F}_{p^r}$. To get a better overall view of the set of numbers ν_r, we consider the generating function

$$P_X(t) = \sum_{r=1}^{\infty} \nu_r t^r.$$

A deep general theorem asserts that this function is always a rational function of t (for a fairly elementary proof, see Koblitz [47], Chap. V). In this way the function $P_X(t)$ gives an expression in finite terms for the infinite sequence of numbers ν_r.

The function $P_X(t)$ associated with the closed set X has some properties analogous to those of the Riemann zeta function. To express these, note that if $x \in X$ is a point whose coordinates are in $\mathbb{F}_{p^r}$ and generate this field, then X contains all the points $\varphi^i(x)$ for $i = 1, \ldots, r$, and these are all distinct. We call a set $\xi = \{\varphi^i(x)\}$ of this form a *cycle*, and the number r of points of ξ the *degree* of ξ, denoted $\deg \xi$. Now we can group together all the ν_r points $x \in X$ with coordinates in $\mathbb{F}_{p^r}$ into cycles. The coordinates of any of these points generate some subfield $\mathbb{F}_{p^d} \subset \mathbb{F}_{p^r}$, and it is known that $d \mid r$ (see for example van der Waerden [73], 6.7, Ex. 6.23 (§43, Ex. 1)). We get a formula

$$\nu_r = \sum_{d \mid r} d\mu_d,$$

where μ_d is the number of cycles of degree d, hence

$$P_X(t) = \sum_{r=1}^{\infty}\sum_{d \mid r} d\mu_d t^r = \sum_{d=1}^{\infty} d\mu_d \sum_{m=1}^{\infty} t^{md} = \sum_{d=1}^{\infty} \mu_d \frac{dt^d}{1-t^d}. \qquad (1)$$

We introduce the function

$$Z_X(t) = \prod_{\xi} \frac{1}{1 - t^{\deg \xi}}, \qquad (2)$$

where the product runs over all cycles ξ. Then the formula (1) can obviously be rewritten as

$$P_X(t) = \frac{Z'_X(t)}{Z_X(t)} t.$$

(2) is analogous to the Euler product for the Riemann zeta function. To emphasise this analogy we set $p^{\deg \xi} = N(\xi)$ and $t = p^{-s}$. Then (2) takes the form

$$Z_X(t) = \zeta_X(s) = \prod_\xi \frac{1}{1 - N(\xi)^{-s}}.$$

This function (either $Z_X(t)$ or $\zeta_X(s)$) is called the *zeta function* of X.

We now find out how a regular map acts on the ring of regular functions on a closed set. We start with a remark concerning arbitrary maps between sets. If $f: X \to Y$ is a map from a set X to a set Y then we can associate with every function u on Y (taking values in an arbitrary set Z) a function v on X by setting $v(x) = u(f(x))$. Obviously the map $v: X \to Z$ is the composite of $f: X \to Y$ and $u: Y \to Z$. We set $v = f^*(u)$, and call it the *pullback* of u. We get in this way a map f^* from functions on Y to functions on X. Now suppose that $f: X \to Y$ is a regular map; then f^* takes regular functions on Y into regular functions on X. Indeed, if u is given by a polynomial function $G(T_1, \ldots, T_n)$ and f by polynomials $F_1, \ldots, F_m$ then $v = f^*(u)$ is obtained simply by substituting F_i for T_i in G, so that v is given by the polynomial $G(F_1, \ldots, F_m)$. Moreover, regular maps can be characterised as the maps that take regular functions into regular functions. Indeed, suppose that a map $f: X \to Y$ of closed set has the property that for any regular function u on Y the function $f^*(u)$ on X is again regular. Then this applies in particular to the functions t_i defined by the coordinates T_i on Y for $i = 1, \ldots, m$; thus the functions $f^*(t_i)$ are regular on X. But this just means that f is a regular map.

We have seen that if f is regular then the pullback of functions defines a map $f^*: k[Y] \to k[X]$. It follows easily from the definition of f^* that it is a homomorphism of k-algebras. We show that, conversely, every algebra homomorphism $\varphi: k[Y] \to k[X]$ is of the form $\varphi = f^*$ for some regular map $f: X \to Y$. Let $t_1, \ldots, t_m$ be coordinates in the ambient space $\mathbb{A}^m$ of Y, viewed as functions on Y. Obviously $t_i \in k[Y]$, and hence $\varphi(t_i) \in k[X]$. Set $\varphi(t_i) = s_i$ and consider the map f given by the formula $f(x) = (s_1(x), \ldots, s_m(x))$. This is of course a regular map. We prove that $f(X) \subset Y$. Indeed, if $H \in \mathfrak{A}_Y$ then $H(t_1, \ldots, t_m) = 0$ in $k[Y]$, hence also $\varphi(H) = 0$ on X. Let $x \in X$; then $H(f(x)) = \varphi(H)(x) = 0$, and therefore $f(x) \in Y$.

Definition. A regular map $f: X \to Y$ of closed sets is an *isomorphism* if it has an inverse, that is, if there exists a regular map $g: Y \to X$ such that $f \circ g = 1$ and $g \circ f = 1$. In this case we say that X and Y are *isomorphic*.

An isomorphism is obviously a one-to-one correspondence. It follows from what we have said that if f is an isomorphism then $f^*: k[Y] \to k[X]$ is an isomorphism of algebras. It is easy to see that the converse is also true; in other words, closed sets are isomorphic if and only if their rings of regular functions are isomorphic over k.

The facts we have just proved show that $X \mapsto k[X]$ defines an equivalence of categories between closed subsets of affine spaces (with regular maps between them) and a certain subcategory of the category of commutative algebras over k (with algebra homomorphisms). What is this subcategory, that is, which algebras are of the form $k[X]$?

Theorem 1. *An algebra A over a field k is isomorphic to a coordinate ring $k[X]$ of some closed subset X if and only if A has no nilpotents (that is $f^m = 0$ implies that $f = 0$ for $f \in A$) and is finitely generated as an algebra over k.*

Proof. These conditions are all obviously necessary. If an algebra A is generated by finitely many elements $t_1, \ldots, t_n$ then $A \cong k[T_1, \ldots, T_n]/\mathfrak{A}$, where $\mathfrak{A}$ is an ideal of the polynomial ring $k[T_1, \ldots, T_n]$. Suppose that $\mathfrak{A} = (F_1, \ldots, F_m)$, and consider the closed set $X \subset \mathbb{A}^n$ defined by the equations $F_1 = \cdots = F_m = 0$; we prove that $\mathfrak{A}_X = \mathfrak{A}$, from which it will follow that $k[X] \cong k[T_1, \ldots, T_n]/\mathfrak{A} \cong A$.

If $F \in \mathfrak{A}_X$ then $F^r \in \mathfrak{A}$ for some $r > 0$ by the Nullstellensatz. Since A has no nilpotents, also $F \in \mathfrak{A}$. Thus $\mathfrak{A}_X \subset \mathfrak{A}$, and since obviously $\mathfrak{A} \subset \mathfrak{A}_X$, we have $\mathfrak{A}_X = \mathfrak{A}$. The theorem is proved.

Example 7. The generalised parabola, defined by the equation $y = x^k$ is isomorphic to the line, and the maps $f(x, y) = x$ and $g(t) = (t, t^k)$ define an isomorphism.

Example 8. The projection $f(x, y) = x$ of the hyperbola $xy = 1$ to the x-axis is not an isomorphism, since the map is not a one-to-one correspondence: the hyperbola does not contain any point (x, y) for which $f(x, y) = 0$. Compare also Ex. 4.

Example 9. The map $f(t) = (t^2, t^3)$ of the line to the curve defined by $y^2 = x^3$ is easily seen to be a one-to-one correspondence. However, it is not an isomorphism, since the inverse map is of the form $g(x, y) = y/x$, and the function y/x is not regular at the origin. (See Ex. 5.)

Example 10. Let X and $Y \subset \mathbb{A}^n$ be closed sets. Consider $X \times Y \subset \mathbb{A}^{2n}$ as in 2.1, Example 5, and the linear subspace $\Delta \subset \mathbb{A}^{2n}$ defined by equations $t_1 = u_1, \ldots, t_n = u_n$, called the *diagonal*. Consider the map that sends each point $z \in X \cap Y$ to $\varphi(z) = (z, z) \in \mathbb{A}^{2n}$, which is obviously a point of $X \times Y \cap \Delta$. It is easy to check that the map $\varphi \colon X \cap Y \to X \times Y \cap \Delta$ obtained in this way is an isomorphism from $X \cap Y$ to $X \times Y \cap \Delta$. Using this, we can always reduce the study of the intersection of two closed sets to considering the intersection of a different closed set with a linear subspace.

Example 11. Let X be a closed set and G a finite group of automorphisms of X. Suppose that the characteristic of the field k does not divide the order N of G. Set $A = k[X]$, and let A^G be the subalgebra of invariants of G in A, that is, $A^G = \{f \in A \mid g^*(f) = f \text{ for all } g \in G\}$. According to Appendix, §4, Proposition 1, the algebra A^G is finitely generated over k. From Theorem 1 it follows that there exists a closed set Y such that $A^G \cong k[Y]$, and a regular map $\varphi \colon X \to Y$ such that $\varphi^*(k[Y]) = A^G$. This set Y is called the *quotient variety* or *quotient space* of X by the action of G, and is written X/G.

Given two points $x_1, x_2 \in X$, there exists $g \in G$ such that $x_2 = g(x_1)$ if and only if $\varphi(x_1) = \varphi(x_2)$. Indeed, if $x_2 = g(x_1)$ then $f(x_2) = f(x_1)$ for every $f \in k[X]^G = k[Y]$, and hence $\varphi(x_1) = \varphi(x_2)$. Conversely, if $x_2 \neq g(x_1)$ then we must take a function $f \in k[X]$ such that $f(g(x_2)) = 1$, $f(g(x_1)) = 0$ for all $g \in G$. Then the symmetrised function $S(f)$ (see Appendix, §4) is G-invariant and satisfies $S(f)(x_2) = 1$ and $S(f)(x_1) = 0$, and hence $\varphi(x_2) \neq \varphi(x_1)$. Thus X/G parametrises the orbits $\{g(x) \mid g \in G\}$ of G acting on X.

In what follows we will mainly be interested in notions and properties of closed sets invariant under isomorphism. The system of equations defining a set is clearly not a notion of this kind; two sets X and Y can be isomorphic although given by different systems of equations in different spaces $\mathbb{A}^n$. Thus it would be natural to try to give an intrinsic definition of a closed set independent of its realisation in some affine space; a definition of this kind will be given in Chap. V–VI in connection with the notion of a scheme.

Now we determine when a homomorphism $f^* \colon k[Y] \to k[X]$ corresponding to a regular map $f \colon X \to Y$ has no kernel, that is, when f^* is an isomorphic inclusion $k[Y] \hookrightarrow k[X]$. For $u \in k[Y]$, let's see when $f^*(u) = 0$. This means that $u(f(x)) = 0$ for all $x \in X$. In other words, u vanishes at all points of the image $f(X)$ of X. The points $y \in Y$ for which $u(y) = 0$ obviously form a closed set, and hence if this contains $f(X)$, it also contains the closure $\overline{f(X)}$. Repeating the same arguments backwards, we see that $f^*(u) = 0$ if and only if u vanishes on $\overline{f(X)}$, or equivalently, $u \in \mathfrak{a}_{\overline{f(X)}}$. It follows in particular that the kernel of f^* is zero if and only if $\overline{f(X)} = Y$, that is, $f(X)$ is dense in Y.

This is certainly the case if $f(X) = Y$, but cases with $\overline{f(X)} = Y$ but $f(X) \neq Y$ are possible (see Example 3).

In what follows we will be concerned mainly with algebraic varieties in projective space. But closed subsets of affine space have a geometry with a specific flavour, which is often quite nontrivial. As an example we give the following theorem due to Abhyankar and Moh:

Theorem. *A curve $X \subset \mathbb{A}^2$ is isomorphic to $\mathbb{A}^1$ if and only if there exists an automorphism of $\mathbb{A}^2$ that takes X to a line. (Here an automorphism is an isomorphism from $\mathbb{A}^2$ to itself.)* $\square$

The group $\operatorname{Aut} \mathbb{A}^2$ of automorphisms of the plane is an extremely interesting object. Some examples of automorphisms are very simply to construct: the affine linear maps, and maps of the form

$$x' = \alpha x, \qquad y' = \beta y + f(x), \tag{3}$$

where α, $\beta \neq 0$ are constants, and f a polynomial. It is known that the whole group $\operatorname{Aut} \mathbf{A}^2$ is generated by these automorphisms. Moreover, the expression of an element $g \in \operatorname{Aut} \mathbf{A}^2$ as a word in affine linear maps and maps of the form (3) is almost unique: the only relations in $\operatorname{Aut} \mathbf{A}^2$ between maps of these two classes are those expressing the fact that the two classes have a subset in common, namely maps of the form (3) with f a linear polynomial. In the language of abstract group theory, $\operatorname{Aut} \mathbf{A}^2$ is the free product (or amalgamation) of two subgroups, the maps of the form (3) and the affine maps, over their common subgroup (see Kurosh [51], Vol. II, Chap. IX, §35 and Ex. 10).

A famous unsolved problem related to automorphisms of $\mathbf{A}^2$ is the Jacobian conjecture. This asserts that, if the ground field k has characteristic 0, a map given by

$$x' = f(x, y), \quad y' = g(x, y)$$

with f, $g \in k[x, y]$ is an automorphism of $\mathbf{A}^2$ if and only if the Jacobian determinant $\frac{\partial(f,g)}{\partial(x,y)}$ is a nonzero constant. At present this conjecture is proved when the degrees of f and g are not too large (the order of 100). There is a similar conjecture for the n-dimensional affine space $\mathbf{A}^n$.

Exercises to §2

1. The set $X \subset \mathbf{A}^2$ is defined by the equation $f \colon x^2 + y^2 = 1$ and $g \colon x = 1$. Find the ideal $\mathfrak{A}_X$. Is it true that $\mathfrak{A}_X = (f, g)$?

2. Let $X \subset \mathbf{A}^2$ be the algebraic plane curve defined by $y^2 = x^3$. Prove that an element of $k[X]$ can be written uniquely in the form $P(x) + Q(x)y$ with $P(x)$, $Q(x)$ polynomials.

3. Let X be the curve of Ex. 2 and $f(t) = (t^2, t^3)$ the regular map $\mathbf{A}^1 \to X$. Prove that f is not an isomorphism. [Hint: Try to construct the inverse of f as a regular map, using the result of Ex. 2.]

4. Let X be the curve defined by the equation $y^2 = x^2 + x^3$ and $f \colon \mathbf{A}^1 \to X$ the map defined by $f(t) = (t^2 - 1, t(t^2 - 1))$. Prove that the corresponding homomorphism f^* maps $k[X]$ isomorphically to the subring of the polynomial ring $k[t]$ consisting of polynomials $g(t)$ such that $g(1) = g(-1)$. (Assume that char $k \neq 2$.)

5. Prove that the hyperbola defined by $xy = 1$ and the line $\mathbf{A}^1$ are not isomorphic.

6. Consider the regular map $f \colon \mathbf{A}^2 \to \mathbf{A}^2$ defined by $f(x, y) = (x, xy)$. Find the image $f(\mathbf{A}^2)$; is it open in $\mathbf{A}^2$? Is it dense? Is it closed?

7. The same question as in Ex. 6 for the map $f \colon \mathbf{A}^3 \to \mathbf{A}^3$ defined by $f(x, y, z) = (x, xy, xyz)$.

8. An isomorphism $f: X \to X$ of a closed set X to itself is called an *automorphism*. Prove that all automorphisms of the line $\mathbf{A}^1$ are of the form $f(x) = ax + b$ with $a \neq 0$.

9. Prove that the map $f(x, y) = (\alpha x, \beta y + P(x))$ is an automorphism of $\mathbf{A}^2$, where α, $\beta \in k$ are nonzero elements, and $P(x)$ is a polynomial. Prove that maps of this type form a group B.

10. Let A be the group of affine linear transformations of the plane $\mathbf{A}^2$, B the group described in Ex. 9 and $C = A \cap B$. Choose a system of representatives $\overline{A}$ for the cosets of $C \backslash A$ and $\overline{B}$ for the cosets of $C \backslash B$, with the elements of C itself deleted. Prove that for any $a \in C$, $\overline{a}_i \in \overline{A}$ and $\overline{b}_i \in \overline{B}$, the product $a\overline{b}_1\overline{a}_1\overline{b}_2\overline{a}_2\overline{b}_3 \cdots \neq e$. [Hint: Set $s_{2i-1} = a\overline{b}_1 \cdots \overline{b}_i$ and $s_{2i} = a\overline{b}_1 \cdots \overline{a}_i$. Assume that s_j is of the form $(x, y) \mapsto (f_j(x, y), g_j(x, y))$. Verify that $\deg f_{2i} = \deg g_{2i} = \deg f_{2i+1} < \deg g_{2i+1}$, so that the number $\max(\deg f_j, \deg g_j)$ either stays the same, or increases.] Deduce from this that an element $g \in \operatorname{Aut} \mathbf{A}^2$ has a unique expression in the form $a\overline{b}_1\overline{a}_1\overline{b}_2\overline{a}_2\overline{b}_3 \cdots$.

11. Prove that if $f(x_1, \ldots, x_n) = \left(P_1(x_1, \ldots, x_n), \ldots, P_n(x_1, \ldots, x_n)\right)$ is an automorphism of $\mathbf{A}^n$ then the Jacobian $J(f) = \det \left|\frac{\partial P_i}{\partial x_j}\right| \in k$. Prove that $f \mapsto J(f)$ is a homomorphism from the group of automorphisms of $\mathbf{A}^n$ into the multiplicative group of nonzero elements of k.

12. Suppose that X consists of two points. Prove that the coordinate ring $k[X]$ is isomorphic to the direct sum of two copies of k.

13. Let $f: X \to Y$ be a regular map. The subset $\Gamma_f \subset X \times Y$ consisting of all points of the form $(x, f(x))$ is called the *graph* of f. Prove that (a) $\Gamma_f \subset X \times Y$ is a closed subset, and (b) Γ_f is isomorphic to X.

14. The map $p_Y: X \times Y \to Y$ defined by $p_Y(x, y) = y$ is called the *projection* to Y or the *second projection*. Prove that if $Z \subset X$ and $f: X \to Y$ is a regular map then $f(Z) = p_Y((Z \times Y) \cap \Gamma_f)$, where Γ_f is the graph of f and $Z \times Y \subset X \times Y$ is the subset of (z, y) with $z \in Z$.

15. Prove that for any regular map $f: X \to Y$ there exists a regular map $g: X \to X \times Y$ that is an isomorphism of X with a closed subset of $X \times Y$ and such that $f = p_Y \circ g$. In other words, any map is the composite of an embedding and a projection.

16. Prove that if $X = \bigcup U_\alpha$ is any covering of a closed set X by open subsets U_α then there exists a finite number $U_{\alpha_1}, \ldots, U_{\alpha_r}$ of the U_α such that $X = U_{\alpha_1} \cup \cdots \cup U_{\alpha_r}$.

17. Prove that the Frobenius map φ (see 2.3, Example 6) is a one-to-one correspondence. Is it an isomorphism, for example if $X = \mathbf{A}^1$?

18. Find the zeta function $Z_X(t)$ for $X = \mathbf{A}^n$.

19. Determine $Z_X(t)$ for X a nonsingular conic in $\mathbf{A}^2$.

3. Rational Functions

3.1. Irreducible Algebraic Subsets

In 1.1 we introduced the notion of an irreducible algebraic curve in the plane. Here we formulate the analogous notion in general.

Definition. A closed algebraic set X is *reducible* if there exist proper closed subsets $X_1, X_2 \subsetneqq X$ such that $X = X_1 \cup X_2$. Otherwise X is *irreducible*.

Theorem 1. *Any closed set X is a finite union of irreducible closed sets.*

Proof. Suppose that the theorem fails for a set X. Then X is reducible, $X = X_1 \cup X_1'$, and the theorem must fail either for X_1 or for X_1'. If X_1, then it is reducible, and again it is made up of closed sets one of which is reducible. In this way we construct an infinite strictly decreasing chain of closed subsets $X \supsetneqq X_1 \supsetneqq X_2 \cdots$. We prove that there cannot be such a chain. Indeed, the ideals corresponding to the X_i would form an increasing chain $\mathfrak{A}_X \subsetneqq \mathfrak{A}_{X_1} \subsetneqq \mathfrak{A}_{X_2} \subsetneqq \cdots$. But such an infinite strictly increasing chain cannot exist, since every ideal of the polynomial ring has a finite basis, and hence an increasing chain of ideals terminates. The theorem is proved.

If $X = \bigcup X_i$ is an expression of X as a finite union of irreducible closed sets, and if $X_i \subset X_j$ for some $i \neq j$ then we can delete X_i from the expression. Repeating this several times, we arrive at a representation $X = \bigcup X_i$ in which $X_i \not\subset X_j$ for all $i \neq j$. We say that such a representation is *irredundant*, and the X_i are the *irreducible components* of X.

Theorem 2. *The irredundant representation of X as a finite union of irreducible closed sets is unique.*

Proof. Let $X = \bigcup_i X_i = \bigcup_j Y_j$ be two irredundant representations. Then

$$X_i = X_i \cap X = X_i \cap \bigcup_j Y_j = \bigcup_j (X_i \cap Y_j).$$

Since by assumption X_i is irreducible, we have $X_i \cap Y_j = X_i$ for some j, that is, $X_i \subset Y_j$. Repeating the argument with the X_i and Y_j interchanged gives $Y_j \subset X_{i'}$ for some i'. Hence $X_i \subset Y_j \subset X_{i'}$, so that by the irredundancy of the representation, $i = i'$ and $Y_j = X_i$. The theorem is proved.

We now restate the condition that a closed set X is irreducible in terms of its coordinate ring $k[X]$. If $X = X_1 \cup X_2$ is reducible then since $X \supsetneqq X_1$ there exists a polynomial F_1 that is 0 on X_1 but not 0 on X, and a similar polynomial F_2 for X_2. Then the product $F_1 F_2$ is 0 on both X_1 and X_2, hence on X. The corresponding regular functions f_1, $f_2 \in k[X]$ have the property that f_1, $f_2 \neq 0$, but $f_1 f_2 = 0$. In other words, f_1 and f_2 are zerodivisors in $k[X]$. Conversely, suppose that $k[X]$ has zerodivisors f_1, $f_2 \neq 0$, with $f_1 f_2 = 0$. Write X_1, X_2 for the closed subsets of X corresponding to the ideals (f_1) and (f_2) of $k[X]$. In other words, X_i consists of the points $x \in X$ such that $f_i(x) = 0$, for $i = 1$ or 2. Obviously both $X_i \subsetneqq X$, since $f_i \neq 0$ on X, and $X = X_1 \cup X_2$ since $f_1 f_2 = 0$ on X, so that at each point $x \in X$ either $f_1(x) = 0$ or $f_2(x) = 0$. Therefore, a closed set X is irreducible if and only if its coordinate ring $k[X]$ has no zerodivisors. This in turn is equivalent to $\mathfrak{A}_X$ being a prime ideal.

If a closed subset Y is contained in X then obviously so are its irreducible components. In terms of the ring $k[X]$ the irreducibility of a closed subset $Y \subset X$ is reflected in $\mathfrak{a}_Y \subset k[X]$ being a prime ideal.

A hypersurface $X \subset \mathbf{A}^n$ with equation $f = 0$ is irreducible if and only if the polynomial f is irreducible. Thus our terminology is compatible with that used in §1 in the case of plane curves.

Theorem 3. *A product of irreducible closed sets is irreducible.*

Proof. Suppose that X and Y are irreducible, but $X \times Y = Z_1 \cup Z_2$, with $Z_i \subsetneqq X \times Y$ for $i = 1, 2$. For any point $x \in X$, the closed set $x \times Y$, consisting of points (x, y) with $y \in Y$, is isomorphic to Y, and is therefore irreducible. Since

$$x \times Y = \big((x \times Y) \cap Z_1 \big) \cup \big((x \times Y) \cap Z_2 \big),$$

either $x \times Y \subset Z_1$ or $x \times Y \subset Z_2$. Consider the subset $X_1 \subset X$ consisting of points $x \in X$ such that $x \times Y \subset Z_1$; we now prove that X_1 is a closed set. Indeed, for any point $y \in Y$, the set X_y of points $x \in X$ such that $x \times y \in Z_1$ is closed: it is characterised by $(X \times y) \cap Z_1 = X_y \times y$, and the left-hand side is closed as an intersection of closed sets; now $X_1 = \bigcap_{y \in Y} X_y$ is closed. In the same way, the set X_2 consisting of all points $x \in X$ such that $x \times Y \subset Z_2$ is also closed. We see that $X_1 \cup X_2 = X$, and since X is irreducible it follows from this that $X_1 = X$ or $X_2 = X$. In the first case $X \times Y = Z_1$, and in the second $X \times Y = Z_2$. This contradiction proves the theorem.

3.2. Rational Functions

It is known that any ring without zerodivisors can be embedded into a field, its field of fractions.

Definition. If a closed set X is irreducible then the field of fractions of the coordinate ring $k[X]$ is the *function field* or *field of rational functions* of X; it is denoted by $k(X)$.

Recalling the definition of the field of fractions, we can say that the function field $k(X)$ consists of rational functions $F(T)/G(T)$ such that

$G(T) \notin \mathfrak{A}_X$, and $F/G = F_1/G_1$ if $FG_1 - F_1G \in \mathfrak{A}_X$. This means that the field $k(X)$ can be constructed as follows. Consider the subring $\mathcal{O}_X \subset k(T_1, \ldots, T_n)$ of rational functions $f = P/Q$ with $P, Q \in k[T]$ and $Q \notin \mathfrak{A}_X$. The functions f with $P \in \mathfrak{A}_X$ form an ideal M_X and $k(X) = \mathcal{O}_X/M_X$.

In contrast to regular functions, a rational function on a closed set X does not necessarily have well-defined values at every point of X; for example, the function $1/x$ at $x = 0$ or x/y at $(0,0)$. We now find out when this is possible.

Definition. A rational function $\varphi \in k(X)$ is *regular* at $x \in X$ if it can be written in the form $\varphi = f/g$ with $f, g \in k[X]$ and $g(x) \neq 0$. In this case we say that the element $f(x)/g(x) \in k$ is the *value* of φ at x, and denote it by $\varphi(x)$.

Theorem 4. *A rational function φ that is regular at all points of a closed subset X is a regular function on X.*

Proof. Suppose $\varphi \in k(X)$ is regular at every point $x \in X$. This means that for every $x \in X$ there exists $f_x, g_x \in k[X]$ with $g_x(x) \neq 0$ such that $\varphi = f_x/g_x$. Consider the ideal $\mathfrak{a}$ generated by all the functions g_x for $x \in X$. This has a finite basis, so that there are a finite number of points $x_1, \ldots, x_N$ such that $\mathfrak{a} = (g_{x_1}, \ldots, g_{x_N})$. The functions g_{x_i} do not have a common zero $x \in X$, since then all functions in $\mathfrak{a}$ would vanish at x, but $g_x(x) \neq 0$. From the analogue of the Nullstellensatz it follows that $\mathfrak{a} = (1)$, and hence there exist functions $u_1, \ldots, u_N \in k[X]$ such that $\sum_{i=1}^{N} u_i g_{x_i} = 1$. Multiplying both sides of this equality by φ and using the fact that $\varphi = f_{x_i}/g_{x_i}$, we get that $\varphi = \sum_{i=1}^{N} u_i f_{x_i}$, that is, $\varphi \in k[X]$. The theorem is proved.

If φ is a rational function on a closed set X, the set of points at which φ is regular is nonempty and open. The first assertion follows since φ can be written $\varphi = f/g$ with $f, g \in k[X]$ and $g \neq 0$; hence $g(x) \neq 0$ for some $x \in X$, and obviously φ is regular at this point. To prove the second assertion, consider all possible representations $\varphi = f_i/g_i$. For any regular function g_i the set $Y_i \subset X$ of points $x \in X$ for which $g_i(x) = 0$ is obviously closed, and hence $U_i = X \setminus Y_i$ is open. The set U of points at which φ is regular is by definition $U = \bigcup U_i$, and is therefore open. This open set is called the *domain of definition* of φ. For any finite system $\varphi_1, \ldots, \varphi_m$ of rational functions, the set of points $x \in X$ at which they are all regular is again open and nonempty. The first assertion follows since the intersection of a finite number of open sets is open, and the second from the following useful proposition: the intersection of a finite number of nonempty open sets of an irreducible closed set is nonempty. Indeed, let $U_i = X \setminus Y_i$ for $i = 1, \ldots, m$ be such that $\bigcap U_i = \emptyset$. Then $Y_i \neq X$ and $\bigcup Y_i = X$; but the Y_i are closed sets, and this contradicts the irreducibility of X.

Thus for any finite set of rational functions, there is some nonempty open set on which they are all defined and can be compared. This remark is useful because a rational function $\varphi \in k(X)$ is uniquely determined if it is specified on some nonempty open subset $U \subset X$. Indeed, if $\varphi(x) = 0$ for all $x \in U$ and $\varphi \neq 0$ on X then any expression $\varphi = f/g$ with $f, g \in k[X]$ gives a representation of X as a union $X = X_1 \cup X_2$ of two closed sets, where $X_1 = X - U$ and X_2 is defined by $f = 0$. This contradicts the irreducibility of X.

3.3. Rational Maps

Let $X \subset A^n$ be an irreducible closed set. A rational map $\varphi \colon X \to A^m$ is a map given by an arbitrary m-tuple of rational functions $\varphi_1, \ldots, \varphi_m \in k(X)$. Thus a rational map φ is not a map defined on the whole set X to the set A^m, but it clearly defines a map of some nonempty open set $U \subset X$ to A^m. Working with functions and maps that are not defined at all points is an essential difference between algebraic geometry and other branches of geometry, for example, topology.

We now define the notion of rational map $\varphi \colon X \to Y$ to a closed subset $Y \subset A^m$.

Definition. A *rational map* $\varphi \colon X \to Y \subset A^m$ is an m-tuple of rational functions $\varphi_1, \ldots, \varphi_m \in k(X)$ such that, for all points $x \in X$ at which all the φ_i are regular, $\varphi(x) = (\varphi_1(x), \ldots, \varphi_m(x)) \in Y$; we say that φ is *regular* at such a point x, and $\varphi(x) \in Y$ is the *image* of x. The *image* of X under a rational map φ is the set of points

$$\varphi(X) = \{\varphi(x) \mid x \in X \text{ and } \varphi \text{ is regular at } x\}.$$

As we proved at the end of 3.2, there exists a nonempty open set $U \subset X$ on which all the rational functions φ_i are defined, hence also the rational map $\varphi = (\varphi_1, \ldots, \varphi_m)$. Thus we can view rational maps as maps defined on open subsets; but we have to bear in mind that different maps may have different domains of definition. The same of course also applies to rational functions.

To check that rational functions $\varphi_1, \ldots, \varphi_m \in k(X)$ define a rational map $\varphi \colon X \to Y$ we need to check that $\varphi_1, \ldots, \varphi_m$, as elements of $k(X)$, satisfy all the equations of Y. Indeed, if this property holds then for any polynomial $u(T_1, \ldots, T_m) \in \mathfrak{A}_Y$ the function $u(\varphi_1, \ldots, \varphi_m) = 0$ on X. Then at each point x at which all the φ_i are regular, we have $u(\varphi_1(x), \ldots, \varphi_m(x)) = 0$ for all $u \in \mathfrak{A}_Y$, that is, $(\varphi_1(x), \ldots, \varphi_m(x)) \in Y$. Conversely, if $\varphi \colon X \to Y$ is a rational map, then for every $u \in \mathfrak{A}_Y$ the function $u(\varphi_1, \ldots, \varphi_m) \in k(X)$ vanishes on some nonempty open set $U \subset X$, and so is 0 on the whole of X. It follows from this that $u(\varphi_1, \ldots, \varphi_m) = 0$ in $k(X)$.

We now study how rational maps act on rational functions on a closed set. Let $\varphi \colon X \to Y$ be a rational map and assume that $\varphi(X)$ is dense in Y. Consider φ as a map $U \to \varphi(X) \subset Y$, where U is the domain of definition of φ, and construct the map φ^* on functions corresponding to it. For any function $f \in k[Y]$ the function $\varphi^*(f)$ is a rational function on X. Indeed, if $Y \subset \mathbf{A}^m$, and f is given by a polynomial $u(T_1, \ldots, T_m)$, then $\varphi^*(f)$ is given by the rational function $u(\varphi_1, \ldots, \varphi_m)$. Thus we have a map $\varphi^*\colon k[Y] \to k(X)$ which is obviously a ring homomorphism of the ring $k[Y]$ to the field $k(X)$. This homomorphism is even an isomorphic inclusion $k[Y] \hookrightarrow k(X)$. Indeed, if $\varphi^*(u) = 0$ for $u \in k(Y)$ then $u = 0$ on $\varphi(X)$. But if $u \neq 0$ on Y then the equality $u = 0$ defines a closed subset $V(u) \subsetneqq Y$. Then $\varphi(X) \subset V(u)$, but this contradicts the assumption that $\varphi(X)$ is dense in Y. The inclusion $\varphi^*\colon k[Y] \hookrightarrow k(X)$ can be extended in an obvious way to an isomorphic inclusion of the field of fractions $k(Y)$ into $k(X)$. Thus if $\varphi(X)$ is dense in Y, the rational map φ defines an isomorphic inclusion $\varphi^*\colon k(Y) \hookrightarrow k(X)$.

Given two rational maps $\varphi\colon X \to Y$ and $\psi\colon Y \to Z$ such that $\varphi(X)$ is dense in Y then it is easy to see that we can define a composite $\psi \circ \varphi\colon X \to Z$; if in addition $\psi(Y)$ is dense in Z then so is $(\psi \circ \varphi)(X)$. Then the inclusions of fields satisfy the relation $(\psi \circ \varphi)^* = \varphi^* \circ \psi^*$.

Definition. A rational map $\varphi\colon X \to Y$ is *birational* or is a *birational equivalence* if φ has an inverse rational map $\psi\colon Y \to X$, that is, $\varphi(X)$ is dense in Y and $\psi(Y)$ in X, and $\psi \circ \varphi = 1$, $\varphi \circ \psi = 1$ (where defined). In this case we say that X and Y are *birational* or *birationally equivalent*.

Obviously if $\varphi\colon X \to Y$ is a birational map then the inclusion of fields $\varphi^*\colon k(Y) \to k(X)$ is an isomorphism. It is easy to see that the converse is also true (for algebraic plane curves this was done in 1.4). Thus closed sets X and Y are birational if and only if the fields $k(X)$ and $k(Y)$ are isomorphic over k.

Examples. In §1 we treated a series of examples of birational maps between algebraic plane curves. Isomorphic closed sets are obviously birational. The regular maps in 2.3, Examples 8–9, although not isomorphisms, are birational maps.

A closed set that is birational to an affine space $\mathbf{A}^n$ is said to be *rational*. Rational algebraic curves were discussed in §1. We now give some other examples of rational closed sets.

Example 1. An irreducible quadric $X \subset \mathbf{A}^n$ defined by a quadratic equation $F(T_1, \ldots, T_n) = 0$ is rational. The proof given in 1.2 for the case $n = 2$ works in general. The corresponding map can once again be interpreted as the projection of X from some point $x \in X$ to a hyperplane $L \subset \mathbf{A}^n$ not passing through x (*stereographic projection*). We need only choose x so that it is not a vertex of X, that is, so that $\partial F/\partial T_i(x) \neq 0$ for at least one value of $i = 1, \ldots, n$.

Example 2. Consider the hypersurface $X \subset \mathbf{A}^3$ defined by the 3rd degree equation $x^3 + y^3 + z^3 = 1$. We suppose that the characteristic of the ground field k is different from 3. The surface X contains several lines, for example the two skew lines L_1 and L_2 defined by

$$L_1: x + y = 0, z = 1, \quad \text{and} \quad L_2: x + \varepsilon y = 0, z = \varepsilon,$$

where $\varepsilon \neq 1$ is a cube root of 1.

We give a geometric description of a rational map of X to the plane, and leave the reader to write out the formulas, and also to check that it is birational. Choose some plane $E \subset \mathbf{A}^3$ not containing L_1 or L_2. For $x \in X \setminus (L_1 \cup L_2)$, it is easy to verify that there is a unique line L passing through x and intersecting L_1 and L_2. Write $f(x)$ for the point of intersection $L \cap E$; then $x \mapsto f(x)$ is the required rational map $X \to E$.

This argument obviously applies to any cubic surface in $\mathbf{A}^3$ containing two skew lines.

In algebraic geometry we work with two different equivalence relations between closed sets, isomorphism and birational equivalence. Birational equivalence is clearly a coarser equivalence relation than isomorphism; in other words, two closed sets can be birational without being isomorphic. Thus it often turns out that the classification of closed sets up to birational equivalence is simpler and more transparent than the classification up to isomorphism. Since it is defined at every point, isomorphism is closer to geometric notions such as homeomorphism and diffeomorphism, and so more convenient. Understanding the relation between these two equivalence relations is an important problem; the question is to understand how much coarser birational equivalence is compared to isomorphism, or in other words, how many closed sets are distinct from the point of view of isomorphism but the same from that of birational equivalence. This problem will reappear frequently later in this book.

We conclude this section by proving one result that illustrates the notion of birational equivalence.

Theorem 5. *Any irreducible closed set X is birational to a hypersurface of some affine space $\mathbf{A}^m$.*

Proof. $k(X)$ is generated over k by a finite number of elements, for example the coordinates $t_1, \ldots, t_n$ in $\mathbf{A}^n$, viewed as functions on X.

Suppose that d is the maximal number of the t_i that are algebraically independent over k. According to Appendix, §5, Proposition 1, the field $k(X)$ can be written in the form $k(z_1, \ldots, z_{d+1})$, where $z_1, \ldots, z_d$ are algebraically independent over k and

$$f(z_1, \ldots, z_{d+1}) = 0, \tag{1}$$

with f irreducible over k and $f'_{T_{d+1}} \neq 0$. The function field $k(Y)$ of the closed set Y defined by equation (1) is obviously isomorphic to $k(X)$. This means that X and Y are birational. The theorem is proved.

Remark 1. According to Appendix, §5, Proposition 1, the element z_{d+1} is separable over the field $k(z_1, \ldots, z_d)$. Hence the $k(z_1, \ldots, z_d) \subset k(X)$ is a finite separable field extension.

Remark 2. It follows from the proof of Appendix, §5, Proposition 1 and the primitive element theorem of Galois theory that $z_1, \ldots, z_{d+1}$ can be chosen as linear combinations of the original coordinates $x_1, \ldots, x_n$, that is, of the form $z_i = \sum_{j=1}^{n} c_{ij} x_j$ for $i = 1, \ldots, d+1$. The map $(x_1, \ldots, x_n) \mapsto (z_1, \ldots, z_{d+1})$ given by these formulas is a projection of the space $\mathbb{A}^n$ parallel to the linear subspace defined by $\sum_{j=1}^{n} c_{ij} x_j = 0$ for $i = 1, \ldots, d+1$. This shows the geometric meaning of the birational map whose existence is established in Theorem 5.

Exercises to §3

1. Suppose that k is a field of characteristic $\neq 2$. Decompose into irreducible components the closed set $X \subset \mathbb{A}^3$ defined by $x^2 + y^2 + z^2 = 0$, $x^2 - y^2 - z^2 + 1 = 0$.

2. Prove that if X is the closed set of §2, Ex. 4 then the elements of the field $k(X)$ can be expressed in a unique way in the form $u(x) + v(x)y$ where $u(x)$ and $v(x)$ are arbitrary rational functions of x.

3. Prove that the maps f of §2, Ex. 3, 4 and 6 are birational.

4. Decompose into irreducible components the closed set $X \subset \mathbb{A}^3$ defined by $y^2 = xz$, $z^2 = y^3$. Prove that all its components are birational to $\mathbb{A}^1$.

5. Let $X \subset \mathbb{A}^n$ be the hypersurface defined by an equation $f_{n-1}(T_1, \ldots, T_n) + f_n(T_1, \ldots, T_n) = 0$, where f_{n-1} and f_n are homogeneous polynomials of degrees $n-1$ and n. (A hypersurface of this form is called a *monoid*.) Prove that if X is irreducible then it is birational to $\mathbb{A}^{n-1}$. (Compare the case of plane curves treated in 1.4.)

6. At what points of the circle given by $x^2 + y^2 = 1$ is the rational function $(1-y)/x$ regular?

7. At which points of the curve X defined by $y^2 = x^2 + x^3$ is the rational function $t = y/x$ regular? Prove that $y/x \notin k[X]$.

4. Quasiprojective Varieties

4.1. Closed Subsets of Projective Space

Let V be a vector space of dimension $n + 1$ over the field k. The set of lines (that is, 1-dimensional vector subspaces) of V is called the n-*dimensional projective space*, and denoted by $\mathbb{P}(V)$ or $\mathbb{P}^n$. If we introduce coordinates $\xi_0, \ldots, \xi_n$ in V then a point $\xi \in \mathbb{P}^n$ is given by $n + 1$ elements $(\xi_0 : \cdots : \xi_n)$ of the field k, not all equal to 0; and two points $(\xi_0 : \cdots : \xi_n)$ and $(\eta_0 : \cdots : \eta_n)$ are considered to be equal in $\mathbb{P}^n$ if and only if there exists $\lambda \neq 0$ such that $\eta_i = \lambda \xi_i$ for $i = 0, \ldots, n$. Any set $(\xi_0 : \cdots : \xi_n)$ defining the point ξ is called a set of *homogeneous coordinates* for ξ (compare 1.6).

We say that a polynomial $f(S) \in k[S_0, \ldots, S_n]$ vanishes at $\xi \in \mathbb{P}^n$ if $f(\xi_0, \ldots, \xi_n) = 0$ for any choice of the coordinates $(\xi_0, \ldots, \xi_n)$ of ξ. Obviously, then also $f(\lambda \xi_0, \ldots, \lambda \xi_n) = 0$ for all $\lambda \in k$ with $\lambda \neq 0$. Write f in the form $f = f_0 + f_1 + \cdots + f_r$, where f_i is the sum of all terms of degree i in f. Then

$$f(\lambda \xi_0, \ldots, \lambda \xi_n) = f_0(\xi_0, \ldots, \xi_n)$$
$$+ \lambda f_1(\xi_0, \ldots, \xi_n) + \cdots + \lambda^r f_r(\xi_0, \ldots, \xi_n).$$

Since k is an infinite field, the equality $f(\lambda \xi_0, \ldots, \lambda \xi_n) = 0$ for all $\lambda \in k$ with $\lambda \neq 0$ implies that $f_i(\lambda \xi_0, \ldots, \lambda \xi_n) = 0$. Thus if f vanishes at a point ξ then all of its homogeneous components f_i also vanish at ξ.

Definition. $X \subset \mathbb{P}^n$ is a *closed subset* if it consists of all points at which a finite number of polynomials with coefficients in k vanish. A closed subset defined by one homogeneous equation $F = 0$ is call a *hypersurface*, as in the affine case. The degree of the polynomial is the *degree* of the hypersurface. A hypersurface of degree 2 is called a *quadric*.

The set of all polynomials $f \in k[S_0, \ldots, S_n]$ that vanish at all points $x \in X$ forms an ideal of $k[S]$, called the ideal of the closed set X, and denoted by $\mathfrak{A}_X$. By what we said above, the ideal $\mathfrak{A}_X$ has the property that whenever it contains an element f it also contains all the homogeneous components of f. An ideal with this property is said to be *homogeneous* or *graded*. Thus the ideal of a closed set X of projective space is homogeneous. It follows from this that it has a basis consisting of homogeneous polynomials: we need only start from any basis and take the system of homogeneous components of polynomials of the basis. In particular, any closed set can be defined by a system of homogeneous equations.

Thus to each closed subset $X \subset \mathbb{P}^n$ there is a corresponding homogeneous ideal $\mathfrak{A}_X \subset k[S_0, \ldots, S_n]$. Conversely, any homogeneous ideal $\mathfrak{A} \subset k[S]$ defines a closed subset $X \subset \mathbb{P}^n$. That is, if $F_1, \ldots, F_m$ is a homogeneous basis of $\mathfrak{A}$ then X is defined by the system of equation $F_1 = \cdots = F_m = 0$. If this system of equations has no other solutions in the vector space V other than 0 then it is natural to take X to be the empty set.

Examples of closed subsets of projective space

Example 1. The Grassmannian. The projective space $\mathbb{P}(V)$ parametrises the 1-dimensional vector subspaces $L^1 \subset V$ of a vector space V. The *Grassmannian* or *Grassmann variety* $\mathrm{Grass}(r, V)$ plays the same role for r-dimensional vector subspaces $L^r \subset V$. To define this, consider the rth exterior power $\bigwedge^r V$ of V, and send a basis $f_1, \ldots, f_r$ of a vector subspace L into the element $f_1 \wedge \cdots \wedge f_r \in \bigwedge^r V$. On passing to another basis of the same vector subspace this element is multiplied by a nonzero element $\alpha \in k$, the determinant of the matrix of the coordinate change, and hence the corresponding point of the projective space $\mathbb{P}(\bigwedge^r V)$ is uniquely determined by the subspace L. Write $P(L)$ for this point. It is easy to see that it determines the subspace L uniquely. If $\{e_i\}$ is a basis of V then $\{e_{i_1} \wedge \cdots \wedge e_{i_r}\}$ is a basis of $\bigwedge^r V$ and $P(L) = \sum_{i_1 < \cdots < i_r} p_{i_1 \ldots i_r}(e_{i_1} \wedge \cdots \wedge e_{i_r})$. The homogeneous coordinates $p_{i_1 \ldots i_r}$ of $P(L)$ are called the *Plücker coordinates* of L.

Except for the trivial cases of subspaces having dimension or codimension 1, not every point $P \in \mathbb{P}(\bigwedge^r V)$ is of the form $P(L)$, or in other words, not every element $x \in \bigwedge^r V$ is of the form $f_1 \wedge \cdots \wedge f_r$ with $f_i \in V$. Necessary and sufficient conditions for this use the notion of *convolution*. Let $u \in V^*$ be a vector of the dual vector space. For $x \in \bigwedge^1 V = V$ the convolution $u \lrcorner x$ is an element of k, and is just the scalar product (u, x) or the value $u(x)$. For $x \in \bigwedge^0 V = k$ we set $u \lrcorner x = 0$. For any $x \in \bigwedge^r V$ the convolution $u \lrcorner x = 0$ can be extended in a unique way from $x \in \bigwedge^1 V$ if we require the property

$$u \lrcorner (x \wedge y) = (u \lrcorner x) \wedge y + (-1)^a (x \wedge (u \lrcorner y)) \qquad \text{for } x \in \bigwedge^a V. \tag{1}$$

Here $u \lrcorner \bigwedge^r V \subset \bigwedge^{r-1} V$. The element $u \lrcorner x$ for $u \in V^*$ and $x \in \bigwedge^r V$ is called the *convolution* of u and x. Finally, for $u_1, \ldots, u_s \in V^*$ the element $u_1 \lrcorner (u_2 \lrcorner \cdots \lrcorner (u_s \lrcorner x) \cdots)$ depends only on x and $y = u_1 \wedge \cdots \wedge u_s \in \bigwedge^s V^*$, and is denoted by $y \lrcorner x$. Here $y \lrcorner x \in \bigwedge^{r-s} V$ if $r \geq s$ and $y \lrcorner x = 0$ if $r < s$.

Necessary and sufficient conditions for $x \in \bigwedge^r V$ to be of the form $x = f_1 \wedge \cdots \wedge f_r$ are given by

$$(y \lrcorner x) \wedge x = 0 \qquad \text{for all } y \in \bigwedge^{r-1} V^*. \tag{2}$$

It is obviously enough to check the conditions (2) for $y = u_{i_1} \wedge \cdots \wedge u_{i_{r-1}}$, where $\{u_i\}$ is a basis of V^*; in particular, if we take $\{u_i\}$ to be the basis dual to the basis $\{e_i\}$ of V then (2) can be written in coordinates. They take the form

$$\sum_{t=1}^{r+1} (-1)^t p_{i_1 \ldots i_{r-1} j_t} p_{j_1 \ldots \widehat{j_t} \ldots j_{r+1}} = 0 \tag{3}$$

for all sequences $i_1, \ldots, i_{r-1}$ and $j_1, \ldots, j_{r+1}$.

The variety defined in $\mathbb{P}(\bigwedge^r V)$ by the relations (2) or (3) is called the *Grassmannian*, and denoted by $\mathrm{Grass}(r, V)$ or $\mathrm{Grass}(r, n)$ where $n = \dim V$.

We need a method of reconstructing a vector subspace L explicitly from its Plücker coordinates $p_{i_1 \ldots i_r}$ satisfying (3). Suppose for example that $p_{1 \ldots r} \neq 0$. If $p = (p_{i_1 \ldots i_r}) = P(L)$ then L has a basis of the form

$$f_i = e_i + \sum_{k > r} a_{ik} e_k \qquad \text{for } i = 1, \ldots, r.$$

It follows easily from this that $p_{1 \ldots \hat{k} \ldots rk} = (-1)^k a_{ik}$, from which we get $a_{ik} = (-1)^k p_{1 \ldots \hat{k} \ldots rk}$, where we have set $p_{1 \ldots r} = 1$ for convenience.

Thus the open affine sets $p_{i_1 \ldots i_r} \neq 0$ of $\mathrm{Grass}(r, V)$ are all isomorphic to the affine space $\mathbb{A}^{r(n-r)}$ with coordinates a_{ik} (for $i = 1, \ldots, r$ and $k = r+1, \ldots, n$). We can see, for example, that in the open set $p_{1 \ldots r} \neq 0$ the equations (3) can be solved explicitly with the coordinates $p_{1 \ldots r} \neq 0$ and $p_{1 \ldots \hat{k} \ldots rk}$ as free parameters. That is, if $m \geq 2$ of the subscripts $i_1, \ldots, i_r$ are $> r$ then

$$p_{i_1 \ldots i_r} = \frac{F(\ldots, p_{1 \ldots \hat{k} \ldots rk}, \ldots)}{(p_{1 \ldots r})^m},$$

where F is a form of degree m in $p_{1 \ldots r} \neq 0$ and $p_{1 \ldots \hat{i} \ldots rk}$ with $i \leq r$ and $k > r$. A detailed treatment of Grassmannians is contained, for example, in the survey article Kleiman and Laksov [45].

The first nontrivial case of this theory is when $r = 2$. Then by (1)

$$(u \lrcorner x) \wedge x = \frac{1}{2}(u \lrcorner (x \wedge x)) \qquad \text{for } u \in V^* \text{ and } x \in \bigwedge^2 V.$$

Hence (2) reduces to $u \lrcorner (x \wedge x) = 0$ for all $u \in V^*$, that is, simply

$$x \wedge x = 0. \tag{4}$$

Finally, when $n = 4$ we have $\dim \bigwedge^4 V = 1$, so that (4) reduces to a single equation in the Plücker coordinates $p_{12}, p_{13}, p_{14}, p_{23}, p_{24}, p_{34}$:

$$p_{12} p_{34} - p_{13} p_{24} + p_{14} p_{23} = 0. \tag{5}$$

Planes $L \subset V$ in a 4-dimensional vector space V correspond to lines $\ell \subset \mathbb{P}(V)$ in projective 3-space. In this case, coordinates in V are denoted by x_0, x_1, x_2, x_3 and the Plücker coordinates $p_{01}, p_{02}, p_{03}, p_{12}, p_{13}, p_{23}$, and (5) takes the form

$$p_{01} p_{23} - p_{02} p_{13} + p_{03} p_{12} = 0. \tag{6}$$

This is a quadric in projective 5-space $\mathbb{P}(\bigwedge^2 V)$.

Example 2. The variety of associative algebras. Let A be an associative algebra over a field k of rank n. Then after a choice of basis, A is determined by its multiplication table

$$e_i e_j = \sum c_{ij}^l e_l$$

with structure constants $c_{ij}^l \in k$. The associative condition for multiplication in A takes the form

$$\sum_l c_{ij}^l c_{lk}^m = \sum_l c_{il}^m c_{jk}^l \qquad \text{for } i, j, k, m = 1, \ldots, n. \qquad (7)$$

this is again a system of quadratic equation in the structure constants c_{ij}^l. Multiplying all the basis elements e_i by a nonzero element $\alpha^{-1} \in k$ has the effect of multiplying all the c_{ij}^l by α. Thus if we discard the algebra with zero multiplication, all algebras are described by points of the closed set in the projective space $\mathbb{P}^{n^3-1}$ defined by the equations (7).

To be more precise, points of this set correspond to associative multiplication laws written out in terms of a chosen basis $e_1, \ldots, e_n$. The change to a different basis is given by a nondegenerate $n \times n$ matrix. Thus the set of associative algebras of rank n over a field k, up to isomorphism, is parametrised by the quotient of the set defined by (7) by the group of nondegenerate $n \times n$ matrixes. The extent to which this type of quotient can be identified with an algebraic variety is an extremely delicate question.

Example 3. Determinantal varieties. Quadratic forms in n variables form a vector space V of dimension $\binom{n+1}{2} = (1/2)n(n+1)$. Quadrics in an $(n-1)$-dimensional projective space are parametrised by points of the projective space $\mathbb{P}(V)$. Among these, the degenerate quadrics are characterised by $\det(f) = 0$, where f is the corresponding quadratic form. This is a hypersurface $X_1 \subset \mathbb{P}(V)$. The quadrics of rank $\leq n - k$ correspond to points of a set X_k defined by setting all $(n - k + 1) \times (n - k + 1)$ minors of the matrix of f to 0. A set of this type is called a *determinantal variety*. Another type of determinantal variety M_k is defined in the space $\mathbb{P}(V)$, where V is the space of $n \times m$ matrixes, by the condition that a matrix has rank $\leq k$.

In the case of closed subsets of affine space, an ideal $\mathfrak{A} \subset k[T]$ defines the empty set only if $\mathfrak{A} = (1)$; this is the assertion of the Nullstellensatz. For closed subsets of projective spaces this is not the case: for example, the ideal $(S_0, \ldots, S_n)$ also defines the empty set. Write I_s for the ideal of $k[S]$ consisting of polynomials having only terms of degree $\geq s$. Obviously I_s also defines the empty set: it contains, for example, the monomials S_i^s for $i = 0, \ldots, n$, which have a common zero only at the origin.

Lemma. *A homogeneous ideal $\mathfrak{A} \subset k[S]$ defines the empty set if and only if it contains the ideal I_s for some $s > 0$.*

Proof. We have already seen that the ideal I_s defines the empty set, and the same holds a fortiori for any ideal containing I_s. Suppose that a homogeneous ideal $\mathfrak{A} \subset k[S]$ defines the empty set. Let $F_1, \ldots, F_r$ be a homogeneous basis of the ideal $\mathfrak{A}$ and set $\deg F_i = m_i$. Then from the assumption, it follows that the polynomials $F_i(1, T_1, \ldots, T_n)$ have no common root, where $T_j = S_j/S_0$. Indeed, a common root $(\alpha_1, \ldots, \alpha_n)$ would give a common root $(1, \alpha_1, \ldots, \alpha_n)$ of $F_1, \ldots, F_r$. By the Nullstellensatz there must exist polynomials $G_i(T_1, \ldots, T_n)$ such that $\sum_i F_i(1, T_1, \ldots, T_n) G_i(T_1, \ldots, T_n) = 1$. Setting $T_j = S_j/S_0$ in this equality and multiplying through by a common denominator of the form $S_0^{l_0}$ we get $S_0^{l_0} \in \mathfrak{A}$. In the same way, for each $i = 1, \ldots, n$ there exists a number $l_i > 0$ such that $S_i^{l_i} \in \mathfrak{A}$. If now $l = \max(l_0, \ldots, l_n)$ and $s = (l-1)(n+1) + 1$ then in any term $S_0^{a_0} \cdots S_n^{a_n}$ of degree $a_0 + \cdots + a_n \geq s$ we must have at least one term S_i with exponent $a_i \geq l \geq l_i$, and since $S_i^{l_i} \in \mathfrak{A}$, this term is contained in $\mathfrak{A}$. This proves that $I_s \subset \mathfrak{A}$. The lemma is proved.

From now on we consider closed subsets of affine and projective spaces at one and the same time. We again call these affine and projective closed sets. For projective closed sets, we use the same terminology as for affine sets; that is, if $Y \subset X$ are two closed sets then we say that $X \setminus Y$ is an *open set* in X. As before, a union of an arbitrary number of open sets, and an intersection of finitely many open sets is again open. The set $\mathbb{A}_0^n \subset \mathbb{P}^n$ of points $\xi = (\xi_0 : \cdots : \xi_n)$ for which $\xi_0 \neq 0$ is obviously open. Its points can be put in one-to-one correspondence with the points of an n-dimensional affine space by setting $\alpha_i = \xi_i/\xi_0$ for $i = 1, \ldots, n$, and sending $\xi \in \mathbb{A}_0^n$ to $(\alpha_1, \ldots, \alpha_n) \in \mathbb{A}^n$. Thus we call the set $\mathbb{A}_0^n$ an *affine piece* of $\mathbb{P}^n$. In the same way, for $i = 0, \ldots, n$, the set $\mathbb{A}_i^n$ consists of points for which $\xi_i \neq 0$. Obviously $\mathbb{P}^n = \bigcup_i \mathbb{A}_i^n$.

For any projective closed set $X \subset \mathbb{P}^n$, and any $i = 0, \ldots, n$, the set $U_i = X \cap \mathbb{A}_i^n$ is open in X. It is closed as a subset of $\mathbb{A}_i^n$. Indeed, if X is given by a system of homogeneous equations $F_0 = \cdots = F_m = 0$ and $\deg F_j = n_j$ then, for example, U_0 is given by the system

$$S_0^{-n_j} F_j = F_j(1, T_1, \ldots, T_n) = 0 \qquad \text{for } j = 1, \ldots, m,$$

where $T_i = S_i/S_0$ for $i = 1, \ldots, n$. We call U_i the *affine pieces* of X; obviously $X = \bigcup U_i$. A closed subset $U \subset \mathbb{A}_0^n$ defines a closed projective set $\overline{U}$ called its *projective completion*; $\overline{U}$ is the intersection of all projective closed sets containing U. It is easy to check that the homogeneous equations of $\overline{U}$ are obtained by a process inverse to that just described. If $F(T_1, \ldots, T_n)$ is any polynomial in the ideal $\mathfrak{A}$ of U of degree $\deg F = k$, then the equations of $\overline{U}$ are of the form $S_0^k F(S_1/S_0, \ldots, S_n/S_0)$. It follows from this that

$$U = \overline{U} \cap \mathbb{A}_0^n. \tag{8}$$

Up to now we have considered two classes of objects that could claim to be called algebraic varieties; affine and projective closed sets. It is natural to try to introduce a unified notion of which both of these types will be particular cases. This will be done most systematically in Chap. V–VI in connection with the notion of scheme. For the moment we introduce a more particular notion, that unifies projective and affine closed sets.

Definition. A *quasiprojective variety* is an open subset of a closed projective set.

A closed projective set is obviously a quasiprojective variety. For affine closed sets this follows from (8). A *closed subset* of a quasiprojective variety is its intersection with a closed set of projective space. Open set and neighbourhood of a point are defined similarly. The notion of irreducible variety and the theorem on decomposing a variety as a union of irreducible components carries over word-for-word from the case of affine sets.

From now on we use *subvariety* Y of a quasiprojective variety $X \subset \mathbb{P}^n$ to mean any subset $Y \subset X$ which is itself a quasiprojective variety in $\mathbb{P}^n$. This is obviously equivalent to saying that $Y = Z \setminus Z_1$ with Z and $Z_1 \subset X$ closed subsets.

4.2. Regular Functions

We proceed to considering functions on quasiprojective varieties, and start with the projective space $\mathbb{P}^n$ itself. Here we meet an important distinction between functions of homogeneous and inhomogeneous coordinates: a rational function of the homogeneous coordinates

$$f(S_0, \ldots, S_n) = \frac{P(S_0, \ldots, S_n)}{Q(S_0, \ldots, S_n)} \tag{1}$$

cannot be viewed as a function of $x \in \mathbb{P}^n$, even when $Q(x) \neq 0$, since the value $f(\alpha_0, \ldots, \alpha_n)$ in general changes when all the α_i are multiplied through by a common factor. However, when f is a homogeneous function of degree 0, that is, when P and Q are homogeneous of the same degree, then f can be viewed as a function of $x \in \mathbb{P}^n$.

If $X \subset \mathbb{P}^n$ is a quasiprojective variety, $x \in X$ and $f = P/Q$ is a homogeneous function of degree 0 with $Q(x) \neq 0$, then f defines a function on a neighbourhood of x in X with values in k. We say that f is *regular* in a neighbourhood of x, or simply at x. A function on X that is regular at all points $x \in X$ is a *regular function* on X. All regular functions on X form a ring, that we denote by $k[X]$.

Let's prove that for a closed subset X of an affine space, our definition of regular function here is the same as that in 2.2 (after an obvious passage to inhomogeneous coordinates). For X irreducible, this is stated in 3.2, Theorem 4. In the general case we only need to change slightly the arguments used to prove this theorem. In this proof we let f be a regular function in the affine sense of 2.2.

By assumption, each point $x \in X$ has a neighbourhood U_x with $q_x \neq 0$ on U_x in which $f = p_x/q_x$, where p_x, q_x are regular functions on X and $q_x \neq 0$ on U_x. Hence

$$q_x f = p_x \tag{2}$$

on U_x. But we can assume that (2) holds over the whole of X. To achieve this, we multiply both p_x and q_x by a regular function equal to 0 on $X \setminus U_x$ and nonzero at x; then (2) holds also on $X \setminus U_x$, since both sides are 0 there. As in the proof of 3.2, Theorem 4, we can find points $x_1, \ldots, x_N \in X$ and regular functions $h_1, \ldots, h_N$ such that $\sum_{i=1}^{N} q_{x_i} h_i = 1$. Multiply (2) for $x = x_i$ by h_i and add, to get

$$f = \sum_{i=1}^{N} p_{x_i} h_i,$$

that is, f is a regular function.

In contrast to the case of closed affine sets, the ring $k[X]$ may consist only of constants. We will prove later (5.2, Theorem 3, Corollary 1) that this is always the case if X is an irreducible closed projective set. This is easy to prove directly if $X = \mathbb{P}^n$: indeed, if $f = P/Q$, with P and Q forms of the same degree, we can assume that P and Q have no common factors; then f is not regular at points x where $Q(x) = 0$. On the other hand, when X is only quasiprojective, $k[X]$ may turn out to be an unexpectedly large ring. If X is an affine closed set then as we have seen $k[X]$ is finitely generated as an algebra over k. However, Rees and Nagata constructed examples of quasiprojective varieties for which $k[X]$ is not finitely generated. This shows that $k[X]$ is only a reasonable invariant when X is an affine closed set.

We pass to maps. Any map of a quasiprojective variety X to an affine space $\mathbb{A}^n$ is given by n functions on X with values in k. If these functions are regular then we say the map is *regular*.

Definition. Let $f \colon X \to Y$ be a map between quasiprojective varieties, with $Y \subset \mathbb{P}^m$. This map is *regular* if for every point $x \in X$ and for some affine piece $\mathbb{A}_i^m$ containing $f(x)$ there exists a neighbourhood $U \ni x$ such that $f(U) \subset \mathbb{A}_i^m$ and the map $f \colon U \to \mathbb{A}_i^m$ is regular.

We check that the regularity property is independent of the choice of affine piece $\mathbb{A}_i^m$ containing $f(x)$. If $f(x) = (y_0, \ldots, \widehat{1}, \ldots, y_m) \in \mathbb{A}_i^m$ (where $\widehat{1}$ in the ith place means that this coordinate is discarded) is also contained in $\mathbb{A}_j^m$, then $y_j \neq 0$, and the coordinates of this point in $\mathbb{A}_j^m$ are $(y_0/y_j, \ldots, 1/y_j, \ldots, \widehat{1}, \ldots y_m/y_j)$, with $1/y_j$ in the ith place and $\widehat{1}$ discarded from the jth. Therefore if $f \colon U \to \mathbb{A}_i^m$ is given by functions $(f_0, \ldots, \widehat{1}, \ldots, f_m)$, the map f to $\mathbb{A}_j^m$ is given by

$$(f_0/f_j, \ldots, 1/f_j, \ldots, \widehat{1}, \ldots f_m/f_j).$$

By assumption $f_j(x) \neq 0$, and the subset $U' \subset U$ of points at which $f_j \neq 0$ is open. The functions $f_0/f_j, \ldots, 1/f_j, \ldots, f_m/f_j$ are regular on U', and hence $f: U' \to \mathbb{A}_j^m$ is regular.

In the same way as for affine closed sets, a regular map $f: X \to Y$ defines a homomorphism $f^*: k[Y] \to k[X]$.

The question of how to write down formulas defining a regular map on an irreducible variety is solved in complete analogy with the case $n = 2$ treated in 1.6. Suppose for example that $f(x) \in \mathbb{A}_0^m$, and the map $f: U \to \mathbb{A}_0^m$ is given by regular functions $f_1, \ldots, f_m$. By definition $f_i = P_i/Q_i$ where P_i, Q_i are forms of the same degree in the homogeneous coordinates of x and $Q_i(x) \neq 0$. Putting these fractions over a common denominator gives $f_i = F_i/F_0$, where $F_0, \ldots, F_m$ are forms of the same degree and $F_0(x) \neq 0$. In other words, $f(x) = (F_0(x) : \cdots : F_m(x)) \in \mathbb{P}^m$. In this process, we must bear in mind that the representation of a regular function as a ratio of two forms is not unique. Hence two different formulas

$$f(x) = (F_0(x) : \cdots : F_m(x)) \quad \text{and} \quad g(x) = (G_0(x) : \cdots : G_m(x)) \quad (3)$$

may define the same map; this happens if and only if

$$F_i G_j = F_j G_i \text{ on } X \qquad \text{for } 0 \leq i, j \leq m. \quad (4)$$

This brings us to a second form of the definition of a regular map:

Definition. A *regular map* $f: X \to \mathbb{P}^m$ of an irreducible quasiprojective variety X to projective space $\mathbb{P}^m$ is given by an $(m+1)$-tuple of forms

$$(F_0 : \cdots : F_m) \quad (5)$$

of the same degree in the homogeneous coordinates of $x \in \mathbb{P}^n$. We require that for every $x \in X$ there exists an expression (5) for f such that $F_i(x) \neq 0$ for at least one i; then we write $f(x)$ to denote the point $(F_0(x) : \cdots : F_m(x))$. Two maps (3) are considered equal if (4) holds.

Now we have a definition of regular maps between quasiprojective varieties, it is natural to define an *isomorphism* to be a regular map having an inverse regular map.

A quasiprojective variety X' isomorphic to a closed subset of an affine space will be called an *affine variety*. It can happen that X is given as a subset $X \subset \mathbb{A}^n$, but is not closed in $\mathbb{A}^n$. For example, the set $X = \mathbb{A}^1 \setminus 0$ is not closed in $\mathbb{A}^1$, although it is quasiprojective, and is isomorphic to the hyperbola $xy = 1$ (2.3, Example 3), which is a closed set of $\mathbb{A}^2$. Thus the notion of a closed affine set is not invariant under isomorphism, while that of affine variety is invariant by definition.

In the same way, a quasiprojective variety isomorphic to a closed projective set will be called a *projective variety*. We will prove in 5.2, Theorem 2 that if $X \subset \mathbb{P}^n$ is a projective variety then it is closed in $\mathbb{P}^n$, so that the notions of closed projective set and projective variety coincide and are both invariant under isomorphism.

There are quasiprojective varieties that are neither affine nor projective (see Ex. 5 and §5, Ex. 4–6).

In what follows, we will meet some properties of varieties X that need only be verified for some neighbourhood U of any point $x \in X$. In other words, if $X = \bigcup U_\alpha$, with U_α any open sets, then it is enough to verify the property for each of the U_α. We say that properties of this type are *local properties*. We give some example of local properties.

Lemma 1. *The property that a subset $Y \subset X$ is closed in a quasiprojective variety X is a local property.*

Proof. The assertion means that if $X = \bigcup U_\alpha$ with open sets U_α, and $Y \cap U_\alpha$ is closed in each U_α then Y is closed in X. By definition of open sets, $U_\alpha = X \setminus Z_\alpha$ where the Z_α are closed, and by definition of closed sets, $U_\alpha \cap Y = U_\alpha \cap T_\alpha$ where the $T_\alpha \subset X$ are closed.

We check that $Y = \bigcap (Z_\alpha \cup T_\alpha)$, from which it follows of course that Y is closed. If $y \in Y$ and $y \in U_\alpha$ then $y \in U_\alpha \cap Y \subset T_\alpha$, and if $y \notin U_\alpha$ then $y \in X \setminus U_\alpha = Z_\alpha$, so that $y \in Z_\alpha \cup T_\alpha$ for every α. Conversely, suppose that $x \in Z_\alpha \cup T_\alpha$ for every α. Since $X = \bigcup U_\alpha$ it follows that $x \in U_\beta$ for some β. Then $x \notin Z_\beta$, and hence $x \in T_\beta$, so that $x \in T_\beta \cap U_\beta \subset Y$. The lemma is proved.

In studying local properties we can restrict ourselves to affine varieties in view of the following result.

Lemma 2. *Every point $x \in X$ has a neighbourhood isomorphic to an affine variety.*

Proof. By assumption $X \subset \mathbb{P}^n$. If $x \in \mathbb{A}_0^n$ (that is, if the coordinate u_0 of x is nonzero) then $x \in X \cap \mathbb{A}_0^n$, and by definition of a quasiprojective variety $X \cap \mathbb{A}_0^n = Y \setminus Y_1$ where Y and $Y_1 \subset Y$ are closed subsets of $\mathbb{A}_0^n$. Since $x \in Y \setminus Y_1$, there exists a polynomial F of the coordinates of $\mathbb{A}_0^n$ such that $F = 0$ on Y_1 and $F(x) \neq 0$. Write $V(F)$ for the set of points of Y where $F = 0$. Obviously $D(F) = Y \setminus V(F)$ is a neighbourhood of x. We prove that this neighbourhood is isomorphic to an affine variety. Suppose that $G_1 = \cdots = G_m = 0$ are the equations of Y in $\mathbb{A}_0^n$. Define a variety $Z \subset \mathbb{A}^{n+1}$ by the equations

$$G_1(T_1, \ldots, T_n) = \cdots = G_m(T_1, \ldots, T_n) = 0,$$
$$F(T_1, \ldots, T_n) \cdot T_{n+1} = 1. \tag{6}$$

The map $\varphi \colon (x_1, \ldots, x_{n+1}) \mapsto (x_1, \ldots, x_n)$ obviously defines a regular map $Z \to D(F)$ and $\psi \colon (x_1, \ldots, x_n) \mapsto (x_1, \ldots, x_n, F(x_1, \ldots, x_n)^{-1})$ a regular map $D(F) \to Z$ inverse to φ. This proves the lemma.

If $Y = \mathbb{A}^1$, $F = T$ then the isomorphism just constructed is the map considered in 2.3, Example 3.

Definition. An open set $D(f) = X \setminus V(f)$ consisting of the points of an affine variety X such that $f(x) \neq 0$ is called a *principal open set*.

The significance of these sets is that they are affine, as we have seen, and the ring $k[D(f)]$ of regular function on them can be easily determined. Namely, by construction $f \neq 0$ on $D(f)$, so that $f^{-1} \in k[D(f)]$, and 3.2, Theorem 4 together with (6) shows that $k[D(f)] = k[X][f^{-1}]$.

Lemmas 1–2 show for example that closed subsets map to closed subsets under isomorphisms. We prove in addition that the inverse image $f^{-1}(Z)$ under any regular map $f \colon X \to Y$ of any closed subset $Z \subset Y$ is closed in X.

By definition of a regular map $f \colon X \to Y$, for any point $x \in X$ there are neighbourhoods U of x in X and V of $f(x)$ in Y such that $f(U) \subset V \subset \mathbb{A}^m$ and the map $f \colon U \to V$ is regular. By Lemma 2 we can assume that U is an affine variety. By Lemma 1, it is enough to check that $f^{-1}(Z) \cap U = f^{-1}(Z \cap V)$ is closed in U. Since $Z \cap V$ is closed in V, it is defined by equations $g_1 = \cdots = g_m = 0$, where the g_i are regular functions on V. But then $f^{-1}(Z \cap V)$ is defined by the equations $f^*(g_1) = \cdots = f^*(g_m) = 0$, and is hence also closed.

It follows also from what we have just proved that the inverse image of an open set is again open. It is easy to check that a regular map can be defined as a map $f \colon X \to Y$ such that the inverse image of any open set is open (that is, f is "continuous"), and for any point $x \in X$ and any function φ regular in a neighbourhood of $f(x) \in Y$, the function $f^*(\varphi)$ is regular in a neighbourhood of x.

4.3. Rational Functions

In discussing the definition of rational functions on quasiprojective varieties, we met a distinction of substance between the case of affine varieties and the general case. Namely, we defined rational functions on an affine variety X as ratios of functions that are regular on the whole of X. But in the general case, as we have said, it can happen that there are no everywhere regular functions except for the constants, so that if we used the same definition there would also be no rational functions except for the constants. For this reason we define rational functions on a quasiprojective variety $X \subset \mathbb{P}^n$ to be functions defined on X by homogeneous functions on $\mathbb{P}^n$ (as in 1.6 for $n = 2$).

More precisely, consider an irreducible quasiprojective variety $X \subset \mathbb{P}^n$ and (by analogy with 3.2) write $\mathcal{O}_X$ for the set of rational functions $f = P/Q$ in the homogeneous coordinates $S_0, \ldots, S_n$ such that P, Q are forms of the same degree and $Q \notin \mathfrak{A}_X$. As for affine varieties, from the fact that X is irreducible it follows that $\mathcal{O}_X$ is a ring. Write M_X for the set of functions

$f \in \mathcal{O}_X$ with $P \in \mathfrak{A}_X$. Obviously the quotient ring $\mathcal{O}_X/M_X$ is a field, called the *function field* of X, and denoted by $k(X)$. If U is an open subset of an irreducible quasiprojective variety X then, since a form vanishes on X if and only if it vanishes on U, we have $k(X) = k(U)$. In particular, $k(X) = k(\overline{X})$, where $\overline{X}$ is the projective closure of X in $\mathbb{P}^n$. Thus in discussing function fields we can restrict to affine or projective varieties if we want to.

It is easy to check that if X is an affine variety then the definition just given coincides with that given in 3.2. Indeed, dividing the numerator and denominator of a rational function $f = P/Q$ with $\deg P = \deg Q = m$ by S_0^m, we can write it as a rational function in $T_i = S_i/S_0$ for $i = 1, \ldots, n$. By doing this, we establish an isomorphism of the field of homogeneous rational functions of degree 0 in $S_0, \ldots, S_n$ with the field $k(T_1, \ldots, T_n)$. An obvious verification shows that the subring and ideal of $k(T_1, \ldots, T_n)$ denoted in 3.2 by $\mathcal{O}_X$ and M_X correspond to the objects denoted here by the same letters.

In 4.2 we have already used rational functions on $\mathbb{P}^n$ to define regular functions. As there, we say that $f \in k(X)$ is *regular* at a point $x \in X$ if it can be written in the form $f = F/G$, with F and G homogeneous of the same degree and $G(x) \neq 0$. Then $f(x) = F(x)/G(x)$ is the *value* of f at x. As in the case of affine varieties, the set of points at which a given rational function f is regular is a nonempty open set U of X, called the *domain of definition* of f. Obviously a rational function can also be defined as a function regular on some open set $U \subset X$.

A rational map $f \colon X \to \mathbb{P}^m$ is defined (as in the second definition of regular map in 4.2) by giving $m + 1$ forms $(F_0 : \cdots : F_m)$ of the same degree in the $n + 1$ homogeneous coordinates of $\mathbb{P}^n$ containing X. Here at least one of the forms must not vanish on X. Two maps $(F_0 : \cdots : F_m)$ and $(G_0 : \cdots : G_m)$ are equal if $F_i G_j = F_j G_i$ on X for all i, j. If we divide through all the forms F_i by one of them (nonzero on X), we can define a rational map by $m + 1$ rational functions on X, with the same notion of equality of maps. If a rational map f can be defined by functions $(f_0 : \cdots : f_m)$ such that all the f_i are regular at $x \in X$ and not all zero at x, then f is regular at x. It then defines a regular map of some neighbourhood of the point x to $\mathbb{P}^m$.

The set of points at which a rational map is regular is open. Hence we can also define a rational map to be a regular map of some open set $U \subset X$. If $Y \subset \mathbb{P}^m$ is a quasiprojective variety and $f \colon X \to \mathbb{P}^m$ a rational map, we say that f maps X to Y if there exists an open set $U \subset X$ on which f is regular and $f(U) \subset Y$. The union $\widetilde{U}$ of all such open sets is called the *domain of definition* of f, and $f(\widetilde{U}) \subset Y$ the *image* of X in Y.

As in the case of affine varieties, if the image of a rational map $f \colon X \to Y$ is dense in Y then f defines an inclusion of fields $f^* \colon k(Y) \hookrightarrow k(X)$. If a rational map $f \colon X \to Y$ has an inverse rational map then f is *birational* or is a *birational equivalence*, and X and Y are birational. In this case the inclusion of fields $f^* \colon k(Y) \hookrightarrow k(X)$ is an isomorphism.

We can now clarify the relation between the notions of isomorphism and birational equivalence.

Proposition. *Two irreducible varieties X and Y are birational if and only if they contain isomorphic open subsets $U \subset X$ and $V \subset Y$.*

Proof. Indeed, suppose that $f: X \to Y$ is birational, and let $g = f^{-1}: Y \to X$ be the inverse rational map. Write $U_1 \subset X$ and $V_1 \subset Y$ for the domain of definition of f and g. Then by assumption $f(U_1)$ is dense in Y, so that $f^{-1}(V_1) \cap U_1$ is nonempty, and as proved in 4.2, is open. Set $U = f^{-1}(V_1) \cap U_1$ and $V = g^{-1}(U_1) \cap V_1$. A simple check shows that $f(U) = V$, $g(V) = U$ and $fg = 1$, $gf = 1$, that is, U and V are isomorphic.

4.4. Examples of Regular Maps

Example 1. Projection. Let E be a d-dimensional linear subspace of $\mathbb{P}^n$ defined by $n - d$ linearly independent linear equations $L_1 = \cdots = L_{n-d} = 0$, with L_i linear forms. The *projection* with centre E is the rational map $\pi(x) = (L_1(x) : \cdots : L_{n-d}(x))$. This map is regular on $\mathbb{P}^n \setminus E$, since at every point of this set one of the forms L_i does not vanish. Hence if X is any closed subvariety of $\mathbb{P}^n$ disjoint from E, the restriction of π defines a regular map $\pi: X \to \mathbb{P}^{n-d-1}$. The geometric meaning of projection is as follows: as a model of $\mathbb{P}^{n-d-1}$ take any $(n - d - 1)$-dimensional linear subspace $H \subset \mathbb{P}^n$ disjoint from E. Then there is a unique $(d + 1)$-dimensional linear subspace $\langle E, x \rangle$ passing through E and any point $x \in \mathbb{P}^n \setminus E$. This subspace intersects H in a unique point, which is $\pi(x)$. If X intersects E, but is not contained in it, then projection from E is a rational map on X. The case $d = 0$, a projection from a point, has already appeared several times.

Example 2. The Veronese embedding. Consider all the homogeneous polynomials F of degree m in variables $S_0, \ldots, S_n$. These form a vector space, whose dimension is easy to compute: it is the binomial coefficient $\binom{n+m}{m}$.

Consider the hypersurfaces of degree m in $\mathbb{P}^n$. Since polynomials define the same hypersurface if and only if they are proportional, hypersurfaces correspond to points of the projective space $\mathbb{P}^N$ of dimension $N = \nu_{n,m} = \binom{n+m}{m} - 1$. Write $v_{i_0 \ldots i_n}$ for homogeneous coordinates of $\mathbb{P}^N$, where $i_0, \ldots, i_n \geq 0$ are any nonnegative integers such that $i_0 + \cdots + i_n = m$. Consider the map $v_m: \mathbb{P}^n \to \mathbb{P}^N$ defined by

$$v_{i_0 \ldots i_n} = u_0^{i_0} \cdots u_n^{i_n} \qquad \text{for } i_0 + \cdots + i_n = m. \qquad (1)$$

This is obviously a regular map, since the monomials on the right-hand side of (1) include in particular the elements u_i^m, which vanish only if all $u_i = 0$. The map v_m is called the mth *Veronese embedding* of $\mathbb{P}^n$, and the image $v_m(\mathbb{P}^n) \subset \mathbb{P}^N$ the *Veronese variety*. It follows from (1) that the relations

$$v_{i_0...i_n} v_{j_0...j_n} = v_{k_0...k_n} v_{l_0...l_n} \qquad (2)$$

hold on $v_m(\mathbb{P}^n)$ whenever $i_0 + j_0 = k_0 + l_0, \ldots, i_n + j_n = k_n + l_n$. Conversely, it's easy to deduce from (2) that at least one of the coordinates $v_{0...m...0}$ corresponding to the monomial u_i^m is nonzero, and that, for example, on the open set $v_{m0...0} \neq 0$, the map

$$u_0 = v_{m0...0}, \quad u_i = v_{m-1,0...1...0} \quad \text{for } i \geq 2$$

is a regular inverse of v_m. Hence $v_m(\mathbb{P}^n)$ is defined by the equations (2), and v_m is an isomorphic embedding $\mathbb{P}^n \hookrightarrow \mathbb{P}^N$.

The significance of the Veronese embedding is that if

$$F = \sum a_{i_0...i_n} u_0^{i_0} \cdots u_n^{i_n}$$

is a form of degree m in the homogeneous coordinates of $\mathbb{P}^n$ and $H \subset \mathbb{P}^n$ is the hypersurface defined by $F = 0$, then $v_m(H) \subset v_m(\mathbb{P}^n) \subset \mathbb{P}^N$ is the intersection of $v_m(\mathbb{P}^n)$ with the hyperplane of $\mathbb{P}^N$ with equation $\sum a_{i_0...i_n} v_{i_0...i_n}$. Thus the Veronese embedding allows us to reduce the study of some problems concerning hypersurfaces of degree m to the case of hyperplanes.

The mth Veronese image of the projective line $v_m(\mathbb{P}^1) \subset \mathbb{P}^m$ is called the Veronese curve, the twisted m-ic curve, or the *rational normal curve* of degree m.

Exercises to §4

1. Prove that an affine variety U is irreducible if and only if its projective closure $\overline{U}$ is irreducible.

2. Associate with any affine variety $U \subset \mathbb{A}_0^n$ its projective closure $\overline{U}$ in $\mathbb{P}^n$. Prove that this defines a one-to-one correspondence between the affine subvarieties of $\mathbb{A}_0^n$ and the projective subvarieties of $\mathbb{P}^n$ with no components contained in the hyperplane $S_0 = 0$.

3. Prove that the variety $X = \mathbb{A}^2 \setminus (0,0)$ is not isomorphic to an affine variety. [Hint: Compute the ring $k[X]$ of regular functions on X, and use the fact that if Y is an affine variety, every proper ideal $\mathfrak{A} \subsetneq k[Y]$ defines a nonempty set.]

4. Prove that any quasiprojective variety is open in its projective closure.

5. Prove that every rational map $\varphi \colon \mathbb{P}^1 \to \mathbb{P}^n$ is regular.

6. Prove that any regular map $\varphi \colon \mathbb{P}^1 \to \mathbb{A}^n$ maps $\mathbb{P}^1$ to a point.

7. Define a birational map f from an irreducible quadric hypersurface $X \subset \mathbb{P}^3$ to the plane $\mathbb{P}^2$ by analogy with the stereographic projection of 3.3, Example 1. At which points is f not regular? At which points is f^{-1} not regular?

8. In Ex. 7, find the open subsets $U \subset X$ and $V \subset \mathbb{P}^2$ that are isomorphic.

9. Prove that the map $y_0 = x_1 x_2$, $y_1 = x_0 x_2$, $y_2 = x_0 x_1$ defines a birational map of $\mathbb{P}^2$ to itself. At which points are f and f^{-1} not regular? What are the open sets mapped isomorphically by f? (Compare Chap. IV, 3.5.)

10. Prove that the Veronese image $v_m(\mathbb{P}^n) \subset \mathbb{P}^N$ is not contained in any linear subspace of $\mathbb{P}^N$.

11. Prove that the variety $\mathbb{P}^2 \setminus X$, where X is a plane conic, is affine. [Hint: Use the Veronese embedding.]

5. Products and Maps of Quasiprojective Varieties

5.1. Products

The definition of the product of affine varieties (2.1, Example 5) was so natural as not to require any comment. For general quasiprojective varieties, things are somewhat more complicated. Because of this, we first consider quasiprojective subvarieties of affine spaces. If $X \subset \mathbb{A}^n$ and $Y \subset \mathbb{A}^m$ are varieties of this type then $X \times Y = \{(x,y) \mid x \in X, y \in Y\}$ is a quasiprojective variety in $\mathbb{A}^n \times \mathbb{A}^m$. Indeed, if $X = X_1 \setminus X_0$ and $Y = Y_1 \setminus Y_0$ where $X_1, X_0 \subset \mathbb{A}^n$, and $Y_1, Y_0 \subset \mathbb{A}^m$ are closed subvarieties, then writing

$$X \times Y = X_1 \times Y_1 \setminus \left((X_1 \times Y_0) \cup (X_0 \times Y_1) \right)$$

shows that $X \times Y$ is quasiprojective. This quasiprojective variety is the *product* of X and Y. At this point, we should check that if X and Y are replaced by isomorphic varieties then so is $X \times Y$. This is easy to see. Suppose that $\varphi \colon X \to X' \subset \mathbb{A}^p$ and $\psi \colon Y \to Y' \subset \mathbb{A}^q$ are isomorphisms. Then $\varphi \times \psi \colon X \times Y \to X' \times Y'$ defined by $(\varphi \times \psi)(x,y) = (\varphi(x), \psi(y))$ is a regular map, with regular inverse $\varphi^{-1} \times \psi^{-1}$.

We return to quasiprojective varieties, and decide what properties we want the notion of product to have. Let $X \subset \mathbb{P}^n$ and $Y \subset \mathbb{P}^m$ be two quasiprojective varieties. Write $X \times Y$ for the set of pairs (x,y) with $x \in X$ and $y \in Y$. We want to consider this set as a quasiprojective variety, and for this, we have to produce an embedding φ of $X \times Y$ into a projective space $\mathbb{P}^N$ in such a way that the image $\varphi(X \times Y) \subset \mathbb{P}^N$ is a quasiprojective subvariety. At the same time, it is reasonable to require that the definition is local, in the sense that for any points $x \in X$ and $y \in Y$ there exist affine neighbourhoods $X \supset U \ni x$ and $Y \supset V \ni y$ such that $\varphi(U \times V)$ is open in $\varphi(X \times Y)$, and φ

defines an isomorphism of the product of the affine varieties U and V, whose definition we already know, to the subvariety $\varphi(U \times V) \subset \varphi(X \times Y)$.

It is easy to see that the local property of φ determines it uniquely; more precisely, if $\psi \colon X \times Y \to \mathbb{P}^M$ is another embedding of the same kind, then $\psi \circ \varphi^{-1}$ defines an isomorphism between $\varphi(X \times Y)$ and $\psi(X \times Y)$. Indeed, for this, it is enough to prove that for any $x \in X$ and $y \in Y$, there exist neighbourhoods $\varphi(X \times Y) \supset W_1 \ni \varphi(x, y)$ and $\psi(X \times Y) \supset W_2 \ni \psi(x, y)$ such that $\psi \circ \varphi^{-1} \colon W_1 \to W_2$ is an isomorphism. Consider affine neighbourhoods $X \supset U \ni x$ and $Y \supset V \ni y$ the existence of which is provided by the local property; passing if necessary to smaller affine neighbourhoods, we can assume that $U \times V$ is isomorphic to both $\varphi(U \times V)$ and $\psi(U \times V)$. Then $\varphi(U \times V) = W_1$ and $\psi(U \times V) = W_2$ are the affine neighbourhoods we need, since both are isomorphic to the product $U \times V$ of the affine varieties U and V.

We now proceed to construct an embedding φ with the required properties. For this, we can at once restrict to the case $X = \mathbb{P}^n$, $Y = \mathbb{P}^m$; for once an embedding $\varphi \colon \mathbb{P}^n \times \mathbb{P}^m \hookrightarrow \mathbb{P}^N$ is constructed, it is easy to check that its restriction to $X \times Y \subset \mathbb{P}^n \times \mathbb{P}^m$ has all the required properties.

To construct the embedding φ, consider the projective space $\mathbb{P}^N$ with homogeneous coordinates w_{ij} having two subscripts $i = 0, \ldots, n$ and $j = 0, \ldots, m$; thus $N = (n+1)(m+1) - 1$. If $x = (u_0 : \cdots : u_n) \in \mathbb{P}^n$ and $y = (v_0 : \cdots : v_m) \in \mathbb{P}^m$ then we set

$$\varphi(x, y) = (w_{ij}), \quad \text{with } w_{ij} = u_i v_j \quad \text{for } 0 \le i \le n \text{ and } 0 \le j \le m. \tag{1}$$

Multiplying the homogeneous coordinates of x or y by a common scalar obviously does not change the point $\varphi(x, y) \in \mathbb{P}^N$. To prove that $\varphi(\mathbb{P}^n \times \mathbb{P}^m)$ is a closed set of $\mathbb{P}^N$, we write out its defining equations:

$$w_{ij} w_{kl} = w_{kj} w_{il} \quad \text{for } 0 \le i, k \le n \text{ and } 0 \le j, l \le m. \tag{2}$$

Substituting the w_{ij} given by (1) shows at once that they satisfy (2). Conversely, if w_{ij} satisfy (2), and, say, $w_{00} \ne 0$, then setting $k, l = 0$ in (2) gives that $(w_{ij}) = \varphi(x, y)$, where

$$x = (w_{00} : \cdots : w_{n0}) \quad \text{and} \quad y = (w_{00} : \cdots : w_{0m}).$$

This argument proves at the same time that $\varphi(x, y)$ determines x and y uniquely, that is, φ is an embedding $\mathbb{P}^n \times \mathbb{P}^m \hookrightarrow \mathbb{P}^N$ with image the subvariety $W \subset \mathbb{P}^N$ defined by (2). Consider the open sets $\mathbb{A}_0^n \subset \mathbb{P}^n$ given by $u_0 \ne 0$, $\mathbb{A}_0^m \subset \mathbb{P}^m$ by $v_0 \ne 0$, and $\mathbb{A}_{00}^N \subset \mathbb{P}^N$ by $w_{00} \ne 0$, having inhomogeneous coordinates $x_i = u_i/u_0$, $y_j = v_j/v_0$ and $z_{ij} = w_{ij}/w_{00}$ respectively. Then obviously $\varphi(\mathbb{A}_0^n \times \mathbb{A}_0^m) = W \cap \mathbb{A}_{00}^N = W_{00}$. As we have just seen, on W_{00} we have $z_{i0} = x_i$, $z_{0j} = y_j$ and $z_{ij} = x_i y_j = z_{i0} z_{0j}$ for $i, j > 0$. It follows from this that $\varphi(\mathbb{P}^n \times \mathbb{P}^m) \cap \mathbb{A}_{00}^N = W_{00}$ is isomorphic to $\mathbb{A}^{n+m}$ with coordinates $(x_1, \ldots, x_n, y_1, \ldots, y_m)$, and φ defines an isomorphism $\mathbb{A}_0^n \times \mathbb{A}_0^m \to W_{00}$. This proves that φ satisfies the local requirement of our construction. The embedding $\varphi \colon \mathbb{P}^n \times \mathbb{P}^m \hookrightarrow \mathbb{P}^N$ with $N = (n+1)(m+1) - 1$ just constructed is called the *Segre embedding*, and the image $\mathbb{P}^n \times \mathbb{P}^m \subset \mathbb{P}^N$ the Segre variety.

Remark 1. The point (w_{ij}) can be interpreted as an $(n+1) \times (m+1)$ matrix, and equations (2) express the vanishing of the 2×2 minors:

$$\det \begin{vmatrix} w_{ij} & w_{il} \\ w_{kj} & w_{kl} \end{vmatrix} = 0.$$

That is, they express the condition that the matrix (w_{ij}) has rank 1, and equation (1) shows that such a matrix is a product of a $1 \times (n+1)$ column matrix and a $(m+1) \times 1$ row matrix. Thus $\varphi(\mathbb{P}^n \times \mathbb{P}^m)$ is a determinantal variety (see 4.1, Example 3).

Remark 2. The simplest case $n = m = 1$ has a simple geometric interpretation: in this case, (2) is the single equation $w_{11}w_{00} = w_{01}w_{10}$, so that $\varphi(\mathbb{P}^1 \times \mathbb{P}^1)$ is just a nondegenerate quadric surface $Q \subset \mathbb{P}^3$. For $\alpha = (\alpha_0, \alpha_1) \in \mathbb{P}^1$, the set $\varphi(\alpha \times \mathbb{P}^1)$ is the line in $\mathbb{P}^3$ given by $\alpha_1 w_{00} = \alpha_0 w_{10}$, $\alpha_1 w_{01} = \alpha_0 w_{11}$. As α runs through $\mathbb{P}^1$, these lines give all the generators of one of the two families of lines of Q. Similarly the set $\varphi(\mathbb{P}^1 \times \beta)$ is a line of $\mathbb{P}^3$, and as β runs through $\mathbb{P}^1$, these lines give the generators of the other family.

It is convenient, now that we have defined the product $X \times Y$ of quasi-projective varieties using the embedding $\varphi \colon \mathbb{P}^n \times \mathbb{P}^m \hookrightarrow \mathbb{P}^N$, with $N = (n+1)(m+1) - 1$, to explain some ideas of algebraic geometry that are originally defined in terms of $\mathbb{P}^n \times \mathbb{P}^m$ and of this embedding.

Let us determine, for example, what are the subsets of $\mathbb{P}^n \times \mathbb{P}^m$ that are mapped by φ to algebraic subvarieties of $\mathbb{P}^N$; these will then be the closed algebraic subvarieties of the product $\mathbb{P}^n \times \mathbb{P}^m$. A subvariety $V \subset \mathbb{P}^N$ is defined by equations $F_k(w_{00} : \cdots : w_{nm}) = 0$, where the F_k are homogeneous polynomials in the w_{ij}. After making the substitution (1), we can write these in the coordinates u_i and v_j as equations

$$G_k(u_0 : \cdots : u_n; v_0 : \cdots : v_m) = 0,$$

where the G_k are homogeneous in each set of variables $u_0, \ldots, u_n$ and $v_0, \ldots, v_m$, and of the same degree in both. Conversely, it is easy to see that a polynomial with this bihomogeneity property can always be written as a polynomial in the products $u_i v_j$. However, equations that are bihomogeneous in u_i and v_j always define an algebraic subvariety of $\mathbb{P}^n \times \mathbb{P}^m$ even if the degrees of homogeneity in the two sets of variables are different. For if $G(u_0 : \cdots : u_n; v_0 : \cdots : v_m)$ has degree r in u_i and s in v_j, and, say, $r > s$, then $G = 0$ is equivalent to the system of equations $v_i^{r-s} G = 0$ for $i = 0, \ldots, m$, and we know that these define an algebraic variety.

In what follows, we also need to answer the same question for the product $\mathbb{P}^n \times \mathbb{A}^m$. Suppose that $\mathbb{A}^m = \mathbb{A}_0^m \subset \mathbb{P}^m$ is given by $v_0 \neq 0$. The equations of a closed subset of $\mathbb{P}^n \times \mathbb{P}^m$ are $G_k(u_0 : \cdots : u_n; v_0 : \cdots : v_m) = 0$. Suppose that G_k is homogeneous of degree r_k in $v_0 : \cdots : v_m$. Dividing the equation by $v_0^{r_k}$ and setting $y_j = v_j/v_0$ gives equations $g_k(u_0 : \cdots : u_n; y_1 : \cdots : y_m) = 0$ that are homogeneous in the u_i, and (in general) inhomogeneous in the y_j. This proves the following result:

Theorem 1. *A subset $X \subset \mathbb{P}^n \times \mathbb{P}^m$ is a closed algebraic subvariety if and only if it is given by a system of equations*

$$G_k(u_0 : \cdots : u_n; v_0 : \cdots : v_m) = 0 \qquad \text{for } k = 1, \ldots, t,$$

homogeneous separately in each set of variables u_i and v_j. Every closed algebraic subvariety of $\mathbb{P}^n \times \mathbb{A}^m$ is given by a system of equations

$$g_k(u_0 : \cdots : u_n; y_1 : \cdots : y_m) = 0 \qquad \text{for } k = 1, \ldots, t \qquad (3)$$

that are homogeneous in $u_0, \ldots, u_n$). □

Of course, the same kind of thing holds for a product of any number of spaces. For example, a subvariety of $\mathbb{P}^{n_1} \times \cdots \times \mathbb{P}^{n_k}$ is given by a system of equations homogeneous in each of the k sets of variables.

5.2. The Image of a Projective Variety is Closed

The image of an affine variety under a regular map does not have to be a closed set; this is illustrated in 2.3, Examples 3–4 for a map from an affine variety to an affine variety. For maps from an affine variety to a projective variety it is even more obvious: an example is given by the embedding of $\mathbb{A}^n$ into $\mathbb{P}^n$ as the open subset $\mathbb{A}_0^n$. In this respect, projective varieties are fundamentally different from affine varieties.

Theorem 2. *The image of a projective variety under a regular map is closed.*

The proof uses a notion that will occur later. Let $f : X \to Y$ be a regular map between arbitrary quasiprojective varieties. The subset Γ_f of $X \times Y$ consisting of pairs $(x, f(x))$ is called the *graph* of f.

Lemma 1. *The graph of a regular map is closed in $X \times Y$.*

Proof. First of all, it is enough to assume that Y is projective space. Indeed, if $Y \subset \mathbb{P}^m$ then $X \times Y \subset X \times \mathbb{P}^m$, and f defines a map $\bar{f} : X \to \mathbb{P}^m$ with $\Gamma_f = \Gamma_{\bar{f}} \subset X \times Y \subset X \times \mathbb{P}^m$. Thus set $Y = \mathbb{P}^m$. Let ι be the identity map from $\mathbb{P}^m$ to itself. Consider the regular map $(f, \iota) : X \times \mathbb{P}^m \to \mathbb{P}^m \times \mathbb{P}^m$ given by $(f, \iota)(x, y) = (f(x), y)$. Obviously Γ_f is the inverse image under the regular map (f, ι) of the graph Γ_ι of ι. We proved in 4.2 that the inverse image of a closed set under a regular map is closed. Hence everything reduces to proving that $\Gamma_\iota \subset \mathbb{P}^m \times \mathbb{P}^m$ is closed. But Γ_ι consists of points $(x, y) \in \mathbb{P}^m \times \mathbb{P}^m$ such that $x = y$. If $x = (u_0 : \cdots : u_m)$ and $y = (v_0 : \cdots : v_m)$ then the condition is that $(u_0 : \cdots : u_m)$ and $(v_0 : \cdots : v_m)$ are proportional; this condition can be expressed $u_i v_j = u_j v_i$, that is, $w_{ij} = w_{ji}$ for $i, j = 0, \ldots, m$. This proves that Γ_ι is closed, and therefore the lemma.

We return to the proof of the theorem. Let Γ_f be the graph of f, and $p: X \times Y \to Y$ the second projection, defined by $p(x, y) = y$. Obviously $f(X) = p(\Gamma_f)$. In view of Lemma 1, Theorem 2 follows from the following more general assertion.

Theorem 3. *If X is a projective variety, and Y a quasiprojective variety, the second projection $p: X \times Y \to Y$ takes closed sets to closed sets.*

Proof. The proof of this theorem can be reduced to a simple particular case. First of all, if $X \subset \mathbb{P}^n$ is a closed subset then the theorem for X follows from the theorem for $\mathbb{P}^n$: for $X \times Y$ is closed in $\mathbb{P}^n \times Y$, so that if Z is closed in $X \times Y$, it is also closed in $\mathbb{P}^n \times Y$. Thus we can assume that $X = \mathbb{P}^n$. Secondly, since closed is a local property, it is enough to cover Y by affine open sets U_i and prove the theorem for each of these. Hence we can assume that Y is an affine variety. Finally if $Y \subset \mathbb{A}^m$ then $\mathbb{P}^n \times Y$ is closed in $\mathbb{P}^n \times \mathbb{A}^m$, and hence it is enough to prove the theorem in the particular case $X = \mathbb{P}^n$ and $Y = \mathbb{A}^m$.

Let's see what the theorem means in this case. According to Theorem 1, any closed subvariety $Z \subset \mathbb{P}^n \times \mathbb{A}^m$ is defined by equations 5.1, (3), that we write in the form $g_i(u; y) = 0$ for $i = 1, \ldots, t$. Write $p: Z \to \mathbb{A}^m$ for the restriction of the second projection. Obviously the inverse image $p^{-1}(y_0)$ of $y_0 \in \mathbb{A}^m$ consists of all nonzero solutions of the system $g_i(u, y_0) = 0$, and hence $y_0 \in p(Z)$ if and only if the system of equations $g_i(u; y_0) = 0$ has a nonzero solution in $(u_0, \ldots, u_n)$. Thus Theorem 3 asserts that for any system of equations 5.1, (3), the subset T of $y_0 \in \mathbb{A}^m$ for which $g_i(u; y_0) = 0$ has a nonzero solution is closed.

Now in view of 4.1, Lemma 1, $g_i(u; y_0) = 0$ has a nonzero solution if and only if

$$\big(g_1(u, y_0), \ldots, g_t(u, y_0)\big) \not\supset I_s \qquad \text{for all } s = 1, 2, \ldots$$

We now show that for given $s \geq 1$, the set of points $y_0 \in \mathbb{A}^m$ for which $\big(g_1(u, y_0), \ldots, g_t(u, y_0)\big) \not\supset I_s$ is a closed set T_s. Then $T = \bigcap T_s$, and T is also closed. Write k_i for the degree of the homogeneous polynomial $g_i(u, y)$ in the variables $u_0, \ldots, u_n$. Let $\{M^\alpha\}_\alpha$ be the monomials of degree s in $u_0, \ldots, u_n$ written out in some order. The condition $\big(g_1(u, y_0), \ldots, g_t(u, y_0)\big) \supset I_s$ means that each monomial M^α can be expressed in the form

$$M^\alpha = \sum_{i=1}^{t} g_i(u, y_0) F_{i,\alpha}(u). \tag{1}$$

Comparing the homogeneous components of degree s shows that there must also be an expression (1) for M^α with $\deg F_{i,\alpha} = s - k_i$, or $F_{i,\alpha} = 0$ if $k_i > s$. Let $\{N_i^\beta\}_\beta$ be the monomials of degree $s - k_i$ written out in some order. We see that the conditions (1) hold if and only if every monomial M^α is a linear combination of the polynomials $g_i(u, y_0) N_i^\beta$. This, in turn, is equivalent to the condition that the polynomials $g_i(u, y_0) N_i^\beta$ span the entire vector space S of homogeneous polynomials of degree s in $u_0, \ldots, u_n$. Conversely, $(g_1(u, y_0), \ldots, g_t(u, y_0)) \not\supset I_s$ means that $g_i(u, y_0) N_i^\beta$ do not span S. To turn this condition into equations for T_s, write out the coefficients of the M^α appearing in all the polynomials $g_i(u, y_0) N_i^\beta$ as a rectangular matrix $\{a_{\alpha\beta}\}$, and set to zero all of its $\sigma \times \sigma$ minors, where $\sigma = \dim S$. These minors are obviously polynomials in the coefficients of the polynomials $g_i(u, y_0)$, and are therefore polynomials in the coordinates of the point y_0; they give the equations of the set T_s. Theorem 3 is proved, and with it Theorem 2.

Remark. One sees from the proof that Theorem 2 generalises to a wider class of maps $f: X \to Y$ between quasiprojective varieties, namely those that factor as a composite of a *closed embedding* $\iota: X \hookrightarrow \mathbb{P}^n \times Y$ (that is, an isomorphism of X with a closed subvariety) and the projection $p: \mathbb{P}^n \times Y \to Y$. Such maps are said to be *proper*. For example, if $f: X \to Y$ is a regular map of projective varieties then the restriction $f: f^{-1}(U) \to U$ to an open subset $U \subset Y$ is proper. Obviously if $f: X \to Y$ is a proper map the inverse image $f^{-1}(y)$ of a point $y \in Y$ is a projective variety.

Corollary 1. *If φ is a regular function on an irreducible projective variety then $\varphi \in k$, that is, φ is constant.*

Proof. We can view φ as a map $f: X \to \mathbb{A}^1$, and hence as a map $\overline{f}: X \to \mathbb{P}^1$. Since φ is a regular function, f is a regular map, and hence so is $\overline{f}$; by Theorem 2 its image $\overline{f}(X) \subset \mathbb{P}^1$ is closed. But since f itself is regular, $f(X) = \overline{f}(X)$, and therefore $\overline{f}(X)$ is a closed subset of $\mathbb{P}^1$ and is contained in $\mathbb{A}^1$, that is, it does not contain the point at infinity $x_\infty \in \mathbb{P}^1$. It follows from this that either $f(X) = \mathbb{A}^1$ or $f(X)$ is a finite set $S \subset \mathbb{A}^1$ (see 2.1, Example 3). The first case is impossible, since $f(X)$ is also supposed to be closed in $\mathbb{P}^1$, and $\mathbb{A}^1$ is not. Hence $f(X) = S$. If S consists of finitely mány points $\alpha_1, \ldots, \alpha_t$ then $X = \bigcup f^{-1}(\alpha_i)$, and $t > 1$ would contradict the irreducibility of X. Hence S consists of one point only, and so φ is constant. The corollary is proved.

Corollary 1 and 3.2, Theorem 4 provide an example of affine and projective varieties having diametrically opposite properties. On an affine variety there is a host of regular functions (they make up the whole coordinate ring $k[X]$), but on an irreducible projective variety, only the constants. The next result is a second example of affine and projective varieties being opposites.

Corollary 2. *A regular map $f: X \to Y$ from an irreducible projective variety X to an affine variety Y maps X to a point.*

Proof. Suppose that $Y \subset \mathbf{A}^m$. Then f is given by m functions $f(x) = (\varphi_1(x), \ldots, \varphi_m(x))$. Each of the functions φ_i is constant by Corollary 1, that is $\varphi_i = \alpha_i \in k$. Hence $f(X) = (\alpha_1, \ldots, \alpha_m)$. The corollary is proved.

We give another example of an application of Theorem 2. For this, we use the representation of forms of degree m in $n + 1$ variables by points of the projective space $\mathbb{P}^N$ with $N = \nu_{n,m} = \binom{m+n}{m} - 1$, as in 4.4, Example 2.

Proposition. *Points $\xi \in \mathbb{P}^N$ corresponding to reducible homogeneous polynomials F form a closed set.*

Remark 1. The proposition asserts that the condition for a homogeneous polynomial to be reducible can be written as polynomial conditions on its coefficients. For curves of degree 2, that is, the case $m = n = 2$, this relation is well known from coordinate geometry: if $F = \sum_{i=0}^{2} a_{ij} U_i U_j$ then F is irreducible if and only if $\det |a_{ij}| = 0$.

Remark 2. Passing to inhomogeneous coordinates, we see that in the vector space of all polynomial of degree $\leq m$, the reducible polynomials together with the polynomials of degree $< m$ form a closed set.

Proof. Proceeding to the proof of the proposition, we write $X \subset \mathbb{P}^N$ for the set of points ξ corresponding to reducible polynomials, and X_k for the set of points corresponding to polynomials F that split as a product of two polynomials of degrees k and $m - k$ (for $k = 1, \ldots, m$). Obviously $X = \bigcup X_k$, and we need only prove that each X_k is closed.

Consider the projective space $\mathbb{P}^{\nu_{n,k}}$ and $\mathbb{P}^{\nu_{n,m-k}}$ of forms of degree k and $m - k$, where $\nu_{n,k} = \binom{n+k}{k} - 1$ is as in 4.4, Example 2. Multiplying polynomials of degree k and $m - k$ defines a map $f: \mathbb{P}^{\nu_{n,k}} \times \mathbb{P}^{\nu_{n,m-k}} \to \mathbb{P}^N$, and it is easy to see that f is regular. Obviously $X_k = f(\mathbb{P}^{\nu_{n,k}} \times \mathbb{P}^{\nu_{n,m-k}})$. We saw in 5.1 that the product of two projective spaces is a projective variety, and hence X_k closed follows by Theorem 2. The proposition is proved.

5.3. Finite Maps

The projection map introduced in 4.4 has an important property; in order to state this, we first recall some notions from algebra. Let B be a ring, and A a subring containing the identity element 1_B. We say that an element $b \in B$ is *integral* over A if it satisfies an equation

$$b^k + a_1 b^{k-1} + \cdots + a_k = 0 \qquad \text{with } a_i \in A.$$

B is integral over A if every element $b \in B$ is integral over A. It is easy to prove (see for example Atiyah and Macdonald [7], Chap. 5, Proposition 5.1 and Corollary 5.2) that a ring B that is finitely generated as an A-algebra is integral over A if and only if it is finite as a module over A.

Let X and Y be affine varieties and $f \colon X \to Y$ a regular map such that $f(X)$ is dense in Y. Then f^* defines an isomorphic inclusion $k[Y] \hookrightarrow k[X]$. We view $k[Y]$ as a subring of $k[X]$ by means of f^*.

Definition 1. f is a *finite map* if $k[X]$ is integral over $k[Y]$.

From the properties of integral rings recalled above it follows that the composite of two finite maps is again finite. A typical example of a map that is not finite is 2.3, Example 3.

Example 1. Let X be an affine algebraic variety, G a finite group of automorphisms of X and $Y = X/G$ the quotient space (see 2.3, Example 11). Then the map $\varphi \colon X \to Y$ is finite. Indeed, the proof of Appendix, §4, Proposition 1 shows that the generators u_i of the algebra $k[X]$ are integral over the algebra $k[X]^G = k[Y]$. It follows from this that $k[X]$ is integral over $k[Y]$.

If f is a finite map then any point $y \in Y$ has at most a finite number of inverse images. Indeed, suppose that $X \subset \mathbf{A}^n$ and let $t_1, \ldots, t_n$ be the coordinates of $\mathbf{A}^n$ viewed as functions on X. It is enough to prove that any coordinate t_i takes only a finite number of values on the set $f^{-1}(y)$. By definition t_i satisfies an equation $t_i^k + a_1 t_i^{k-1} + \cdots + a_k = 0$ with $a_i \in k[Y]$. For $y \in Y$ and $x \in f^{-1}(y)$, we get an equation

$$t_i(x)^k + a_1(y) t_i(x)^{k-1} + \cdots + a_k(y) = 0, \tag{1}$$

which has only a finite number of roots.

The meaning of the finite condition is that as y moves in Y, none of the roots of (1) tends to infinity, since the coefficient 1 of the leading term does not vanish on Y. Thus as y moves in Y, points of $f^{-1}(y)$ can merge together, but cannot disappear. We make this remark more precise in the following result.

Theorem 4. *A finite map is surjective.*

Proof. Let X and Y be affine varieties, $f\colon X \to Y$ a finite map, and $y \in Y$. Write $\mathfrak{m}_y$ for the ideal of $k[Y]$ consisting of functions that take the value 0 at y. If $t_1, \ldots, t_n$ are the coordinate functions on Y and $y = (\alpha_1, \ldots, \alpha_n)$ then $\mathfrak{m}_y = (t_1 - \alpha_1, \ldots, t_n - \alpha_n)$. The equations of the variety $f^{-1}(y)$ then have the form $f^*(t_1) = \alpha_1, \ldots, f^*(t_n) = \alpha_n$, and by the Nullstellensatz $f^{-1}(y) = \emptyset$ if and only if the elements $f^*(t_i) - \alpha_i$ generate the trivial ideal:

$$(f^*(t_1) - \alpha_1, \ldots, f^*(t_n) - \alpha_n) = k[X].$$

From now on we view $k[Y]$ as a subring of $k[X]$, and do not distinguish between a function $u \in k[Y]$ and $f^*(u) \in k[X]$. Then the above condition is of the form $(t_1 - \alpha_1, \ldots, t_n - \alpha_n) = k[X]$, that is, $\mathfrak{m}_y k[X] = k[X]$. Since $k[X]$ is integral over $k[Y]$ it follows that it is a finite $k[Y]$-module; Theorem 4 follows from this and the following purely algebraic assertion:

Lemma. *If a ring B is a finite A-module where $A \subset B$ is a subring containing 1_B, then for an ideal $\mathfrak{a}$ of A,*

$$\mathfrak{a} \subsetneqq A \implies \mathfrak{a}B \subsetneqq B.$$

See Appendix, §6, Proposition 3, Corollary 1 for the proof. $\quad\square$

This completes the proof of Theorem 4.

Corollary. *A finite map takes closed sets to closed sets.*

Proof. It is enough to check this for an irreducible closed set $Z \subset X$. We apply Theorem 4 to the restriction of f to Z, that is $\overline{f}\colon Z \to \overline{f(Z)}$. This is clearly a finite map between affine varieties, hence $f(Z) = \overline{f(Z)}$ by Theorem 4, that is, $f(Z)$ is closed. The corollary is proved.

Finiteness is a local property:

Theorem 5. *If $f\colon X \to Y$ is a regular map of affine varieties, and every point $x \in Y$ has an affine neighbourhood $U \ni x$ such that $V = f^{-1}(U)$ is affine and $f\colon V \to U$ is finite, then f itself is finite.*

Proof. Set $k[X] = B$, $k[Y] = A$. Principal open sets were defined in 4.2. We can take a neighbourhood U of any point of Y such that U is a principal open set and satisfies the assumption of the theorem (see Ex. 11). Let $D(g_\alpha)$ be a family of such open sets, which we can take to be finite. Then $Y = \bigcup D(g_\alpha)$, that is, the ideal generated by the g_α is the whole of A. In our case $V_\alpha = f^{-1}(D(g_\alpha)) = D(f^*(g_\alpha))$ and $k[D(g_\alpha)] = A[1/g_\alpha]$, $k[V_\alpha] = B[1/g_\alpha]$. By assumption $B[1/g_\alpha]$ has a finite basis $\omega_{i,\alpha}$ over $A[1/g_\alpha]$. We can assume that $\omega_{i,\alpha} \in B$, since if the basis consisted of elements $\omega_{i,\alpha}/g_\alpha^{m_i}$ with $\omega_{i,\alpha} \in B$ then the elements $\omega_{i,\alpha}$ would also be a basis. We take the union of all the bases $\omega_{i,\alpha}$ and prove that they form a basis of B over A.

An element $b \in B$ has an expression

$$b = \sum_i \frac{a_{i,\alpha}}{g_\alpha^{n_\alpha}} w_{i,\alpha}$$

for each α. Since the $g_\alpha^{n_\alpha}$ generate the unit ideal of A, there exist $h_\alpha \in A$ such that $\sum_\alpha g_\alpha^{n_\alpha} h_\alpha = 1$. Hence

$$b = b \sum_\alpha g_\alpha^{n_\alpha} h_\alpha = \sum_i \sum_\alpha a_{i,\alpha} h_\alpha w_{i,\alpha},$$

which proves the theorem.

Definition 2. A regular map $f : X \to Y$ of quasiprojective varieties is *finite* if any point $y \in Y$ has an affine neighbourhood V such that the set $U = f^{-1}V$ is affine and $f : U \to V$ is a finite map between affine varieties.

Obviously, for a finite map f the set $f^{-1}(y)$ is finite for every $y \in Y$. It follows from Theorem 4 that any finite map is surjective. This property has important consequences, that relate to arbitrary maps.

Theorem 6. *If $f : X \to Y$ is a regular map and $f(X)$ is dense in Y then $f(X)$ contains an open set of Y.*

Proof. The assertion of the theorem reduces at once to the case that both X and Y are irreducible and affine, and we assume this in what follows. Then $k[Y] \subset k[X]$. We write r for the transcendence degree of the field extension $k(X)/k(Y)$, and choose r elements $u_1, \ldots, u_r \in k[X]$ that are algebraically independent over $k(Y)$. Then

$$k[X] \supset k[Y][u_1, \ldots, u_r] \supset k[Y] \quad \text{and} \quad k[Y][u_1, \ldots, u_r] = k[Y \times \mathbf{A}^r].$$

This represents f as the composite $f = g \circ h$ of two maps $h : X \to Y \times \mathbf{A}^r$ and $g : Y \times \mathbf{A}^r \to Y$, where g is simply the projection to the first factor. Any element $v \in k[X]$ is algebraic over $k[Y \times \mathbf{A}^r]$, hence there exists an element $a \in k[Y \times \mathbf{A}^r]$ such that av is integral over $k[Y \times \mathbf{A}^r]$. Let $v_1, \ldots \; v_m$ be a system of generators of $k[X]$, and $a_1, \ldots, a_m \in k[Y \times \mathbf{A}^r]$ elements such that each $a_i v_i$ is integral over $k[Y \times \mathbf{A}^r]$, and set $F = a_1 \cdots a_m$. Since all the functions a_i are invertible on the principal open set $D(F) \subset Y \times \mathbf{A}^r$, the functions v_i on $D(h^*(F)) \subset X$ are integral over $k[Y \times \mathbf{A}^r][1/F]$, that is, the restricted map

$$h : D(h^*(F)) \to D(F)$$

is finite. Thus $h(D(h^*(F))) = D(F)$ by Theorem 4, so that $D(F) \subset h(X)$. It remains to prove that $g(D(F))$ contains an open set of Y. Suppose that

$$F = F(y, T) = \sum F_\alpha(y) T^\alpha,$$

where T^α are monomials in the variables $T_1, \ldots, T_r$, the coordinates of $\mathbb{A}^r$. For points $y \in Y$ at which not all $F_\alpha(y) = 0$, there exist values $T_i = \tau_i$ for which $F(y, \tau) \neq 0$. Hence $g(D(F)) \supset \bigcup D(F_\alpha)$. This proves Theorem 6.

Theorem 6 shows one respect in which regular maps of algebraic varieties are simpler than continuous or differentiable maps. The famous example of an everywhere dense line in the torus $T = \mathbb{R}^2/\mathbb{Z}^2$, a map such as

$$f \colon \mathbb{R}^1 \to T \qquad \text{given by} \qquad f(x) = f(x, \sqrt{2}x) \bmod \mathbb{Z}^2$$

is an example of a situation that cannot happen for algebraic varieties, by Theorem 6.

Theorem 7. *If $X \subset \mathbb{P}^n$ is a closed subvariety disjoint from a d-dimensional linear subspace $E \subset \mathbb{P}^n$ then the projection $\pi \colon X \to \mathbb{P}^{n-d-1}$ with centre E (see 4.4, Example 1) defines a finite map $X \to \pi(X)$.*

Proof. Let $y_0, \ldots, y_{n-d-1}$ be homogeneous coordinates on $\mathbb{P}^{n-d-1}$, and suppose that π is given by $y_j = L_j(x)$ for $j = 0, \ldots, n-d-1$, where $x \in X$. Obviously $U_i = \pi^{-1}(\mathbb{A}_i^{n-d-1}) \cap X$ is given by the condition $L_i(x) \neq 0$, and is an affine open subset of X. We prove that $\pi \colon U_i \to \mathbb{A}_i^{n-d-1} \cap \pi(X)$ is a finite map. Any function $g \in k[U_i]$ is of the form $g = G_i(x_0, \ldots, x_n)/L_i^m$, where G_i is a form of degree m. Consider the map $\pi_1 \colon X \to \mathbb{P}^{n-d}$ given by $z_j = L_i^m(x)$ for $j = 0, \ldots, n-d-1$ and $z_{n-d} = G_i(x)$, where $z_0, \ldots, z_{n-d}$ are homogeneous coordinates in $\mathbb{P}^{n-d}$. This is a regular map, and its image $\pi_1(X) \subset \mathbb{P}^{n-d}$ is closed by 5.2, Theorem 2. Suppose that $\pi_1(X)$ is given by equations $F_1 = \cdots = F_s = 0$.

Since X is disjoint from E, the forms L_i for $i = 0, \ldots, n-d-1$ have no common zeros on X. Hence the point $0 = (0 : \cdots : 0 : 1) \in \mathbb{P}^{n-d}$ is not contained in $\pi_1(X)$, or in other words, the equations $z_0 = \cdots = z_{n-d-1} = F_1 = \cdots = F_s = 0$ do not have solutions in $\mathbb{P}^{n-d}$. By 4.1, Lemma 1, it follows from this that $(z_0, \ldots, z_{n-d-1}, F_1, \ldots, F_s) \supset I_k$ for some $k > 0$. In particular, $(z_0, \ldots, z_{n-d-1}, F_1, \ldots, F_s) \ni z_{n-d}^k$. This means that we can write

$$z_{n-d}^k = \sum_{j=0}^{n-d-1} z_j H_j + \sum_{j=1}^{s} F_j P_j,$$

where H_j and P_j are polynomials. Writing $H^{(q)}$ for the homogeneous component of H of degree q, we deduce from this that

$$\Phi(z_0, \ldots, z_{n-d}) = z_{n-d}^k - \sum z_j H_j^{(k-1)} = 0 \quad \text{on } \pi_1(X). \tag{2}$$

The homogeneous polynomial Φ is of degree k and as a polynomial in z_{n-d} it has leading coefficient 1:

$$\Phi = z_{n-d}^k - \sum_{j=0}^{k-1} A_{k-j}(z_0, \ldots, z_{n-d-1}) z_{n-d}^j. \tag{3}$$

If we substitute in (2) the formulas defining the map $\pi_1 \colon X \to \mathbb{P}^{n-d}$, we get that $\Phi(L_0^m, \ldots, L_{n-d-1}^m, G_i) = 0$ on X, with Φ of the form (3). Dividing this relation by L_i^{mk} we get the required relation

$$g^k - \sum_{j=0}^{k-1} A_{k-j}(x_0^m, \ldots, 1, \ldots, x_{n-d-1}^m) g^j = 0,$$

where $x_r = y_r/y_i$ are coordinates in $\mathbb{A}_0^{n-d-1}$. The theorem is proved.

Using the Veronese embedding (4.4, Example 2) allows the following substantial generalisation of Theorem 7.

Theorem 8. *Suppose that $F_0, \ldots, F_s$ are forms of degree m on $\mathbb{P}^n$ having no common zeros on a closed variety $X \subset \mathbb{P}^n$. Then*

$$\varphi(x) = \big(F_0(x) : \cdots : F_s(x)\big)$$

defines a finite map $\varphi \colon X \to \varphi(X)$.

Proof. Let $v_m \colon \mathbb{P}^n \to \mathbb{P}^N$ be the Veronese embedding (with $N = \binom{n+m}{m} - 1$) and L_i the linear forms on $\mathbb{P}^N$ corresponding to the forms F_i on $\mathbb{P}^n$. Then obviously $\varphi = \pi \circ v_m$ where π is the projection defined by the linear forms $L_0, \ldots, L_s$. Since $v_m \colon X \to v_m(X)$ is an isomorphism, Theorem 8 follows from Theorem 7.

5.4. Noether Normalisation

Consider an irreducible projective variety $X \subset \mathbb{P}^n$ distinct from the whole of $\mathbb{P}^n$. Then there exists a point $x \in \mathbb{P}^n \setminus X$, and the map φ obtained by projecting X away from x will be regular. The image $\varphi(X) \subset \mathbb{P}^{n-1}$ is projective by Theorem 2, and the map $\varphi \colon X \to \varphi(X)$ is finite by Theorem 7. If $\varphi(X) \ne \mathbb{P}^{n-1}$ then we can repeat the same argument for it. We finally arrive at a map $X \to \mathbb{P}^m$, which is finite, since it is a composite of finite maps. The result we have proved is called the Noether normalisation theorem:

Theorem 9. *For an irreducible projective variety X there exists a finite map $\varphi \colon X \to \mathbb{P}^m$ to a projective space.* $\square$

The analogous result also holds for affine varieties. To prove this, consider an affine variety $X \subset \mathbb{A}^n$. Embed $\mathbb{A}^n$ as an open $\mathbb{A}^n \subset \mathbb{P}^n$, and write $\overline{X}$ for the projective closure of X in $\mathbb{P}^n$. Suppose that $X \ne \mathbb{A}^n$. We choose a point at infinity $x \in \mathbb{P}^n \setminus \mathbb{A}^n$ with $x \notin \overline{X}$, and consider the projection $\varphi \colon \overline{X} \to \mathbb{P}^{n-1}$ from this point. Here X will map to points in the finite part of $\mathbb{P}^{n-1}$, that is, to points of $\mathbb{A}^{n-1} = \mathbb{P}^{n-1} \cap \mathbb{A}^n$. We can repeat this process as long as $X \neq \mathbb{A}^n$, and as a result we arrive at a projection $\varphi \colon \overline{X} \to \mathbb{P}^m$ for which $\varphi(X) = \mathbb{A}^m$. This proves the following result.

Theorem 10. *For an irreducible affine variety X there exists a finite map $\varphi \colon X \to \mathbb{A}^m$ to an affine space.* $\square$

Theorems 9–10 allow us to reduce the study of certain (very coarse) properties of projective and affine varieties to the case of projective and affine spaces. When $m = 1$ this point of view is due to Riemann, who considered algebraic curves as coverings of the Riemann sphere ($\mathbb{P}^1$ over the complex number field $\mathbb{C}$).

Theorem 10 means that an integral domains A that is finitely generated over the field k is integral over a subring isomorphic to a polynomial ring. This result can also easily be proved directly.

Exercises to §5

1. Prove that the Segre variety $\varphi(\mathbb{P}^n \times \mathbb{P}^m) \subset \mathbb{P}^N$ (where $N = (n+1)(m+1) - 1$) is not contained in any linear subspace strictly smaller than the whole of $\mathbb{P}^N$.

2. Consider the two maps of varieties $\mathbb{P}^1 \times \mathbb{P}^1 \to \mathbb{P}^1$ given by $p_1(x, y) = x$ and $p_2(x, y) = y$. Prove that $p_1(X) = p_2(X) = \mathbb{P}^1$ for any closed irreducible subset $X \subset \mathbb{P}^1 \times \mathbb{P}^1$, unless X is of one of the following types: (a) a point $(x_0, y_0) \in \mathbb{P}^1 \times \mathbb{P}^1$; (b) a line $x_0 \times \mathbb{P}^1$ for $x_0 \in \mathbb{P}^1$ a fixed point; (c) a line $\mathbb{P}^1 \times y_0$.

3. Verify Theorem 2, Corollary 1 directly for the case $X = \mathbb{P}^n$.

4. Let $X = \mathbb{A}^2 \setminus x$ where x is a point. Prove that X is not isomorphic to an affine nor a projective variety (compare §4, Ex. 3).

5. The same question as Ex. 4, for $X = \mathbb{P}^2 \setminus x$.

6. The same question as Ex. 4, for $X = \mathbb{P}^1 \times \mathbb{A}^1$.

7. Is the map $f \colon \mathbb{A}^1 \to X$ finite, where X is given by $y^2 = x^3$, and f by $f(t) = (t^2, t^3)$.

8. Let $X \subset \mathbb{A}^r$ be a hypersurface of $\mathbb{A}^r$ and L a line of $\mathbb{A}^r$ through the origin. Let φ_L be the map projecting X parallel to L to an $(r-1)$-dimensional subspace not containing L. Write S for the set of all lines L such that φ_L is not finite. Prove that S is an algebraic variety.[Hint: Prove that $S = \overline{X} \cap \mathbb{P}_\infty^{r-1}$.] Find S if $r = 2$ and X is given by $xy = 1$.

9. Prove that any intersection of affine open subsets is affine. [Hint: Use 2.3, Example 10.]

10. Prove that forms of degree $m = kl$ in $n+1$ variables that are lth powers of forms correspond to the points of a closed subset of $\mathbb{P}^N$, where $N = \binom{n+m}{m} - 1 = \nu_{n,m}$.

11. Let $f\colon X \to Y$ be a regular map of affine varieties. Prove that the inverse image of a principal affine open set is a principal affine open set.

6. Dimension

6.1. Definition of Dimension

In §2 we saw that closed algebraic subvarieties $X \subset \mathbb{A}^2$ are finite sets of points, algebraic plane curves, and $\mathbb{A}^2$ itself. This division into three cases corresponds to the intuitive notion of dimension, with varieties of dimension 0, 1 and 2. Here we give the definition of the dimension of an arbitrary algebraic variety.

How could we arrive at this definition? First, of course, we take the dimension of $\mathbb{P}^n$ and $\mathbb{A}^n$ to be n. Secondly, if there exists a finite map $X \to Y$ then it is natural to suppose that X and Y have the same dimension. Since by Noether normalisation (5.4, Theorems 9–10), any projective or affine variety X has a finite map to some $\mathbb{P}^m$ or $\mathbb{A}^m$, it is natural to take m as the definition of the dimension of X. However, the question then arises as to whether this is well defined: might there not exist two finite maps $f\colon X \to \mathbb{A}^m$ and $g\colon X \to \mathbb{A}^n$ with $m \neq n$? Suppose that X is irreducible. Then the finiteness of a regular map $f\colon X \to \mathbb{A}^m$ implies that the rational function field $k(X)$ is a finite extension of the field $f^*(k(\mathbb{A}^m))$, which is in turn isomorphic to $k(t_1,\ldots,t_m)$. Hence $k(X)$ has transcendence degree m over k; this gives a characterisation of the number m independent of the choice of the finite map $f\colon X \to \mathbb{A}^m$. This gives some motivation for the definition of dimension.

Definition. The *dimension* of an irreducible quasiprojective variety X is the transcendence degree of the function field $k(X)$; it is denoted by $\dim X$. The dimension of a reducible variety is the maximum of the dimension of its irreducible components. If $Y \subset X$ is a closed subvariety of X then the number $\dim X - \dim Y$ is called the *codimension* of Y in X, and written $\operatorname{codim} Y$ or $\operatorname{codim}_X Y$. Algebraic varieties of dimension 1 and 2 are called *curves* and *surfaces*.[3]

Note that if X is an irreducible variety and $U \subset X$ is open then $k(U) = k(X)$, and hence $\dim U = \dim X$.

---

[3] n-dimensional varieties are often called *n-folds*, for example 3-folds, 4-folds (or threefolds, fourfolds).

Example 1. $\dim \mathbf{A}^n = \dim \mathbb{P}^n = n$, because the field $k(\mathbf{A}^n)$ is the field of rational functions in n variables. Since dimension is by definition invariant under birational equivalence, we see that $\mathbf{A}^n$ and $\mathbf{A}^m$ are not birational if $n \neq m$.

Example 2. An irreducible plane curve is 1-dimensional, as we saw in 1.3.

Example 3. If X consists of a single point then obviously $\dim X = 0$, and thus the same holds if X is a finite set. Conversely, if $\dim X = 0$ then X is a finite set. It is enough to prove this for an irreducible affine variety X. Let $X \subset \mathbf{A}^n$, and write $t_1, \ldots, t_n$ for the coordinates on $\mathbf{A}^n$ as functions on X, that is, as elements of $k[X]$. By assumption the t_i are algebraic over k, and can hence only take finitely many values. It follows from this that X is finite.

Example 4. We prove that if X and Y are irreducible varieties then

$$\dim(X \times Y) = \dim X + \dim Y.$$

We need only consider the case that $X \subset \mathbf{A}^N$ and $Y \subset \mathbf{A}^M$ are affine varieties. Suppose that $\dim X = n$, $\dim Y = m$, and let $t_1, \ldots, t_N$ and $u_1, \ldots, u_M$ be coordinates of $\mathbf{A}^N$ and $\mathbf{A}^M$ considered as functions on X and Y respectively, such that $t_1, \ldots, t_n$ are algebraically independent in $k(X)$ and $u_1, \ldots, u_m$ in $k(Y)$. By definition $k[X \times Y]$ is generated by the elements $t_1, \ldots, t_N, u_1, \ldots, u_M$, and under the current assumptions all of these are algebraically dependent on $t_1, \ldots, t_n, u_1, \ldots, u_m$. Hence it is enough to prove that these elements are algebraically independent. Suppose that there is a relation $F(T, U) = F(T_1, \ldots, T_n, U_1, \ldots, U_m) = 0$ on $X \times Y$. Then for any point $x \in X$ we have $F(x, U_1, \ldots, U_m) = 0$ on Y. Since $u_1, \ldots, u_m$ are algebraically independent in $k(Y)$, every coefficient $a_i(x)$ of the polynomial $F(x, U)$ is zero; this means that the corresponding polynomial $a_i(T_1, \ldots, T_n)$ is 0 on X. Now we use the fact that $t_1, \ldots, t_n$ are algebraically independent in $k(X)$ and deduce from this that $a_i(T_1, \ldots, T_n) = 0$, and hence $F(T, U)$ is identically 0.

Example 5. The Grassmannian $\mathrm{Grass}(r, n)$ (see 4.1, Example 1) is covered by open sets $p_{i_1 \ldots i_r} \neq 0$ isomorphic to the affine space $\mathbf{A}^{r(n-r)}$. Thus $\dim \mathrm{Grass}(r, n) = r(n - r)$. It also follows from this that $\mathrm{Grass}(r, n)$ is rational.

Theorem 1. *If $X \subset Y$ then $\dim X \leq \dim Y$. If Y is irreducible and $X \subset Y$ is a closed subvariety with $\dim X = \dim Y$ then $X = Y$.*

Proof. It is enough to prove the assertions for X and Y irreducible affine varieties.

Suppose $X \subset Y \subset \mathbf{A}^N$ with $\dim Y = n$. Then any $n + 1$ of the coordinate functions $t_1, \ldots, t_N$ are algebraically dependent as elements of $k[Y]$, that is, are connected by a relation $F(t_{i_1}, \ldots, t_{i_{n+1}}) = 0$ on Y. A fortiori this holds on X. This means that the transcendence degree of $k(X)$ is at most n, so that $\dim X \leq \dim Y$.

Now suppose that $\dim X = \dim Y = n$. Then some n of the coordinates $t_1, \ldots, t_N$ are algebraically independent on X; suppose that these are $t_1, \ldots, t_n$. Then a fortiori they are algebraically independent on Y. Let $u \in k[Y]$ with $u \neq 0$ on Y. Then u on Y is algebraically dependent on $t_1, \ldots, t_n$, that is, there is a polynomial $a(t, U) \in k[t_1, \ldots, t_n][U]$ such that the relation

$$a_0(t_1, \ldots, t_n)u^k + \cdots + a_k(t_1, \ldots, t_n) = 0 \tag{1}$$

holds on Y. We can choose $a(t, U)$ to be irreducible, and then $a_k(t_1, \ldots, t_n) \neq 0$ on Y. Relation (1) holds a fortiori on X. Suppose that $u = 0$ on X. Then (1) implies that $a_k(t_1, \ldots, t_n) = 0$ on X. Since by assumption $t_1, \ldots, t_n$ are independent on X, it follows that $a_k(t_1, \ldots, t_n) = 0$ on the whole of $\mathbf{A}^N$. This contradicts $a_k(t_1, \ldots, t_n) \neq 0$ on Y. Thus if $u = 0$ on X then also $u = 0$ on Y, and therefore $X = Y$. The theorem is proved.

We have seen that an irreducible algebraic plane curve is 1-dimensional. The following result is a generalisation.

Theorem 2. *Every irreducible component of a hypersurface in $\mathbf{A}^n$ or $\mathbb{P}^n$ has codimension 1.*

Proof. It is enough to consider the case of a hypersurface in $\mathbf{A}^n$. Suppose that a variety $X \subset \mathbf{A}^n$ is given by an equation $F(T) = 0$. The factorisation $F = F_1 \ldots F_k$ of F into irreducible factors corresponds to an expression $X = X_1 \cup \cdots \cup X_k$, where X_i is defined by $F_i = 0$. It is obviously sufficient to prove the theorem for each variety X_i. Let us prove that X_i is irreducible: if X_i were reducible, there would exist polynomials G and H such that $GH = 0$ on X_i but $G, H \neq 0$ on X. From the Nullstellensatz it follows that $F_i \mid (GH)^l$ for some $l > 0$. Since F_i is irreducible it follows from this that $F_i \mid G$ or $F_i \mid H$, and this contradicts $G \neq 0$, $H \neq 0$ on X_i.

Suppose that the variable T_n actually appears in the polynomial $F_i(T)$, and prove that the coordinates $t_1, \ldots, t_{n-1}$ are algebraically independent on X_i. Indeed, a relation $G(t_1, \ldots, t_{n-1}) = 0$ on X_i would imply that $F_i \mid G^l$ for some $l > 0$, which is impossible since G does not involve T_n. Thus $\dim X_i \geq n - 1$; since $X \neq \mathbf{A}^n$, it follows from Theorem 1 that $\dim X_i = n - 1$. Theorem 2 is proved.

Theorem 3. *Let $X \subset \mathbf{A}^n$ be a variety, and suppose that all the components of X have dimension $n - 1$. Then X is a hypersurface and the ideal $\mathfrak{A}_X$ is principal.*

Proof. We only need consider the case that X is irreducible. Since $X \neq \mathbf{A}^n$ (because $\dim X = n - 1$), there exists a nonzero polynomial F which is zero on X. Since X is irreducible, some irreducible factor H of F is also zero on X. Write $Y \subset A^n$ for the hypersurface defined by $H = 0$; we saw in the proof of Theorem 2 that Y is irreducible. Then $X \subset Y$, so that $X = Y$ by Theorem 1. If $G \in \mathfrak{A}_X$ then by the Nullstellensatz $H \mid G^l$ for some $l > 0$, and then $G \in (H)$ by the irreducibility of H, that is $\mathfrak{A}_X = (H)$.

Theorem 3 is proved.

The following analogue of Theorem 3 is proved similarly:

Theorem 3'. *Let $X \subset \mathbb{P}^{n_1} \times \cdots \times \mathbb{P}^{n_k}$ be a variety, and suppose that all the components of X have dimension $n_1 + \cdots + n_k - 1$. Then X is defined by one equation that is homogeneous in each of the k sets of variables.*

Proof. We need only replace the unique factorisation of polynomials used in the proof of Theorem 3 by the unique factorisation of polynomials that are homogeneous in each of the k groups of variables into irreducible polynomials of the same type. This comes from the fact that if $F(x_0, \ldots, x_{n_1}, y_0, \ldots, y_{n_2}, \ldots, u_0, \ldots, u_{n_k})$ is homogeneous in each of the k sets of variables $\{x_0, \ldots, x_{n_1}\}, \ldots, \{u_0, \ldots, u_{n_k}\}$ and F factorises as $F = G \cdot H$, then G and H have the same homogeneity property. Theorem 3' is proved.

6.2. Dimension of Intersection with a Hypersurface

If we try to study varieties defined by more than one equation, we come up at once against the question of the dimension of intersection of a variety with a hypersurface. We study this question first for projective varieties. If X is closed in $\mathbb{P}^N$ and a form F is not zero on X then we write X_F for the closed subvariety of X defined by $F = 0$.

For any projective variety $X \subset \mathbb{P}^N$ we can find a form $G(U_0, \ldots, U_N)$ of any specified degree m which does not vanish on any components X_i of X. For this, it is enough to choose one point $x_i \in X_i$ in each irreducible component of X, and find a linear form L not vanishing on any of these; then we can take $G = L^m$ to be the appropriate power of L. Suppose that $X \subset \mathbb{P}^N$ is closed, and that a form F is not zero on any component of X. By Theorem 1 we have $\dim X_F < \dim X$. Set $X_F = X^{(1)}$ and apply the same argument to $X^{(1)}$, finding a form F_1 with $\deg F_1 = \deg F$ not vanishing on any component of $X^{(1)}$. We get a chain of varieties $X^{(i)}$ and forms F_i such that

$$X = X^{(0)} \supset X^{(1)} \supset \cdots, \quad \text{with } X^{(i+1)} = X_{F_i}^{(i)} \text{ and } F_0 = F. \qquad (1)$$

By Theorem 1, $\dim X^{(i+1)} < \dim X^{(i)}$. Hence if $\dim X = n$, then $X^{(n+1)}$ is empty. In other words, the forms $F_0 = F, F_1, \ldots, F_n$ have no common zeros on X.

Suppose now that X is irreducible. Consider the map $\varphi\colon X \to \mathbb{P}^n$ given by

$$\varphi(x) = (F_0(x) : \cdots : F_n(x)). \tag{2}$$

This map satisfies the assumptions of 5.3, Theorem 8, and by this theorem the map $X \to \varphi(X)$ is finite. But if $X \to Y$ is a finite map then, as we have seen, $\dim X = \dim Y$. Hence $\dim \varphi(X) = \dim X = n$, and since $\varphi(X) \subset \mathbb{P}^n$ is closed by 5.2, Theorem 2, we get $\varphi(X) = \mathbb{P}^n$ by 6.1, Theorem 1. Suppose now that $\dim X^{(1)} = \dim X_F < n - 1$. Then in (1), already $X^{(n)}$ is empty. In other words, the forms $F_0, \ldots, F_{n-1}$ have no common zeros on X. This means that the point $(0 : \cdots : 0 : 1)$ is not contained in $\varphi(X)$, which contradicts $\varphi(X) = \mathbb{P}^n$. Thus we have proved the following result.

Theorem 4. *If a form F is not 0 on an irreducible projective variety X then* $\dim X_F = \dim X - 1$. $\square$

Recall that this means that X_F contains one or more irreducible components of dimension $\dim X - 1$.

Corollary 1. *A projective variety X contains subvarieties of any dimension $s < \dim X$.* $\square$

Corollary 2 (Inductive definition of dimension). *If X is an irreducible projective variety then $\dim X = 1 + \sup \dim Y$, where Y runs through all proper subvarieties of X.* $\square$

Corollary 3. *The dimension of a projective variety X can be defined as the maximal integer n for which there exists a strictly decreasing chain $Y_0 \supsetneqq Y_1 \supsetneqq \cdots \supsetneqq Y_n \supsetneqq \emptyset$ of length n of irreducible subvarieties $Y_i \subset X$.* $\square$

Corollary 4. *The dimension n of a projective variety $X \subset \mathbb{P}^N$ can be defined as $N - s - 1$, where s is the maximum dimension of a linear subspace of $\mathbb{P}^N$ disjoint from X.*

Proof. Let $E \subset \mathbb{P}^N$ be a linear subspace of dimension s. If $s \geq N - n$ then E can be defined by $\leq n$ equations, and successive application of Theorem 4 proves that $\dim(X \cap E) \geq 0$, and hence $X \cap E$ is nonempty (the dimension of the empty set is -1!). Setting $m = 1$ in the construction of the chain (1) gives $n + 1$ linear forms $L_0, \ldots, L_n$ with no common zeros on X. If E is the linear subspace defined by these, then $\dim E = N - n - 1$ and $X \cap E$ is empty. Corollary 4 is proved.

Corollary 5. *The variety of common zeros of r forms $F_1, \ldots, F_r$ on an n-dimensional projective variety has dimension $\geq n - r$.*

The proof is by $r - 1$ applications of Theorem 4. Corollary 5 provides a rather strong existence theorem.

Proposition. *If $r \leq n$ then r forms have a common zero on an n-dimensional projective variety. For example, in the case $X = \mathbb{P}^n$, this says that n homogeneous equations in $n + 1$ variables have a nonzero solution.* $\square$

This existence theorem allows us to make a number of important deductions.

Corollary 1. *Any two curves of $\mathbb{P}^2$ intersect.* $\square$

This is clear, since a curve is given by a single homogeneous equation. However, there exist nonintersecting curves on a nonsingular quadric surface $Q \subset \mathbb{P}^3$, for example the lines of one family of generators. Therefore $\mathbb{P}^2$ and Q are not isomorphic. Since they are birational (3.3, Example 1) we get an example of two varieties that are birational but not isomorphic. This example will appear again later (Chap. II, 4.1 and 4.5, Chap. IV, 2.3, Example 2).

Corollary 2. *Theorem 3 fails already for the curves on a nonsingular quadric surface Q: there exist curves $C \subset Q$ that cannot defined by setting to zero a single form on $\mathbb{P}^3$.*

Indeed, if we assume that each of the disjoint curves C_1 and C_2 which we found on Q is defined by one equation $F_1 = 0$ and $F_2 = 0$, we get a contradiction to Corollary 5, according to which the system of equations $G = F_1 = F_2 = 0$ have a common solution (where G is the equation of Q).

Corollary 3. *Any curve of degree ≥ 3 has an inflexion point.*

Proof. We have seen in 1.6 that the inflexion points of an algebraic plane curve with equation $F = 0$ is defined by $H(F) = 0$, where $H(F)$ is the Hessian form of F. If F has degree n then $H(F)$ has degree $3(n - 2)$. Therefore for $n \geq 3$ the system of equations $F = H(F) = 0$ has a nonzero solution; that is, the curve $F = 0$ has an inflexion point. Corollary 3 is proved.

The simplest case is when $n = 3$. We see that every cubic curve in $\mathbb{P}^2$ has an inflexion point. Choose a coordinate system (ξ_0, ξ_1, ξ_2) so that the inflexion point is $(0, 0, 1)$, and the inflexional tangent is the line $\xi_1 = 0$. Setting $u = \xi_0/\xi_2$, $v = \xi_1/\xi_2$, we see easily that our assumption is equivalent to saying that the equation $\varphi(u, v)$ of the curve has no constant term, or term in u or u^2. Changing to coordinates $x = \xi_0/\xi_1$, $y = \xi_2/\xi_1$, so that the inflexion point is at infinity, we find that the equation of our cubic has no term in y^3, $y^2 x$ or $y x^2$, that is, it is of the form $a y^2 + (bx + c)y + g(x) = 0$, where g is a polynomial of degree ≤ 3. If $a = 0$ then the inflexion point is singular. If $a \neq 0$ we can assume that $a = 1$. Assuming that char $k \neq 2$, we can complete the square by setting $y_1 = y + (1/2)(bx + c)$ and reduce the equation to the form $y_1^2 = g_1(x)$, where $g_1(x)$ has degree ≤ 3, and $= 3$ if the cubic curve is nonsingular. Thus the equation of a nonsingular cubic has Weierstrass normal form in some coordinate system. In 1.4 we proved only the weaker statement that a cubic is isomorphic to a curve with equation in Weierstrass normal form.

Corollary 4. (Tsen's theorem). *Let $F(x_1, \ldots, x_n)$ be a form in n variables of degree $m < n$ whose coefficients are polynomials in one variable t. Then the equation $F(x_1, \ldots, x_n) = 0$ has a solution in polynomials $x_i = p_i(t)$.*

Proof. We look for x_i of the form $x_i = \sum_{j=0}^{l} u_{ij} t^j$ with unknown coefficients u_{ij}. Substituting these expressions in the equation $F(x_1, \ldots, x_n) = 0$, we get a polynomial in t all of whose coefficients must be set to 0. If the maximum of the degrees of the coefficients of a polynomial F equals k then the number of equations is at most $ml + k + 1$. The number of indeterminates is $n(l + 1)$. Since by assumption $n > m$, for l sufficiently large, the number of unknowns is greater than the number of equations, and hence the system has a nonzero solution. $\blacksquare$

Example 1. An important particular case of Tsen's theorem is when $n = 3$ and F is a quadratic form. It can be given the following geometric interpretation: suppose that a surface $X \subset \mathbb{P}^2 \times \mathbb{A}^1$ is defined by the equation

$$q(x_0 : x_1 : x_2; t) = \sum_{i,j=0}^{2} a_{ij}(t) x_i x_j \qquad \text{with } a_{ij}(t) \in k[t],$$

where $(x_0 : x_1 : x_2)$ are coordinates in $\mathbb{P}^2$ and t a coordinate on $\mathbb{A}^1$. The fibres of the map $X \to \mathbb{A}^1$ are the conics $q(x_0 : x_1 : x_2; a) = 0$ for $a \in \mathbb{A}^1$, and the surface is called a *conic bundle* or *pencil of conics*. Tsen's theorem proves that a pencil of conics has a section, that is, there exists a regular map $\varphi \colon \mathbb{A}^1 \to X$ such that $\varphi(a)$ is a point of the fibre over a for every $a \in \mathbb{A}^1$.

Another interpretation of this result is as follows. Consider our surface X as the conic C with equation $q(x_0 : x_1 : x_2; t) = \sum_{i,j=0}^{2} a_{ij} x_i x_j = 0$ in $\mathbb{P}^2$ over the algebraically nonclosed field $K = k(t)$. Obviously $K(C) = k(X)$. Then C has a point with coordinates in K.

We assume that the curve C is irreducible for a general point $t \in \mathbb{A}^1$, that is, that $\det |a_{ij}(t)|$ is not identically 0; we say that the pencil of conics is *nondegenerate* in this case. In 1.2 we saw that the conic is then rational, with the birational map to $\mathbb{P}^1$ defined over $K = k(t)$. In other words, the field $K(C)$ is isomorphic over K to the field $K(x)$, and since $K(C) = k(X)$ it follows that $k(X)$ is isomorphic to $K(x) = k(t, x)$. We have proved the next result.

Corollary 5. *A nondegenerate pencil of conics over $\mathbf{A}^1$ is a rational surface.* $\square$

Theorem 5. *Under the assumptions of Theorem 4, every component of X_F has dimension* $\dim X - 1$.

Proof. Consider the finite map $\varphi \colon X \to \mathbb{P}^n$ (with $n = \dim X$) constructed in the proof of Theorem 4, and let $\mathbf{A}_i^n \subset \mathbb{P}^n$ for $i = 0, \ldots, n$ be the affine open sets covering $\mathbb{P}^n$. Then using the Veronese embedding with $m = \deg F$, it is easy to see that $\varphi^{-1}(\mathbf{A}_i^n) = U_i$ are affine open sets of X. It is obviously enough to prove that each component of the affine variety $X_F \cap U_i$ has dimension $n - 1$ for each i. From now on our arguments apply to some fixed U_i, which we denote by U. Obviously $X_F \cap U = V(f)$, where $f = F/F_i$, that is, X_F coincides on U with the set of zeros of the regular function $f \in k[U]$. We constructed above a finite map $\varphi \colon U \to \mathbf{A}^n$, given by n regular functions $f_1, \ldots, f_n$, with $f = f_1$.

To prove that each component of $V(f)$ has dimension $n - 1$, we only need to prove that it has dimension $\geq n - 1$. We prove that the functions $f_2, \ldots, f_n$ are algebraically independent on each component. Let $P \in k[T_2, \ldots, T_n]$. To prove that $R = P(f_2, \ldots, f_n)$ does not vanish on any component of $V(f)$ it is enough to prove that for $Q \in k[U]$,

$$RQ = 0 \text{ on } V(f) \implies Q = 0 \text{ on } V(f).$$

Indeed, if $V(f) = U^{(1)} \cup \cdots \cup U^{(t)}$ is an irredundant decomposition into irreducible components, and $R = 0$ on $U^{(1)}$, then take Q to be any function that vanishes on $U^{(2)} \cup \cdots \cup U^{(t)}$ but not on $U^{(1)}$. Then $RQ = 0$ on $V(f)$ but $Q \neq 0$ on $V(f)$.

By the Nullstellensatz our assertion can be restated as follows: if $f \mid (RQ)^l$ for some $l > 0$ then $f \mid Q^k$ for some $k > 0$. Thus Theorem 5 follows from the following purely algebraic fact:

Lemma. *Set $B = k[T_1, \ldots, T_n]$, and let $A \supset B$ be an integral domain that is integral over B; write $x = T_1$, and let $y = P(T_2, \ldots, T_n) \neq 0$. Then for any $u \in A$,*

$$x \mid (yu)^l \text{ in } A \text{ for some } l > 0 \implies x \mid u^k \text{ for some } k > 0.$$

Proof of the lemma. The only property of x and y that we use is that they are relatively prime in the UFD $k[T_1, \ldots, T_n]$. Note that we can replace y^l by z and u^l by v, and then it is enough to prove that if x and z are relatively prime in $k[T_1, \ldots, T_n]$ then $x \mid zv$ in A implies that $x \mid v^k$ for some $k > 0$. Thus the lemma asserts that the property of polynomials x, $z \in B$ being relatively prime is in a certain sense preserved on passing to a ring A that is integral over B.

Write K for the field of fractions of B. If $t \in A$ is integral over B then it is algebraic over K. Let $F(T) \in K[T]$ be the minimal polynomial of t over K, that is, the polynomial of least degree with leading coefficient 1 such that $F(t) = 0$. Division with remainder shows that any polynomial $G(T) \in K[T]$ with $G(t) = 0$ is divisible by $F(T)$ in $K[T]$. Now from this it follows that t is integral over B if and only if $F[T] \in B[T]$. Indeed, if t is integral and $G(t) = 0$ for $G \in B[T]$ with leading coefficient 1, then $G(T) = F(T)H(T)$ in $K[T]$. But $B = k[T_1, \ldots, T_n]$ is a UFD, so a simple application of Gauss' lemma shows that $F(T), H(T) \in B[T]$.

It is now easy to complete the proof of the lemma. Suppose that $zv = xw$ with v, $w \in A$ and let $F(T) = T^k + b_1 T^{k-1} + \cdots + b_k$ be the minimal polynomial of w. Since w is integral over B, the coefficients b_i of F satisfy $b_i \in B$. It is easy to see that the minimal polynomial $G(T)$ of $v = xw/z$ is given by $(x/z)^k F(zT/x)$. Therefore

$$G(T) = T^k + \frac{xb_1}{z}T^{k-1} + \cdots + \frac{x^k b_k}{z^k},$$

$$\text{and} \quad v^k + \frac{xb_1}{z}v^{k-1} + \cdots + \frac{x^k b_k}{z^k} = 0. \tag{3}$$

Since v is integral over B, also $x^i b_i / z^i \in B$, and because x and z are relatively prime it follows that $z^i \mid b_i$. It then follows from (3) that $x \mid v^k$. The lemma is proved, and with it Theorem 5.

Corollary 1. *If $X \subset \mathbb{P}^N$ is an irreducible quasiprojective variety and F a form that is not identically 0 on X, then every (nonempty) component of X_F has codimension 1. ($X_F = \emptyset$ is of course possible for quasiprojective varieties.)*

Proof. By definition X is open in some closed subset $\overline{X} \subset \mathbb{P}^N$. Since X is irreducible, so is $\overline{X}$, and hence $\dim \overline{X} = \dim X$. By Theorem 5, $(\overline{X})_F = \bigcup Y_i$ with $\dim Y_i = \dim X - 1$. But it is easy to see that $X_F = (\overline{X})_F \cap X$; it follows that $X_F = \bigcup (Y_i \cap X)$, and $Y_i \cap X$ is either empty or is open in Y_i, so that $\dim(Y_i \cap X) = \dim X - 1$. This proves Corollary 1.

The particular case of this lemma that usually turns up is when $X \subset \mathbb{A}^n$ is an affine variety. Let $\mathbb{A}^n \subset \mathbb{P}^n$ be the subset $\mathbb{A}_0^n$ given by $u_0 \neq 0$, and write $m = \deg F$ and $f = F/u_0^m$; then $X_F = V(f)$. In other words, X_F is just the set of zeros of some regular function $f \in k[X]$.

Corollary 2. *Let $X \subset \mathbb{P}^N$ be an irreducible n-dimensional quasiprojective variety, and $Y \subset X$ the set of zeros of m forms on X. Then every (nonempty) component of Y has dimension $\geq n - m$.*

Proof. The proof is by an obvious induction on m. In the case of an affine variety X we can again say that Y is the set of zeros of m regular functions on X. If X is projective and $m \le n$ then by the proposition after Theorem 4, Corollary 5 we can assert that $Y \ne \emptyset$. Corollary 2 is proved.

Theorem 6. *Let $X, Y \subset \mathbb{P}^N$ be irreducible quasiprojective varieties with $\dim X = n$ and $\dim Y = m$. Then any (nonempty) component Z of $X \cap Y$ has $\dim Z \ge n + m - N$.*

Moreover, if X and Y are projective and $n + m \ge N$ then $X \cap Y \ne \emptyset$.

Proof. The theorem is obviously local in nature, and we therefore only need to prove it in the case of affine varieties. Suppose that $X, Y \subset \mathbb{A}^N$. Write $\Delta \subset \mathbb{A}^N \times \mathbb{A}^N = \mathbb{A}^{2N}$ for the diagonal (see 2.3, Example 10). Then $X \cap Y$ is isomorphic to $(X \times Y) \cap \Delta \subset \mathbb{A}^{2N}$. The theorem follows from Theorem 5, Corollary 2, since $\Delta \subset \mathbb{A}^{2N}$ is defined by N equations.

For the final sentence, apply the first part to the affine cone over X and Y. The theorem is proved.

Theorem 6 can be stated in a more symmetric form, in which it generalises at once to the intersection of any number of subvarieties:

$$\operatorname{codim}_X \bigcap_{i=1}^{r} Y_i \le \sum_{i=1}^{r} \operatorname{codim}_X Y_i. \tag{4}$$

6.3. The Theorem on the Dimension of Fibres

For a given regular map $f \colon X \to Y$ of quasiprojective varieties, and $y \in Y$, the set $f^{-1}(y)$ is called the *fibre* of f over y. It is obviously a closed subvariety of X. The idea behind the terminology is that f fibres X as the disjoint union of the fibres over the different points $y \in f(X)$.

Theorem 7. *Let $f \colon X \to Y$ be a regular map between irreducible varieties. Suppose that f is surjective: $f(X) = Y$, and that $\dim X = n$, $\dim Y = m$. Then $m \le n$, and*

(i) $\dim F \ge n - m$ for any $y \in Y$ and for any component F of the fibre $f^{-1}(y)$;

(ii) there exists a nonempty open subset $U \subset Y$ such that $\dim f^{-1}(y) = n - m$ for $y \in U$.

Proof of (i). This property is obviously local over Y, and it is enough to prove it after replacing Y by any open set $U \subset Y$ with $U \ni y$ and X by $f^{-1}(U)$. Hence we can assume that Y is affine. Suppose that $Y \subset \mathbb{A}^N$. In the chain of subvarieties of Y given by 6.2, (1), $Y^{(m)}$ is a finite set $Y^{(m)} = Y \cap Z$, where Z is defined by m equations and $y \in Z$. The open set U can be chosen such that $Z \cap Y \cap U = \{y\}$, and so we can assume that $Z \cap Y = \{y\}$. The subspace Z is defined by m equations $g_1 = \cdots = g_m = 0$. Thus in Y the system of equations $g_1 = \cdots = g_m = 0$ defines the point y. This means that in X the system of equations $f^*(g_1) = \cdots = f^*(g_m) = 0$ defines the subvariety $f^{-1}(y)$. Assertion (i) now follows from Theorem 5, Corollary 2 (the affine case).

Proof of (ii). We can replace Y by an affine open subset W and X by an open affine set $V \subset f^{-1}(W)$. Since V is dense in $f^{-1}(W)$ and f is surjective, $f(V)$ is dense in W. Hence f defines an inclusion $f^*: k[W] \hookrightarrow k[V]$. From now on we take $k[W] \subset k[V]$, therefore $k(W) \subset k(V)$. Write $k[W] = k[w_1, \ldots, w_M]$ and $k[V] = k[v_1, \ldots, v_N]$. Since $\dim W = m$ and $\dim V = n$, the field $k(V)$ has transcendence degree $n - m$ over $k(W)$. Suppose that $v_1, \ldots, v_{n-m}$ are algebraically independent over $k(W)$, and the remaining v_i algebraic over $k(W)[v_1, \ldots, v_{n-m}]$, with relations

$$F_i(v_i; v_1, \ldots, v_{n-m}; w_1, \ldots, w_M) = 0 \quad \text{for } i = n - m + 1, \ldots, N.$$

Write $\bar{v}_i$ for the function v_i restricted to $f^{-1}(y) \cap V$. Then

$$k[f^{-1}(y) \cap V] = k[\bar{v}_1, \ldots, \bar{v}_N]. \tag{1}$$

We now view F_i as a polynomial in $v_i, v_1, \ldots, v_{n-m}$, with coefficients functions of $w_1, \ldots, w_M$, and define Y_i to be the subvariety of W given by the vanishing of the leading term of F_i. Set $E = \bigcup Y_i$ and $U = W \setminus E$. Obviously U is open and nonempty. By construction of E, if $y \in U$ then none of the polynomials $F_i(T_i; T_1, \ldots, T_{n-m}; w_1(y), \ldots, w_M(y))$ is identically zero, and therefore all the $\bar{v}_i$ are algebraically dependent on $\bar{v}_1, \ldots, \bar{v}_{n-m}$. Together with formula (1) this proves that $\dim f^{-1}(y) \leq n - m$, so that (ii) of the proposition follows from (i). The theorem is proved.

It is easy to give examples where (ii) does not hold for every $y \in Y$; (see for example §2, Ex. 6, and the end of 6.4). That is, the dimension of fibres may jump up.

Corollary. *The sets $Y_k = \{y \in Y \mid \dim f^{-1}(y) \geq k\}$ are closed in Y.*

Proof. By Theorem 7, $Y_{n-m} = Y$, and there exists a closed subset $Y' \subsetneq Y$ such that $Y_k \subset Y'$ if $k > n - m$. If Z_i are the irreducible components of Y' and $f_i: f^{-1}(Z_i) \to Z_i$ the restrictions of f, then $\dim Z_i < \dim Y$, and we can prove the corollary by induction on $\dim Y$. The corollary is proved.

Theorem 7 implies a criterion for a variety to be irreducible which is often useful.

Theorem 8. *Let $f: X \to Y$ be a regular map between projective varieties, with $f(X) = Y$. Suppose that Y is irreducible, and that all the fibres $f^{-1}(y)$ for $y \in Y$ are irreducible and of the same dimension. Then X is irreducible.*

Proof. Let $X = \bigcup X_i$ be an irreducible decomposition. By 5.2, Theorem 2, each $f(X_i)$ is closed. Since $Y = \bigcup f(X_i)$ and Y is irreducible, $Y = f(X_i)$ for some i.

Set $\dim f^{-1}(y) = n$. For each i such that $Y = f(X_i)$, by Theorem 7, (ii), there exists a dense open set $U_i \subset Y$ and an integer n_i such that $\dim(f_i^{-1}(y)) = n_i$ for all $y \in U_i$. Extend the definition of U_i to i such that $f(X_i) \neq Y$ by setting $U_i = Y \setminus f(X_i)$. Consider $y \in \bigcap U_i$. Then since $f^{-1}(y)$ is irreducible, we must have $f^{-1}(y) \subset X_i$ for some i, say $i = 0$. Write $f_0 \colon X_0 \to Y$ for the restriction of f. Then $f^{-1}(y) \subset f_0^{-1}(y)$; but the opposite inclusion is trivial, so that $f^{-1}(y) = f_0^{-1}(y)$ and $n = n_0$.

Now since f_0 is surjective, we know that $f_0^{-1}(y) \subset f^{-1}(y)$ is nonempty for every $y \in Y$, and it has dimension $\geq n_0$ by Theorem 7, (i), so that $f_0^{-1}(y) = f^{-1}(y)$. Therefore $X_0 = X$. The theorem is proved.

A very special case of Theorem 8 is the irreducibility of a product of irreducible projective varieties; see 3.1, Theorem 3.

6.4. Lines on Surfaces

It is only natural, after the effort spent on the proof of 6.2, Theorems 4–6 on the dimension of intersections, to look for some applications of these results. As an example, we now treat a simple question on lines on surfaces in $\mathbb{P}^3$.

As a general rule, the notion of dimension is useful in cases when we need to give rigorous meaning to a statement that some set depends on a given number of parameters. For this, we must identify the set with some algebraic variety, and apply the notion of dimension we have introduced.

For example, we have seen in 4.4, Example 2 that hypersurfaces of $\mathbb{P}^n$, defined by equations of degree m, are in one-to-one correspondence with points of a projective space

$$\mathbb{P}^N, \qquad \text{where } N = \nu_{n,m} = \binom{n+m}{m} - 1.$$

We proceed to subvarieties that are not hypersurfaces, the simplest of which are lines in $\mathbb{P}^3$. In 4.1, Example 1, we saw that lines $l \subset \mathbb{P}^3$ are in one-to-one correspondence with points of the quadric hypersurface of $\Pi \subset \mathbb{P}^5$ defined by $p_{01}p_{23} - p_{02}p_{13} + p_{03}p_{12} = 0$. Obviously $\dim \Pi = 4$.

To study lines lying on surfaces, the following result is important.

Lemma. *The conditions that the line l with Plücker coordinates p_{ij} be contained in the surface X with equation $F = 0$ are algebraic relations between the p_{ij} and the coefficients of F, homogeneous in both the p_{ij} and the coefficients of F.*

Proof. We can write a parametric representation of l it terms of its Plücker coordinates: let x and y be a basis of a plane $\mathcal{L} \subset V$, with $\dim \mathcal{L} = 2$, $\dim V = 4$. Then it is easy to check that as f runs through the space of all linear forms on V, the set of vectors of the form

$$xf(y) - yf(x) \tag{1}$$

coincides with $\mathcal{L}$. If f has coordinates $(\alpha_0, \alpha_1, \alpha_2, \alpha_3)$, that is, if $f(x) = \sum \alpha_i x_i$, then the vector (1) has coordinates $z_i = \sum_j \alpha_j p_{ij}$, where $p_{ij} = x_i y_j - x_j y_i$. Hence if l is the line with Plücker coordinates p_{ij}, the points of l are the points with coordinates $\sum_j \alpha_j p_{ij}$ for $j = 0, \ldots, 3$.

On substituting these expressions into the equation $F(u_0, u_1, u_2, u_3) = 0$ and equating to zero the coefficients of all the monomials in α_i, we get the condition that $l \subset X$, as a set of algebraic relations between the coefficients of F and the Plücker coordinates p_{ij}. The lemma is proved.

We proceed to the question we are interested in, the lines lying on surfaces in $\mathbb{P}^3$. For given m, consider the projective space $\mathbb{P}^N$ with $N = \nu_{3,m} = \binom{m+3}{3} - 1$, whose points parametrise surfaces in $\mathbb{P}^3$ of degree m, that is, given by a homogeneous equation of degree m. Write $\Gamma_m \subset \mathbb{P}^N \times \Pi$ for the set of pairs $(\xi, \eta) \in \mathbb{P}^N \times \Pi$ such that the line l corresponding to $\eta \in \Pi$ is contained in the surface X corresponding to $\xi \in \mathbb{P}^N$. By the lemma, Γ_m is a projective variety. Let us determine the dimension of Γ_m. For this, consider the projection maps $\varphi \colon \mathbb{P}^N \times \Pi \to \mathbb{P}^N$ and $\psi \colon \mathbb{P}^N \times \Pi \to \Pi$ given by $\varphi(\xi, \eta) = \xi$ and $\psi(\xi, \eta) = \eta$. Obviously φ and ψ are regular maps. From now on, we only consider their restrictions to Γ_m. Note that $\psi(\Gamma_m) = \Pi$. This simply means that for every line l there is at least one surface of degree m passing through l, possibly reducible.

We determine the dimension of the fibres $\psi^{-1}(\eta)$ of ψ. By a projective transformation we can assume that the line corresponding to η is given by $u_0 = u_1 = 0$. Points $\xi \in \mathbb{P}^N$ such that $(\xi, \eta) \in \psi^{-1}(\eta) \subset \Gamma_m$ correspond to surfaces of degree m passing through this line. Such a surface is given by $F = 0$, where $F = u_0 G + u_1 H$, with G and H arbitrary forms of degree $m-1$. The set of such forms is of course a linear subspace of $\mathbb{P}^N$ whose dimension is easy to calculate. It is equal to

$$\mu = \frac{m(m+1)(m+5)}{6} - 1. \tag{2}$$

Thus

$$\dim \psi^{-1}(\eta) = \frac{m(m+1)(m+5)}{6} - 1 = N - (m+1).$$

It follows from Theorem 8 that Γ_m is irreducible. Applying Theorem 7 we get that

$$\begin{aligned}
\dim \Gamma_m &= \dim \psi(\Gamma_m) + \dim \psi^{-1}(\eta) \\
&= \frac{m(m+1)(m+5)}{6} + 3 \\
&= N + 3 - m.
\end{aligned} \tag{3}$$

Consider now the other projection $\varphi \colon \Gamma_m \to \mathbb{P}^N$. Its image is a closed subset of $\mathbb{P}^N$, by 5.2, Theorem 2. Obviously $\dim \varphi(\Gamma_m) \leq \Gamma_m$. Thus if

$\dim \Gamma_m < N$ then $\varphi(\Gamma_m) \neq \mathbb{P}^N$, or in other words, not every surface of degree m contains a line. By (3), the inequality $\dim \Gamma_m < N$ reduces to $m > 3$. We have obtained the following result.

Theorem 9. For any $m > 3$, there exist surfaces of degree m that do not contain any lines. Moreover, such surfaces correspond to an open set of $\mathbb{P}^N$.

Thus there exist nontrivial algebraic relations between the coefficients of a form $F(u_0, u_1, u_2, u_3)$ of degree $m > 3$ that are necessary and sufficient for the surface given by $F = 0$ to contain a line.

Of the remaining cases $m = 1, 2, 3$, the case $m = 1$ is trivial. We consider the case $m = 2$, although we already know the answer from 3-dimensional coordinate geometry. When $m = 2$ we have $N = 9$ and $\dim \Gamma_m = 10$. It follows from Theorem 7 that $\dim \varphi^{-1}(\xi) \geq 1$. This is the well-known fact that any quadric surface contains infinitely many lines.

We remark in passing, and without details of the proof, that this already provides an example of the phenomenon mentioned in 6.3 of the dimension of fibres jumping up: if the quadric surface corresponding to a point ξ is irreducible then $\dim \varphi^{-1}(\xi) = 1$, whereas if it splits as a pair of planes then of course $\dim \varphi^{-1}(\xi) = 2$.

Now consider the case $m = 3$. In this case, $\dim \Gamma_m = N = 19$. It is easy to construct a cubic surface $X \subset \mathbb{P}^3$ which contains only a finite number of lines. For example, if X is given in inhomogeneous coordinates by

$$T_1 T_2 T_3 = 1, \tag{4}$$

then X does not have a single line contained in $\mathbb{A}^3$. Indeed, if we write the equation of an affine line in the form $T_i = a_i t + b_i$ for $i = 1, 2, 3$ and substitute in (4), we get a contradiction; whereas the intersection of X with the plane at infinity contains 3 lines. Thus there exists a point of $\mathbb{P}^{19}$ for which $\varphi^{-1}(\xi)$ is nonempty and $\dim \varphi^{-1}(\xi) = 0$. By 6.3, Theorem 7, this is only possible if $\dim \varphi(\Gamma_3) = 19$. Using Theorem 1, we see that $\varphi(\Gamma_3) = \mathbb{P}^{19}$. We have proved the following result.

Theorem 10. Every cubic surface contains at least one line. There exists an open subset U of the space $\mathbb{P}^{19}$ parametrising all cubic surfaces such that a surface corresponding to a point of U contains only finitely many lines. □

Cubic surfaces that contain infinitely many lines do exist, for example cubic cones. Thus again the dimension of fibres can jump up. We will see later that most cubic surfaces contain only finitely many lines, and we will determine the number of these.

Exercises to §6

1. Let $L \subset \mathbb{P}^n$ be an $(n-1)$-dimensional linear subspace, $X \subset L$ an irreducible closed variety and y a point in $\mathbb{P}^n \setminus L$. Join y to all points $x \in X$ by lines, and denote by Y the set of points lying on all these lines, that is, the cone over X with vertex y. Prove that Y is an irreducible projective variety and $\dim Y = \dim X + 1$.

2. Let $X \subset \mathbb{A}^3$ be the reducible curve whose components are the 3 coordinate axes. Prove that the ideal $\mathfrak{A}_X$ cannot be generated by 2 elements.

3. Let $X \subset \mathbb{P}^2$ be the reducible 0-dimensional variety consisting of 3 points not lying on a line. Prove that the ideal $\mathfrak{A}_X$ cannot be generated by 2 elements.

4. Prove that any finite set $S \subset \mathbb{A}^2$ can be defined by two equations. [Hint: Choose the coordinates x, y in $\mathbb{A}^2$ in such a way that all points of S have different x coordinates; then show how to define S by the two equations $y = f(x)$, $\prod(x - \alpha_i) = 0$, where $f(x)$ is a polynomial.]

5. Prove that any finite set of points $S \subset \mathbb{P}^2$ can be defined by two equations.

6. Let $X \subset \mathbb{A}^3$ be an algebraic curve, and x, y, z coordinates in $\mathbb{A}^3$; suppose that X does not contain a line parallel to the z-axis. Prove that there exists a nonzero polynomial $f(x, y)$ vanishing at all points of X. Prove that all such polynomials form a principal ideal $(g(x, y))$, and that the curve $g(x, y) = 0$ in $\mathbb{A}^2$ is the closure of the projection of X onto the (x, y)-plane parallel to the z-axis.

7. We use the notation of Ex. 6. Suppose that $h(x, y, z) = g_0(x, y)z^n + \cdots + g_n(x, y)$ is the irreducible polynomial of smallest positive degree in z contained in the ideal $\mathfrak{A}_X$. Prove that if $f \in \mathfrak{A}_X$ has degree m as a polynomial in z, then we can write $fg_0^m = hU + v(x, y)$, where $v(x, y)$ is divisible by $g(x, y)$. Deduce that the equation $h = g = 0$ defines a reducible curve consisting of X together with a finite number of lines parallel to the x-axis, defined by $g_0(x, y) = g(x, y) = 0$.

8. Use Ex. 6–7 to prove that any curve $X \subset \mathbb{A}^3$ can be defined by 3 equations.

9. By analogy with Ex. 6–8, prove that any curve $X \subset \mathbb{P}^3$ can be defined by 3 equations.

10. Let $F_0(x_0, \ldots, x_n), \ldots, F_n(x_0, \ldots, x_n)$ be forms of degree $m_0, \ldots, m_n$ and consider the system of $n+1$ equations in $n+1$ variables $F_0(x) = \cdots = F_n(x) = 0$. Write Γ for the subset of $\prod_{i=0}^n \mathbb{P}^{\nu_{n,m_i}} \times \mathbb{P}^n$ (where $\nu_{n,m} = \binom{n+m}{m} - 1$) defined by

$$\Gamma = \{(F_0, \ldots, F_n, x) \mid F_0(x) = \cdots = F_n(x) = 0\}.$$

By considering the two projection maps $\varphi \colon \Gamma \to \prod_i \mathbb{P}^{\nu_{n,m_i}}$ and $\psi \colon \Gamma \to \mathbb{P}^n$, prove that $\dim \Gamma = \dim \varphi(\Gamma) = \sum_i \nu_{n,m_i} - 1$. Deduce from this that there exists a polynomial $R = R(F_0, \ldots, F_n)$ in the coefficients of the forms $F_0, \ldots, F_n$ such that $R = 0$ is a necessary and sufficient condition for the system of $n+1$ equations in $n+1$ variables to have a nonzero solution. What is the polynomial R if the forms $F_0, \ldots, F_n$ are linear?

11. Prove that the Plücker hypersurface $\Pi \subset \mathbb{P}^5$ contains two systems of 2-dimensional linear subspaces. A plane of the first system is defined by a point $\xi \in \mathbb{P}^3$ and consists of all points of Π corresponding to lines $l \subset \mathbb{P}^3$ through ξ. A plane of the second system is defined by a plane $\Xi \subset \mathbb{P}^3$ and consists of all points of Π corresponding to lines $l \subset \mathbb{P}^3$ contained in Ξ. There are no other planes contained in Π.

12. Let $F(x_0, x_1, x_2, x_3)$ be an arbitrary form of degree 4. Prove that there exists a polynomial Φ in the coefficients of F such that $\Phi(F) = 0$ is a necessary and sufficient condition for the surface $F = 0$ to contain a line.

13. Let $Q \subset \mathbb{P}^3$ be an irreducible quadric surface and $\Lambda_X \subset \Pi$ the set of points on the Plücker hypersurface $\Pi \subset \mathbb{P}^5$ corresponding to lines contained in Q. Prove that Λ_X consists of two disjoint conics.

Chapter II. Local Properties

1. Singular and Nonsingular Points

1.1. The Local Ring of a Point

This chapter investigates local properties of points of algebraic varieties, that is, properties of points $x \in X$ that remain unchanged if X is replaced by any neighbourhood of x. Since any point has an affine neighbourhood, in the study of local properties of points we can restrict ourselves to affine varieties.

The basic local invariant of a point x of a variety is its *local ring* $\mathcal{O}_x$, the ring consisting of all functions, each of which is regular in some neighbourhood of x. This definition requires a little care, however, since each function is regular in a different neighbourhood.

If X is irreducible, $\mathcal{O}_x$ is the subring of the function field $k(X)$ consisting of all functions $f \in k(X)$ that are regular at x. Recalling the definition of $k(X)$ as the field of fractions of the coordinate ring $k[X]$ we see that $\mathcal{O}_x$ consists of fractions f/g with $f, g \in k[X]$ and $g(x) \neq 0$.

This construction becomes clearer if we focus on its general and purely algebraic nature. It can be applied to an arbitrary commutative ring A and prime ideal $\mathfrak{p}$ of A. In this generality there is a new difficulty caused by possible zerodivisors in A.

Consider the set of pairs (f, g) with $f, g \in A$ and $g \notin \mathfrak{p}$; we identify pairs according to the rule

$$(f, g) = (f', g') \iff \exists\, h \in A \setminus \mathfrak{p} \text{ such that } h(fg' - gf') = 0. \tag{1}$$

Algebraic operations are defined on this set as follows:

$$(f, g) + (f', g') = (fg' + gf', gg'), \tag{2}$$
$$(f, g)(f', g') = (ff', gg'). \tag{3}$$

It is easy to check that in this way we get a ring. It is called the *local ring* of A at the prime ideal $\mathfrak{p}$, and denoted by $A_\mathfrak{p}$.

The map $\varphi \colon A \to A_\mathfrak{p}$ given by $\varphi(h) = (h, 1)$ is a homomorphism. The elements $\varphi(g)$ with $g \notin \mathfrak{p}$ are invertible in $A_\mathfrak{p}$, and any element $u \in A_\mathfrak{p}$ can be written $u = \varphi(f)/\varphi(g)$ with $g \notin \mathfrak{p}$; we sometimes use the somewhat imprecise notation $u = f/g$. The elements of the form $\varphi(f)/\varphi(g)$ with $f \in \mathfrak{p}$ and $g \notin \mathfrak{p}$ form an ideal $\mathfrak{m} \subset A_{\mathfrak{p}}$; moreover every element $u \in A_{\mathfrak{p}}$ with $u \notin \mathfrak{m}$ has an inverse. Therefore $\mathfrak{m}$ contains every other ideal of $A_{\mathfrak{p}}$.

We arrive at one of the fundamental notions of commutative algebra: a ring $\mathcal{O}$ is a *local ring* if it has an ideal $\mathfrak{m} \subset \mathcal{O}$ with $\mathfrak{m} \neq \mathcal{O}$ such that $\mathfrak{m}$ contains every other ideal of $\mathcal{O}$.

Lemma. *If A is a Noetherian ring then so is every local ring $A_{\mathfrak{p}}$.*

Proof. Indeed, for any ideal $\mathfrak{a} \subset A_{\mathfrak{p}}$, set $\bar{\mathfrak{a}} = \varphi^{-1}(\mathfrak{a})$. This is an ideal of A, and so by assumption has a finite basis, $\bar{\mathfrak{a}} = (f_1, \ldots, f_r)$. If $u \in \mathfrak{a}$ then $u = \varphi(f)/\varphi(g)$ with $f, g \in A$ and $g \notin \mathfrak{p}$. By the identification rule (1), it follows that there exists $h \in A \setminus \mathfrak{p}$ such that $hf \in \bar{\mathfrak{a}}$, and since $1/\varphi(hg) \in A_{\mathfrak{p}}$, we get $u \in \varphi(\bar{\mathfrak{a}})A_{\mathfrak{p}} = (\varphi(f_1), \ldots, \varphi(f_r))$. Hence $\mathfrak{a} = (\varphi(f_1), \ldots, \varphi(f_r))$, and so has a finite basis. The lemma is proved.

If $A = k[X]$ is the affine coordinate ring of an affine variety X and $\mathfrak{p} = \mathfrak{m}_x$ the maximal ideal of a point $x \in X$ then $A_{\mathfrak{p}}$ is called the *local ring* of x, and denoted by $\mathcal{O}_{X,x}$ or $\mathcal{O}_x$. It is Noetherian by the lemma.

For each pair (f, g) defining an element of $\mathcal{O}_x$ the function f/g is regular in the neighbourhood $D(g)$ of x. The rule (1) means that in $\mathcal{O}_x$ we identify functions f/g and f'/g' that are equal in some neighbourhood of x (in the present case $D(hgg')$). Thus we can also define $\mathcal{O}_x$ as the ring whose elements are regular functions in different neighbourhoods of x, with the identification rule just given. The definition is already independent of the choice of some affine neighbourhood U of x.

We choose, in particular, the variety V so that all its irreducible components pass through x. Then a function f that is 0 on some neighbourhood $U \subset V$ of x will be 0 on the whole of V. Hence the homomorphism $\varphi\colon k[V] \to \mathcal{O}_x$ is an inclusion, and we identify $k[V]$ with a subring of $\mathcal{O}_x$. In this situation, we can get rid of the factor h in the identification rule (1). In other words, $\mathcal{O}_x$ consists of functions on V without any identification, and all functions $\varphi_x \in \mathcal{O}_x$ are of the form f/g, with $f, g \in k[V]$ and $g(x) \neq 0$.

A similar construction is applicable to any irreducible subvariety Y of an affine variety X. Here we need to set $A = k[X]$ and $\mathfrak{p} = \mathfrak{a}_Y$. In this case, the local ring $A_{\mathfrak{p}}$ is called the local ring of X at the irreducible subvariety Y (or along Y), and denoted $\mathcal{O}_{X,Y}$ or $\mathcal{O}_Y$. If X is irreducible then $\mathcal{O}_Y \subset k(X)$ is the ring consisting of all rational functions that are regular at some point of Y (and hence regular on a dense open subset of Y). The maximal ideal $\mathfrak{m}_Y \subset \mathcal{O}_Y$ consists of functions vanishing along Y, and the residue field $\mathcal{O}_Y/\mathfrak{m}_Y = k(Y)$ is the function field of Y.

The passage to the case of an irreducible closed subvariety Y of an arbitrary quasiprojective variety X is just as obvious as when Y was a point. The local ring $\mathcal{O}_Y$ is defined in this case as the local ring of the subvariety

$Y \cap V$, where $V \subset X$ is any open affine variety such that $Y \cap V \neq \emptyset$. The local ring is independent of the choice of V.[4]

1.2. The Tangent Space

We will define the tangent space to an affine variety X at a point x as the set of all lines through x tangent to X. To define tangency of a line $L \subset \mathbf{A}^N$ to a variety $X \subset \mathbf{A}^N$, suppose that the coordinate system in $\mathbf{A}^N$ is chosen so that $x = (0, \ldots, 0) = 0$. Then $L = \{ta \mid t \in k\}$, where $a \neq 0$ is a fixed point. To study the intersection of X with L, suppose that X is given by a system of equations $F_1 = \cdots = F_m = 0$ with $\mathfrak{A}_X = (F_1, \ldots, F_m)$.

The set $X \cap L$ is then given by the equations $F_1(ta) = \cdots = F_m(ta) = 0$. Since we are now dealing with polynomials in one variable t, their common roots are the roots of their highest common factor. Suppose that

$$f(t) = \operatorname{hcf}(F_1(ta), \ldots, F_m(ta)) = c \prod (t - \alpha_i)^{k_i}. \tag{1}$$

The values $t = \alpha_i$ correspond to the points of intersection of L with X. Note that in (1), a root $t = \alpha_i$ has an associated multiplicity k_i, that is naturally interpreted as the multiplicity of intersection of L with X. In particular, since $L \cap X \ni 0$, one of the roots of $f(t)$ in (1) is $t = 0$. We arrive at the following definition.

Definition 1. The *intersection multiplicity* of a line L with a variety X at 0 is the multiplicity of $t = 0$ as a root of the polynomial $f(t) = \operatorname{hcf}(F_1(ta), \ldots, F_m(ta))$.

Thus the intersection multiplicity is the biggest power of t dividing all the $F_i(ta)$. It is ≥ 1 by definition, since $0 \in L \cap X$. If the $F_i(ta)$ are identically 0 then the intersection multiplicity is considered to be $+\infty$.

Obviously, $f(t) = \operatorname{hcf}\{F(ta) \mid F \in \mathfrak{A}_x\}$, and hence the multiplicity of intersection is independent of the choice of the generators F_i of $\mathfrak{A}_X$.

---

[4] Quite generally, $\mathcal{O}_{X,Y}$ is a subring of the direct product of function fields $k(X_i)$ of the irreducible components X_i of X that meet Y, that is,

$$\mathcal{O}_{X,Y} \subset \prod_{Y \cap X_i \neq \emptyset} k(X_i).$$

We can also view it as a quotient of the local ring $\mathcal{O}_{\mathbf{A}^n,Y}$ of rational functions on the ambient space regular on a dense open set of Y, modulo the ideal $\mathfrak{a}_X \mathcal{O}_{\mathbf{A}^n,Y}$ of functions vanishing on X. In discussing rational maps and rational functions as in Chap. I, a point to grasp is that rational functions are defined as *fractions*, and the locus where they are regular is determined subsequently; otherwise you have to worry about when two functions or maps with different domains are equal (for example, is the function z/z with a removable singularity equal to 1?).

Definition 2. A line L is *tangent* to X at 0 if it has intersection multiplicity ≥ 2 with X at 0.

We now write out the conditions for L to be tangent to X. Since $X \ni 0$, each of the polynomials $F_i(T)$ has constant term 0. For $i = 1, \ldots, m$, we write L_i for the linear term, so that $F_i = L_i + G_i$, where G_i has only terms of degree ≥ 2. Then $F_i(at) = tL_i(a) + G_i(ta)$, and $G_i(ta)$ is divisible by t^2. Therefore $F_i(at)$ is divisible by t^2 if and only if $L_i(a) = 0$. Thus the condition for tangency is

$$L_1(a) = \cdots = L_m(a) = 0. \tag{2}$$

Definition 3. The geometric locus of points on lines tangent to X at x is called the *tangent space* to X at x. It is denoted by Θ_x, or by $\Theta_{X,x}$ if we need to specify which variety is intended.

Thus (2) are the equations of the tangent space. They show that Θ_x is a linear subspace of $\mathbf{A}^N$.

Example 1. The tangent space to $\mathbf{A}^n$ at any point is just $\mathbf{A}^n$ itself.

Example 2. Let $X \subset \mathbf{A}^n$ be a hypersurface and $\mathfrak{A}_X = (F)$. If $X \ni 0$ and $F = L + G$ (in the above notation) then Θ_0 is defined by the single equation $L(T_1, \ldots, T_n) = 0$. Hence if $L \neq 0$ then $\dim \Theta_0 = n - 1$ and if $L = 0$ then $\Theta_0 = \mathbf{A}^n$, so that $\dim \Theta_0 = n$. Obviously

$$L = \sum \frac{\partial F}{\partial x_i}(0) x_i,$$

so that for $n = 2$ the definition coincides with that given in Chap. I, 1.5, (3).

Example 3. The tangent space at $(0,0)$ to the curve $y(y - x^2) = 0$ in $\mathbf{A}^2$ is the whole of $\mathbf{A}^2$. (Although both its components have the same tangent line $y = 0$.)

1.3. Intrinsic Nature of the Tangent Space

1.2, Definition 3 was given in terms of the defining equations of a subvariety $X \subset \mathbf{A}^N$. Hence it is not obvious that under an isomorphism $f \colon X \to Y$ the tangent spaces at x and at $f(x)$ are isomorphic (that is, have the same dimension). We now prove this; for this, we reformulate the notion of tangent space so that it depends only on the coordinate ring $k[X]$.

We recall some definitions. If $F(T_1, \ldots, T_N)$ is a polynomial and $x = (x_1, \ldots, x_N)$ a point, then F has a Taylor series expansion

$$F(T) = F(x) + F^{(1)}(T) + \cdots + F^{(k)}(T),$$

where $F^{(i)}$ are homogeneous polynomials of degree i in the variables $T_j - x_j$. The linear form $F^{(1)}$ is the *differential* of F at x, and is denoted by dF or $d_x F$; we have

$$d_x F = \sum_{i=1}^{N} \frac{\partial F}{\partial T_i}(x)(T_i - x_i).$$

It follows from the definition that

$$\begin{aligned} d_x(F + G) &= d_x F + d_x G, \\ d_x(FG) &= F(x)d_x G + G(x)d_x F \end{aligned} \tag{1}$$

Using this notation, we can write the equations 1.2, (2) of the tangent space to X at $x \in X$ in the form

$$d_x F_1 = \cdots = d_x F_m = 0, \tag{2}$$

or

$$\sum_{i=1}^{N} \frac{\partial F_j}{\partial T_i}(x)(T_i - x_i) = 0 \qquad \text{for } j = 1, \ldots, m, \tag{3}$$

where $\mathfrak{A}_X = (F_1, \ldots, F_m)$. Suppose that $g \in k[X]$ is defined as the restriction to X of a polynomial G. If we set $d_x g = d_x G$ then the answer depends on the choice of the polynomial G; more precisely, it would only be defined up to adding a terms $d_x F$ with $F \in \mathfrak{A}_X$. Since $\mathfrak{A}_X = (F_1, \ldots, F_m)$, we have $F = A_1 F_1 + \cdots + A_m F_m$, and by (1) and the fact that $F_i(x) = 0$, we get that $d_x F = A_1(x)d_x F_1 + \cdots + A_m(x)d_x F_m$. Using (2) we see that all the linear forms $d_x F$ for $F \in \mathfrak{A}_X$ are 0 on Θ_x, and hence, if we write $d_x g$ for the restriction to Θ_x of the linear form $d_x G$, that is,

$$d_x g = d_x G|_{\Theta_x}, \tag{4}$$

we get a map that sends any function $g \in k[X]$ into a well-defined linear form $d_x g$ on Θ_x.

Definition. The linear function $d_x g$ defined by (4) is called the *differential* of g at x.

Obviously,

$$d_x(f + g) = d_x f + d_x g, \qquad d_x(fg) = f(x)d_x g + g(x)d_x f. \tag{5}$$

We thus have a homomorphism $d_x \colon k[X] \to \Theta_x^*$, where Θ_x^* is the space of linear forms on Θ_x. Since $d_x \alpha = 0$ for $\alpha \in k$, we can replace the study of this map by that of $d_x \colon \mathfrak{m}_x \to \Theta_x^*$, where $\mathfrak{m}_x = \{f \in k[X] \mid f(x) = 0\}$. Obviously $\mathfrak{m}_x$ is an ideal of $k[X]$.

Theorem 1. *The map d_x defines an isomorphism of the vector spaces $\mathfrak{m}_x/\mathfrak{m}_x^2$ and Θ_x^*.*

Proof. We need to show that $\operatorname{im} d_x = \Theta_x^*$ and $\ker d_x = m_x^2$. The first of these is obvious. Any linear form φ on Θ_x is induced by some linear function f on $\mathbb{A}^N$, and $d_x f = \varphi$. To prove the second assertion, suppose that $x = (0, \ldots, 0)$ and that $g \in m_x$ satisfies $d_x g = 0$. Suppose that g is induced by a polynomial $G \in k[T_1, \ldots, T_N]$. By assumption the linear form $d_x G$ is 0 on Θ_x, and hence is a linear combination of the equations (2) defining this subspace, that is,

$$d_x G = \lambda_1 d_x F_1 + \cdots + \lambda_m d_x F_m.$$

Set $G_1 = G - \lambda_1 F_1 - \cdots - \lambda_m F_m$. We see that G_1 does not have any terms of degree 0 or 1 in $T_1, \ldots, T_N$, and therefore $G_1 \in (T_1, \ldots, T_N)^2$. Furthermore, $G_{1|X} = G_{|X} = g$, and hence $g \in (t_1, \ldots, t_N)^2$, where $t_i = T_{i|X}$. Since obviously $m_x = (t_1, \ldots, t_N)$, this proves the theorem.

As is well known, if L is a vector space and $M = L^*$ is the vector space of all linear forms on L then L can be identified with the vector space of all linear forms on M, that is, $L = M^*$. Applying this in our case gives the following.

Corollary 1. *The tangent space Θ_x at a point x is isomorphic to the vector space of all linear forms on m_x/m_x^2.* $\square$

The vector space m_x/m_x^2 is called the *cotangent space* to X at x.

From this, we make a deduction concerning the behaviour of tangent spaces under a regular map $f \colon X \to Y$ between varieties. Suppose that $x \in X$ and $y = f(x)$. Then f defines a map $f^* \colon k[Y] \to k[X]$, and obviously $f^*(m_y) \subset m_x$ and $f^*(m_y^2) \subset m_x^2$; thus f induces a map $f^* \colon m_y/m_y^2 \to m_x/m_x^2$. Linear functions, like any functions, are contravariant (map in the opposite direction) and since by Corollary 1 the tangent spaces $\Theta_{X,x}$ and $\Theta_{Y,y}$ are isomorphic to the vector space of linear forms on m_x/m_x^2 and m_y/m_y^2 respectively. We get a map $\Theta_{X,x} \to \Theta_{Y,y}$. This map is called the *differential* of f and denoted by $d_x f$.

It is easy to check that if $g \colon Y \to Z$ is another regular map and $z = g(y)$ then the differential $d(g \circ f) \colon \Theta_{X,x} \to \Theta_{Z,z}$ of the composite map is given by $d(g \circ f) = dg \circ df$. If f is the identity map $X \to X$ then for any point $x \in X$ the differential $d_x f$ is also the identity map on the tangent space at any point. These observations imply the following result.

Corollary 2. *Under an isomorphism of varieties, the tangent spaces at corresponding points are isomorphic. In particular the dimension of the tangent space at a point is invariant under isomorphism.* $\square$

Theorem 2. *The tangent space $\Theta_{X,x}$ is a local invariant of a point x of a variety X. Namely, $\Theta_{X,x}$ is the dual vector space of the vector space m_x/m_x^2, where m_x is the maximal ideal of the local ring $\mathcal{O}_x$ of x.*

Proof. We show how to determine Θ_x in terms of the local ring $\mathcal{O}_x$ of x. Recall that the differential of a rational function F/G, where $F, G \in k[T_1, \ldots, T_n]$, at a point where $G(x) \neq 0$, is given by

$$d_x(F/G) = \frac{G(x)d_x F - F(x)d_x G}{G^2(x)}.$$

We can view a function $f \in \mathcal{O}_x$ as the restriction to X of a rational function F/G, and define the differential as $d_x f = d_x(F/G)|_{\Theta_x}$. All the arguments given before Theorem 1 and during its proof go through as before, and we see that d_x defines an isomorphism $d_x \colon m_x/m_x^2 \xrightarrow{\sim} \Theta_x^*$, where now m_x is the maximal ideal $\{f \in \mathcal{O}_x \mid f(x) = 0\}$ of the local ring $\mathcal{O}_x$. This proves Theorem 2.

We define the *tangent space* Θ_x at a point x of any quasiprojective variety x as $(m_x/m_x^2)^*$, where m_x is the maximal ideal of the local ring $\mathcal{O}_x$ of x. By Theorem 2, Θ_x is then also the tangent space at x to any affine neighbourhood of x.

The tangent space is thus defined as an abstract vector space, not realised as a subspace of any ambient space. However, if X is affine and $X \subset \mathbf{A}^N$ then the embedding $i \colon X \hookrightarrow \mathbf{A}^N$ defines an embedding $di \colon \Theta_{X,x} \hookrightarrow \Theta_{\mathbf{A}^N,x}$. Since $\Theta_{\mathbf{A}^N,x}$ may be identified with $\mathbf{A}^N$, we can view $\Theta_{X,x}$ as embedded in $\mathbf{A}^N$, thus returning to the definition given in 1.2.

If $X \subset \mathbf{P}^N$ is a projective variety and $x \in X$ with $x \in \mathbf{A}_i^N$, then $\Theta_{X,x}$ is an affine linear subspace of $\mathbf{A}_i^N$. The closure of $\Theta_{X,x}$ in $\mathbf{P}^N$ does not depend on the choice of the affine piece $\mathbf{A}_i^N$. Despite the ambiguity in using the same term for two different objects, the closure $\overline{\Theta}_{X,x} \subset \mathbf{P}^N$ is sometimes also called the (projective) tangent space to X at x. The usual verification shows that $\overline{\Theta}_{X,x} \subset \mathbf{P}^N$ is defined by the equations

$$\sum_{i=0}^{N} \frac{\partial F_\alpha}{\partial \xi_i}(x)\xi_i = 0,$$

where $\{F_\alpha\}$ is a homogeneous basis for the ideal of X.

The intrinsic nature of the tangent space provides answers to certain questions on embedding varieties in affine spaces. For example, if $x \in X$ is a point such that $\dim \Theta_x = N$, then X is not isomorphic to any subvariety of an affine space $\mathbf{A}^n$ with $n < N$. An isomorphism $f \colon X \xrightarrow{\sim} Y \subset \mathbf{A}^n$ would take Θ_x isomorphically to the subspace $\Theta_{f(x)} \subset \mathbf{A}^n$. From this, for any $n > 1$, one can construct an example of a curve $X \subset \mathbf{A}^n$ not isomorphic to any curve $Y \subset \mathbf{A}^m$ with $m < n$. Namely, take X to be the image of $\mathbf{A}^1$ under the map

$$x_1 = t^n, \quad x_2 = t^{n+1}, \quad \ldots, \quad x_n = t^{2n-1}. \tag{6}$$

It is enough to prove that the tangent space to X as $x = (0, \ldots, 0)$ is the whole of $\mathbf{A}^n$. This means that all polynomials $F \in \mathfrak{A}_X$ do not contain linear terms in $T_1, \ldots, T_n$. Let $F \in \mathfrak{A}_X$ and write

$$F = \sum_{i=1}^{n} a_i T_i + G \quad \text{with } G \in (T_1, \dots, T_n)^2.$$

Substituting (6) in F, we get the following identity in t

$$\sum_{i=1}^{n} a_i t^{n+i-1} + G(t^n, t^{n+1}, \dots, t^{2n-1}) \equiv 0.$$

But if any $a_i \neq 0$ this is impossible, since the terms $a_i t^{n+i-1}$ have degree $\leq 2n - 1$, and terms coming from $G(t^n, \dots, t^{2n-1})$ have degree $\geq 2n$, so that they cannot cancel out.

It follows from the proof just given that no neighbourhood of x in the curve X is isomorphic to a quasiprojective variety in $\mathbf{A}^m$ with $m < n$.

We now consider some examples of tangent spaces. We start by giving an interpretation of the tangent space to a point $q \in \mathbb{P}(V)$ of the projective space corresponding to a vector space V. The tangent space $\Theta_{V,v}$ to V at v can naturally be identified with V, since $\mathfrak{m}_v/\mathfrak{m}_v^2$ is identified with the vector space of linear forms on V, that is, V^*. The map $\pi\colon V \setminus 0 \to \mathbb{P}(V)$ given by $\pi(\xi_0, \dots, \xi_n) = (\xi_0 : \cdots : \xi_n)$ has differential $\mathrm{d}_v\pi\colon \Theta_{V,v} = V \to \Theta_{\mathbb{P}(V),\pi(v)}$. If $\xi_0 \neq 0$ at v then in coordinates $x_i = \xi_i/\xi_0$ a linear form $\varphi \in \Theta_{V,v}$ goes over into the function $\psi = (\mathrm{d}_v\pi)(\varphi)$ on $\mathfrak{m}_{\pi(v)}/\mathfrak{m}_{\pi(v)}^2$ for which

$$\psi(x_i) = \varphi \mathrm{d}_v(\xi_i/\xi_0) = \frac{\xi_i \varphi(\xi_0) - \xi_0 \varphi(\xi_i)}{\xi_0^2}.$$

It follows that the image of $\mathrm{d}_v\pi$ is the whole of $\Theta_{\mathbb{P}(V),\pi(v)}$, and the kernel consists of the vectors $(\eta_0, \dots, \eta_n)$ satisfying $\xi_i \eta_0 = \xi_0 \eta_i$, that is, proportional to $(\xi_0, \dots, \xi_n)$.

Thus for $\xi \in \mathbb{P}(V)$ we have

$$\Theta_{\mathbb{P}(V),\xi} \cong V/l_\xi, \tag{7}$$

where $l_\xi = \pi^{-1}(\xi)$ is the line in V corresponding to a point $\xi \in \mathbb{P}(V)$.

From this, we can say that if $X \subset \mathbb{P}(V)$ is a projective variety defined by a system of homogeneous equations, and $\widetilde{X} \subset V$ the affine cone over X, defined in V by the same equations, then $\Theta_{X,x} \cong \Theta_{\widetilde{X},\widetilde{x}}/l_x$, where $x = \pi(\widetilde{x})$, and l_x is as in (7). We apply this interpretation of the tangent space to a projective variety to the algebraic varieties considered in Chap. I, 4.1, Examples 1–3.

Example 1. The Grassmannian. We consider here only $X = \mathrm{Grass}(2, n)$. It is defined by the equations $x^2 = x \wedge x = 0$ in $\mathbb{P}(\bigwedge^2 V)$. Differentiating these equations, we get that the tangent space to the affine cone $\widetilde{X} \subset \bigwedge^2 V$ at x consists of $y \in \bigwedge^2 V$ such that

$$x \wedge y = 0. \tag{8}$$

Suppose that $x \in \bigwedge^2 L \subset \bigwedge^2 V$ is the point corresponding to a 2-plane $L \subset V$, so that $\bigwedge^2 L = kx$, and let $f \in \operatorname{Hom}(L, V/L)$. Then it is easy to check that for any basis e_1, e_2 of L, the bivector $y = e_1 \wedge f(e_2) - e_2 \wedge f(e_1)$ is uniquely determined in $\bigwedge^2 V/kx$, is independent of the choice of a basis in L up to a scalar multiple, and satisfies (8). Moreover, any solution to (8) is obtained in this way. Thus for any 2-plane $L \subset V$, we see that

$$\Theta_{\operatorname{Grass}(2,V),L} \cong \operatorname{Hom}(L, V/L). \tag{9}$$

We will show in Chap VI, 4.1, Example 3 that a similar relation holds for any $\operatorname{Grass}(r, V)$, and give an interpretation.

Remark. Our starting point for deducing (9) was that $\operatorname{Grass}(2, V)$ is given by the system of equations $x \wedge x = 0$. But in order to apply the definition of the tangent space given in 1.2, we need to know that these equations not only define $X = \operatorname{Grass}(2, V)$ set-theoretically, by also generate the ideal $\mathfrak{A}_X$. At present, we can only assert that if we write out the equations $x \wedge x = 0$ as $F_1 = \cdots = F_m = 0$ then, after restricting to some affine piece of $\mathbb{P}(\bigwedge^2 V)$, the space defined by $\sum (\partial F_i / \partial T_j)(x)(T_i - x_i) = 0$ is isomorphic to $\operatorname{Hom}(L, V/L)$. From this it is already not hard to deduce that $\mathfrak{A}_X = (F_1, \ldots, F_m)$, and hence the relation (9) holds without any reservations (see §3, Ex. 15).

Example 2. Variety of associative algebras. Differentiating the associativity relation Chap. I, 4.1, (7), we see that the tangent space to the variety of associative algebras is defined by the equations

$$\sum_l (\alpha_{ij}^l x_{lk}^m + \alpha_{lk}^m x_{ij}^l) = \sum_l (\alpha_{il}^m x_{jk}^l + \alpha_{jk}^l x_{il}^m). \tag{10}$$

Suppose that $x_{ij}^m = \eta_{ij}^m$ satisfy these equations. Consider the bilinear function $f(x, y)$ with $x, y \in A$ given by $f(e_i, e_j) = \sum_m \eta_{ij}^m e_m$. The relations (10) then take the form

$$x f(y, z) + f(x, yz) = f(xy, z) + f(x, y)z \qquad \text{for all } x, y, z \in A.$$

Functions of this type are called *2-cocycles* on A. Thus the tangent space to the variety of algebras at a point corresponding to an algebra A is isomorphic to the space of 2-cocycles on A.

Remark. As in Example 1, we started from the relations Chap. I, 4.1, (7), that define the variety of associative algebras only in the set-theoretic sense. Whether the left-hand sides of these equations generate the ideal of the variety seems not to be known; it is known that this fails for Lie algebras, and it is plausible that it also fails for associative algebras. Thus the space of 2-cocycles on A is only equal to the tangent space at A to the variety of algebras for those dimensions n for which the equations Chap. I, 4.1, (7) generate the ideal of the variety of algebras, or for A for which these equations generate the ideal locally. However, the associativity relations Chap. I, 4.1, (7) are so natural that any information deduced from them should have some kind of meaning. In particular, for a discussion of the space of 2-cocycles, see Chap. V, 3.4, and Chap. VI, 4.1, Example 4.

Example 3. Variety of quadrics. Let V be the vector space of symmetric $n \times n$ matrixes $A = (x_{ij})$, with $x_{ij} = x_{ji}$, and consider the variety $\Delta \subset \mathbb{P}(V)$ given by $\det A = 0$ for $A \in V$. It is easy to see that $\det A$ is an irreducible polynomial, so that Δ is an irreducible hypersurface. The tangent space to the affine cone $\tilde{\Delta}$ at a matrix A consists of matrixes $B \in V$ such that $\big((d/dt)\det(A + tB)\big)_{|t \,=\, 0} = 0$. Since

$$\frac{\mathrm{d}}{\mathrm{d}t}\det(A + tB)_{|t \,=\, 0} = \det A_1 + \cdots + \det A_n,$$

where A_i is the matrix obtained by replacing the ith row of A by that of B, this expression is 0 if $\operatorname{rank} A < n - 1$. For these points $\Theta_{\Delta,A} = \mathbb{P}(V)$. Suppose that A has rank $n - 1$. Transformations $A \mapsto {}^t C A C$ with C a nondegenerate matrix obviously define automorphisms of Δ. We can use such a transformation to put the quadratic form f corresponding to A in the form $x_1^2 + \cdots + x_{n-1}^2$. Thus we can assume that $f = x_1^2 + \cdots + x_{n-1}^2$, and then the same argument shows that $\big((d/dt)\det(A + tB)\big)_{|t \,=\, 0}$ equals the entry b_{nn} of B. Hence at such points, the tangent space $\Theta_{\Delta,A}$ can be identified with the subspace of matrixes $B \in V$ with $b_{nn} = 0$, that is, the space of quadrics passing through the vertex of the singular quadric $f = 0$.

1.4. Singular Points

We now explain what can be said concerning the dimension of tangent spaces of an irreducible quasiprojective variety X. Our result will be local in nature, so that we restrict ourselves to considering affine varieties.

Let $X \subset \mathbb{A}^N$ be an irreducible variety. Consider the subset Θ of the direct product $\mathbb{A}^N \times X$ consisting of pairs (a, x) with $a \in \mathbb{A}^N$ and $x \in X$ such that $a \in \Theta_x$. The equation 1.3, (2) shows that Θ is closed in $\mathbb{A}^N \times X$. Write $\pi\colon \Theta \to X$ for the second projection, $\pi(a, x) = x$. Obviously $\pi(\Theta) = X$, and $\pi^{-1}(x) = \{(a, x) \mid a \in \Theta_x\}$. Thus Θ is fibred over X with the tangent spaces Θ_x at different points of X as fibres; $\Theta \to X$ is called the *tangent fibre space* to X. We apply to Θ the results on the dimensions of fibres of a map Chap I, 6.3, Theorem 7 and its corollary; then we see that there exists a number s such that $\dim \Theta_x \geq s$ for every $x \in X$, and the points $y \in X$ for which $\dim \Theta_y > s$ form a closed proper subvariety $Y \subsetneq X$, that is, a variety of smaller dimension.

Definition. Let X be an irreducible variety and set $s = \min_{x \in X} \dim \Theta_x$. We say that a point $x \in X$ is *nonsingular*[5] if $\dim \Theta_x = s$; we also say that X is nonsingular at x. A variety X is nonsingular if it is nonsingular at every $x \in X$. If $\dim \Theta_x > s$ then x is a *singular* point of X.

As we have just seen, nonsingular points of X form an open nonempty subvariety, and singular points a closed proper subvariety.

Consider the example of a hypersurface (1.2, Example 2), which contains as the particular case $n = 2$ the case of algebraic plane curves considered in Chap. I, 1.5. If $\mathfrak{A}_X = (F)$ then the equation of the tangent space at x is

$$\sum_{i=1}^{n} \frac{\partial F}{\partial T_i}(x)(T_i - x_i) = 0.$$

We now prove that in this case $s = \min \dim \Theta_x = n - 1$. This is obviously equivalent to saying that the $\partial F / \partial T_i$ are not all identically 0 on X. In characteristic 0 this would mean that F is constant, and in characteristic $p > 0$ that all the indeterminates only appear in F in powers that are multiples of p. But then, as in Chap. I, 1.5, since the field k is algebraically closed, it would follow that $F = F_1^p$, and this contradicts $\mathfrak{A}_X = (F)$.

Thus in our example, nonsingular points $x \in X$ have $\dim \Theta_x = \dim X = n - 1$. We now prove that the same holds for an arbitrary irreducible variety, and that the general case reduces to that of a hypersurface.

Theorem 3. *The dimension of the tangent space at a nonsingular point equals the dimension of the variety.*

Proof. In view of the definition of nonsingular point, the theorem asserts that $\dim \Theta_x \geq \dim X$ for every point x of an irreducible variety X, and that the set of points x with $\dim \Theta_x = \dim X$ is open and nonempty. This is obviously a local assertion, and we need only consider the case of an affine variety. We have seen that there exists an s such that $\dim \Theta_x \geq s$ for every $x \in X$, and the set of points x with $\dim \Theta_x = s$ is open and nonempty. It only remains to prove that $s = \dim X$. We now use Chap. I, 3.3, Theorem 5, which asserts that X is birational to a hypersurface Y.

Let $\varphi \colon X \to Y$ be the birational map of Chap. I, 3.3, Theorem 5. By Chap. I, 4.3, Proposition, there exist nonempty open sets $U \subset X$ and $V \subset Y$ such that φ defines an isomorphism $\varphi \colon U \xrightarrow{\sim} V$. By the remarks made before the statement of the theorem, the set W of nonsingular points of the variety Y is open, and $\dim \Theta_y = \dim Y = \dim X$ for all $y \in W$. The set $W \cap V$ is also open and nonempty, and hence $\varphi^{-1}(W \cap V) \subset U$ is also open and nonempty.

---

[5] The term *smooth* is used interchangeably with nonsingular in the current literature. The first English edition of this book used the archaic term *simple*, which goes back to Zariski, and is a literal translation of the Russian, but is not in current use.

Since the dimension of the tangent space is invariant under isomorphism, $\dim \Theta_x = \dim X$ for $x \in \varphi^{-1}(W \cap V)$. The theorem is proved.

Consider now reducible varieties. Already the inequality $\dim \Theta_x \geq \dim X$ fails for them. For example, if $X = X_1 \cup X_2$ with $\dim X_1 = 1$ and $\dim X_2 = 2$, and if $x \in X_1 \setminus X_2$ is a nonsingular point of X_1 then $\dim \Theta_x = 1$, whereas $\dim X = 2$. This is only to be expected, since a component of X not passing through x contributes to the dimension $\dim X$, but does not affect Θ_x. Hence it is natural to introduce the following notion. The dimension of X at a point x, denoted by $\dim_x X$, is the maximum of the dimensions of the irreducible components of X through x. Obviously $\dim X = \max_{x \in X} \dim_x X$.

Definition. A point x of an affine variety is *nonsingular* if $\dim \Theta_x = \dim_x X$.

It follows from Theorem 3 that $\dim \Theta_x \geq \dim_x X$ for any point $x \in X$. Indeed, if X^i for $i = 1, \ldots, s$ are the irreducible components of X passing through x, and Θ_x^i is the tangent space to X^i at this point then $\dim \Theta_x^i \geq \dim X^i$ and $\Theta_x^i \subset \Theta_x$, so that

$$\dim \Theta_x \geq \max_i \dim \Theta_x^i \geq \max_i \dim X^i = \dim_x X.$$

It follows from Theorem 3 exactly as before that the singular points are contained in a subvariety of X of smaller dimension.

The passage to an arbitrary quasiprojective variety is obvious: a point $x \in X$ is *nonsingular* if it is nonsingular on an affine neighbourhood $U \ni x$. This is equivalent to $\dim \Theta_x = \dim_x X$. A variety is *nonsingular* or *smooth* if all its points are nonsingular.

Examples of singular points of algebraic plane curves appeared in Chap. I, 1.5. We now consider a quadric $Q \subset \mathbb{P}^n$. In a suitable coordinate system, Q has the equation $x_0^2 + \cdots + x_r^2 = 0$ for some $r \leq n$ (here we are assuming that char $k \neq 2$). The singular points of Q are given by $x_0 = \cdots = x_r = 0$, and if $r = n$, there are none. If $r < n$ then the singular points form a $(n - r - 1)$-dimensional linear subspace $L \subset \mathbb{P}^n$. Intersecting Q with the r-dimensional subspace $x_{r+1} = \cdots = x_n = 0$ gives a nonsingular quadric $S \subset \mathbb{P}^r$. For any point $q = (\alpha_0 : \cdots : \alpha_n) \in Q$ the points $s = (\alpha_0 : \cdots : \alpha_r) \in S$, because the equation of Q does not involve the last $n - r$ coordinates $x_{r+1}, \ldots, x_n$. If s is fixed, the points $q \in Q$ with arbitrary $\alpha_{r+1}, \ldots, \alpha_n$ form the $(n-r)$-dimensional linear subspace spanned by s and L. These subspaces sweep out Q. For this reason, we say that Q is a cone with vertex the linear subspace L and base the nonsingular quadric S.

Chap. I, 6.1, Example 5 showed that $\dim \text{Grass}(r, n) = r(n - r)$, and that $\text{Grass}(r, n)$ is nonsingular and rational. In exactly the same way, in the space of quadrics, by 1.3, Example 3, the open set of the determinantal hypersurface Δ consisting of quadrics of rank $n - 1$ is nonsingular. In the case of the variety of associative algebras (Example 2), the situation is more complicated;

there are both nonsingular and singular points, that is, "nonsingular" and "singular" algebras.

1.5. The Tangent Cone

The simplest invariant measuring how far a singular point is from being nonsingular is the dimension of its tangent space. However, there is a much finer invariant, the tangent cone to X at x. We do not need this notion in what follows, and therefore leave the detailed working out of the following arguments to the reader as a (very easy) exercise.

Let X be an irreducible affine variety. The tangent cone to X at $x \in X$ consists of lines through x that we define as the analogue of limiting positions of secants in differential geometry.

Suppose that $X \subset \mathbf{A}^N$ with $x = (0,\ldots,0)$, and that we make $\mathbf{A}^N$ into a vector space using the choice of x as origin. In $\mathbf{A}^{N+1} \cong \mathbf{A}^N \times \mathbf{A}^1$, consider the set $\widetilde{X}$ of pairs (a,t) with $a \in \mathbf{A}^N$ and $t \in \mathbf{A}^1$ such that $at \in X$. Obviously $\widetilde{X}$ is closed in $\mathbf{A}^{N+1}$. We have, as usual, the two projections $\varphi \colon \widetilde{X} \to \mathbf{A}^1$ and $\psi \colon \widetilde{X} \to \mathbf{A}^N$. One sees easily that $\widetilde{X}$ is reducible (if $X \neq \mathbf{A}^N$), and consists of two components: $\widetilde{X} = \widetilde{X}_1 \cup \widetilde{X}_2$, where $\widetilde{X}_2 = \{(a,0) \mid a \in \mathbf{A}^N\}$ and $\widetilde{X}_1$ is the closure in $\widetilde{X}$ of $\varphi^{-1}(\mathbf{A}^1 \setminus 0)$. Write φ_1 and ψ_1 for the restriction to $\widetilde{X}_1$ of the projections φ and ψ. The set $\psi_1(\widetilde{X}_1)$ is the closure of the set of points on all secants of X through x. The set $T_x = \psi_1(\varphi_1^{-1}(0))$ is called the *tangent cone* to X at x.

It is easy to write out the equations of the tangent cone. The equations of $\widetilde{X}$ are of the form

$$F(at) = 0 \qquad \text{for all } F \in \mathfrak{A}_X.$$

Suppose that $F = F_k + F_{k+1} + \cdots + F_l$, where F_j is a form of degree j and $F_k \neq 0$. Then $F(at) = t^k F_k(a) + t^{k+1} F_{k+1}(a) + \cdots + t^l F_l(a)$. Since $F(0) = 0$, we automatically have $k \geq 1$, and the equation of the component $\widetilde{X}_2$ inside $\widetilde{X}$ is $t = 0$. It is easy to see that the equations of T_x are $F_k = 0$ for all $F \in \mathfrak{A}_X$. The form F_k is the *leading form* of F. Thus T_x is defined by setting to 0 the leading forms of all polynomials $F \in \mathfrak{A}_X$. Since T_x is defined by homogeneous polynomials, it is a cone with vertex x. It is easy to see that $T_x \subset \Theta_x$, and that $T_x = \Theta_x$ if x is a nonsingular point.

We consider the example of an algebraic plane curve $X \subset \mathbf{A}^2$. If $\mathfrak{A}_X = (F(x,y))$ and F_k is the leading form of F then T_x has equation $F_k(x,y) = 0$. Since F_k is a form in two variables, and k is algebraically closed, F_k splits as a product of linear forms, $F_k(x,y) = \prod (\alpha_i x + \beta_i y)^{l_i}$. Hence in this case T_x breaks up into several lines $\alpha_i x + \beta_i y = 0$. These lines are called the *tangent lines* to X at x, and l_i their *multiplicities*. If $k > 1$ then $\Theta_x = \mathbf{A}^2$. The number k is called the *multiplicity* of the singular point x. When $k = 2$ or 3 we say that $x \in X$ is a double point or triple point.

For example if $F = x^2 - y^2 + x^3$ and $x = (0,0)$ then T_x consists of the two lines $x + y = 0$ and $x - y = 0$; if $F = x^2y - y^3 + x^4$ and $x = (0,0)$ then T_x consists of 3 lines $y = 0$, $x + y = 0$ and $x - y = 0$. If $F = y^2 - x^3$ and $x = (0,0)$ then $y = 0$ is a tangent line with multiplicity 2.

In exactly the same way as the first definition of tangent space given in 1.2, the above definition of tangent cone uses a notion that is not invariant under isomorphism. However, one can prove that the tangent cone T_x is invariant under isomorphism, and is a local invariant of $x \in X$.

Exercises to §1

1. Prove that the local ring $\mathcal{O}_{X,x}$ of a point x of an irreducible variety X is the union in $k(X)$ of all the rings $k[U]$ for U a neighbourhood of x.

2. The map $\varphi(t) = (t^2, t^3)$ defines a birational map from the line $\mathbf{A}^1$ to the curve $y^2 = x^3$. What are the rational functions in t that correspond to the functions in the local ring $\mathcal{O}_x$ of the point $(0,0)$?

3. The same question for the birational map from $\mathbf{A}^1$ to the curve of Chap. I, 1.2, (1).

4. Prove that the local ring $\mathcal{O}_x$ of the curve $xy = 0$ at $(0,0)$ is isomorphic to the subring $\mathcal{O} \subset \mathcal{O}_1 \oplus \mathcal{O}_2$, where $\mathcal{O}_1$ and $\mathcal{O}_2$ are copies of the local ring of 0 in $\mathbf{A}^1$, consisting of functions f_1, f_2 with $f_1 \in \mathcal{O}_1$ and $f_2 \in \mathcal{O}_2$ such that $f_1(0) = f_2(0)$.

5. Determine the local ring at $(0,0,0)$ of the curve consisting of the three coordinate axes in $\mathbf{A}^3$.

6. Determine the local ring at $(0,0)$ of the curve $xy(x - y) = 0$. Prove that this curve is not isomorphic to that of Ex. 5.

7. Prove that if $x \in X$ and $y \in Y$ are nonsingular points then $(x,y) \in X \times Y$ is nonsingular.

8. Prove that if $X = X_1 \cup X_2$ and $x \in X_1 \cap X_2$, then the tangent spaces $\Theta_{X,x}$, $\Theta_{X_1,x}$ and $\Theta_{X_2,x}$ satisfy
$$\Theta_{X_1,x} + \Theta_{X_2,x} \subset \Theta_{X,x}.$$
Does equality always hold?

9. Prove that a hypersurface of degree 2 with a singular point is a cone.

10. Prove that if a hypersurface X of degree 3 has two singular points then the line joining them is contained in X.

11. Prove that if a plane curve of degree 3 has three singular points then it breaks up as a union of 3 lines.

12. Prove that the singular points of the hypersurface $X \subset \mathbb{P}^n$ defined by $F(x_0, \ldots, x_n) = 0$ are determined by the system of equations

$$F(x_0, \ldots, x_n) = 0, \quad \text{and} \quad \frac{\partial F}{\partial x_i}(x_0, \ldots, x_n) = 0 \quad \text{for } i = 0, \ldots, n.$$

If $\deg F$ is not divisible by the characteristic of the field, then the first equation follows from the others.

13. Prove that if a hypersurface $X \subset \mathbb{P}^n$ contains a linear subspace L of dimension $r \geq n/2$ then X is singular. [Hint: Choose the coordinate system so that L is given by $x_{r+1} = \cdots = x_n = 0$, write out the equation of X and look for singular points contained in L.]

14. For what values of a does the curve $x_0^3 + x_1^3 + x_2^3 + a(x_0 + x_1 + x_2)^3 = 0$ have a singular point? What are its singular points then? Is it reducible?

15. Determine the singular points of the Steiner surface in $\mathbb{P}^3$:

$$x_1^2 x_2^2 + x_2^2 x_0^2 + x_0^2 x_1^2 - x_0 x_1 x_2 x_3 = 0.$$

16. For what values of a does the surface $x_0^4 + x_1^4 + x_2^4 + x_3^4 - a x_1 x_2 x_3 x_4$ have singular points, and what are these points?

17. Let $\mathbb{P}^N$ with $N = \nu_{n,m}$ be as in Chap. I, 4.4, Example 2. Prove that over a field of characteristic 0, the points of the space $\mathbb{P}^N$ for which the corresponding hypersurfaces has a singular point form a hypersurface in $\mathbb{P}^N$. [Hint: Use the result of Chap. I, §6, Ex. 10.]

18. Let $F(x_0, x_1, x_2) = 0$ be the equation of an irreducible curve $X \subset \mathbb{P}^2$ over a field of characteristic 0. Consider the rational map $\varphi \colon X \to \mathbb{P}^2$ given by the formulas $u_i = \partial F / \partial x_i(x_0, x_1, x_2)$ for $i = 0, 1, 2$. Prove (a) $\varphi(X)$ is a point if and only if X is a line; (b) if X is not a line, then φ is regular at $x \in X$ if and only if x is nonsingular. The image $\varphi(X)$ is called the *dual curve* of X.

19. Prove that if X is a conic then so is $\varphi(X)$.

20. Find the dual curve of $x_0^3 + x_1^3 + x_2^3 = 0$.

21. Prove that if $X \subset \mathbb{P}^n$ is a nonsingular hypersurface and not a hyperplane, then as x runs through X, the tangent hyperplanes Θ_x form a hypersurface in the dual space $\mathbb{P}^{n*}$.

22. Let φ be the regular map of a variety $X \subset \mathbb{A}^n$ consisting of the linear projection to some subspace. Determine the map $d\varphi$ on the linear subspaces Θ_x for $x \in X$.

23. Let x be a point of a variety X and $\mathfrak{m}_x \subset \mathcal{O}_x$ the local ring at x and its maximal ideal. Prove that for every integer $t > 0$, the module $\mathfrak{m}_x^t / \mathfrak{m}_x^{t+1}$ is a finite dimensional vector space over k.

2. Power Series Expansions

2.1. Local Parameters at a Point

We study a nonsingular point x of a variety X, with $\dim_x X = n$.

Definition. Functions $u_1, \ldots, u_n \in \mathcal{O}_x$ are *local parameters* at x if each $u_i \in \mathfrak{m}_x$, and the images of $u_1, \ldots, u_n$ form a basis of the vector space $\mathfrak{m}_x/\mathfrak{m}_x^2$.

In view of the isomorphism $d_x \colon \mathfrak{m}_x/\mathfrak{m}_x^2 \to \Theta_x^*$, we see that $u_1, \ldots, u_n \in \mathcal{O}_x$ is a system of local parameters if and only if the n linear forms $d_x u_1, \ldots, d_x u_n$ on Θ_x are linearly independent. Since $\dim \Theta_x = n$, this in turn is equivalent to saying that the system of equations

$$d_x u_1 = \cdots = d_x u_n = 0 \tag{1}$$

has 0 as its only solution in Θ_x.

We can replace X by an affine neighbourhood X' of x on which the $u_1, \ldots, u_n$ are regular functions; set $\mathfrak{A} = \mathfrak{A}_{X'}$. Now write X_i' for the hypersurface in X' defined by $u_i = 0$, and set $\mathfrak{A}_i = \mathfrak{A}_{X_i'}$. Let U_i be a polynomial that defines the function u_i on X'. Then $\mathfrak{A}_i \supset (\mathfrak{A}, U_i)$, and by definition of the tangent space it follows that $\Theta_i \subset L_i$, where Θ_i is the tangent space to X_i' at x and $L_i \subset \Theta_x$ is defined by $d_x U_i = 0$. From the assumption that the system (1) has 0 as its only solution in Θ_x it follows that $L_i \neq \Theta_x$, that is, $\dim L_i = n - 1$, and from the theorem on dimension of intersection and the inequality $\dim \Theta_i \geq \dim X_i'$ it follows that $\dim \Theta_i \geq n - 1$. Hence $\dim \Theta_i = n - 1$, and it follows that x is a nonsingular point of X_i'. In some neighbourhood of x, the intersection of the varieties X_i' is exactly x: for if some component Y of $\bigcap X_i'$ with $\dim Y > 0$ passed through x, the tangent space to Y at x would be contained in all the Θ_i, and this again contradicts the assumption that (1) has 0 as its only solution.

We have thus proved the following assertion.

Theorem 1. *If $u_1, \ldots, u_n$ are local parameters at x such that the u_i are regular on X, and $X_i = V(u_i)$, then x is a nonsingular point on each of the X_i and $\bigcap \Theta_i = 0$, where Θ_i is the tangent space to X_i at x.*

Here we meet a general property of subvarieties that will appear frequently in what follows.

Definition. Subvarieties $Y_1, \ldots, Y_r$ of a nonsingular variety X are *transversal* at a point $x \in \bigcap Y_i$ if

$$\operatorname{codim}_{\Theta_{X,x}}\left(\bigcap_{i=1}^r \Theta_{Y_i,x}\right) = \sum_{i=1}^r \operatorname{codim}_X Y_i. \tag{2}$$

For example, two curves on a nonsingular surface are transversal at a point of intersection if they are both nonsingular and their tangent lines are different (Figure 7).

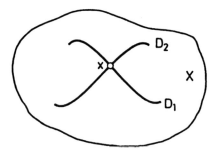

Figure 7. Transversal Curves on a Surface

Using the inequality Chap I, 6.2, (4) for the subspaces $\Theta_{Y_i,x} \subset \Theta_{X,x}$, and the inequalities codim $\Theta_{Y_i,x} \leq \text{codim}_X Y_i$ we see that (2) implies the equality

$$\dim \Theta_{Y_i,x} = \dim Y_i,$$

so that each Y_i is nonsingular at x, and the equality

$$\text{codim}_{\Theta_{X,x}} \left(\bigcap_{i=1}^{r} \Theta_{Y_i,x} \right) = \sum_{i=1}^{r} \text{codim} \, \Theta_{Y_i,x},$$

so that the vector subspaces $\Theta_{Y_i,x} \subset \Theta_{X,x}$ are transversal, in the sense that their intersection is as small as possible for their dimensions. From the inclusion $\bigcap_{i=1}^{r} \Theta_{Y_i,x} \supset \Theta_{Y,x}$, where $Y = \bigcap Y_i$, we deduce in the same way that Y is nonsingular at x.

Thus Theorem 1 asserts that the subvarieties $V(u_i)$ are transversal.

Let X' be an affine neighbourhood of x in which $\bigcap X_i = x = (0,\ldots,0)$. If $X' \subset \mathbb{A}^N$ and t_i are coordinates in $\mathbb{A}^N$, then x is defined by equations $t_1 = \cdots = t_N = 0$, and $\bigcap X_i$ is defined in X' by $u_1 = \cdots = u_n = 0$. By the Nullstellensatz it follows that $(t_1,\ldots,t_N)^k \subset (u_1,\ldots,u_n)$ for some $k > 0$, where $(t_1,\ldots,t_N)$ and $(u_1,\ldots,u_n)$ denote ideals of $k[X']$. A fortiori the same holds for the ideals $(t_1,\ldots,t_N)$ and $(u_1,\ldots,u_n)$ in $\mathcal{O}_x$. Note that $(t_1,\ldots,t_N) = \mathfrak{m}_x$, so that $\mathfrak{m}_x^k \subset (u_1,\ldots,u_n)$. In fact a more precise statement holds.

Theorem 2. *Local parameters at x generate the maximal ideal $\mathfrak{m}_x$ of $\mathcal{O}_x$.*

Proof. This is an immediate consequence of Nakayama's lemma (Appendix, §6, Proposition 3) applied to the maximal ideal $\mathfrak{m}_x$ as an $\mathcal{O}_x$-module. By 1.1, Lemma, $\mathfrak{m}_x$ is a finite $\mathcal{O}_x$-module. Since local parameters generate $\mathfrak{m}_x/\mathfrak{m}_x^2$, they generate $\mathfrak{m}_x$ by Nakayama's lemma. The theorem is proved.

Example. Let X be a nonsingular affine variety and G a finite group of auto-morphisms of X, as in Chap. I, 2.3, Example 11. Suppose that G acts freely on X, that is, $g(x) = x$ implies that $g = e$ for any $g \in G$ and $x \in X$, where e is the identity map. We prove that under these assumptions the quotient variety X/G is again nonsingular. Let $X \to Y = X/G$ be the quotient map constructed in Chap. I, 2.3, Example 11, and set $n = \dim X = \dim Y$. Choose $x \in X$, set $y = f(x) \in Y$ and let $u_1, \ldots, u_n$ be local parameters at x with $u_i \in k[X]$. Then $u_1, \ldots, u_n$ generate $\mathfrak{m}_x$. For each u_i, we construct a function $\overline{u}_i \in k[X]$ such that $\overline{u}_i \equiv u_i \bmod \mathfrak{m}_x^2$ and $\overline{u}_i \in \mathfrak{m}_{g(x)}^2$ for all $g \in G$ with $g \neq e$. For this, we need only multiply u_i by the square of an element $h \in k[X]$ with $h(x) = 1$ and $h(g(x)) = 0$ for all $g \neq e$.

Set $v_i = S(\overline{u}_i)$, where the averaging operator S is as in Appendix, §4, Proposition 1 and Chap. I, 2.3, Example 11. Since $g^*\overline{u}_i \in \mathfrak{m}_x^2$ for $g \neq e$, we have $v_i \equiv u_i \bmod \mathfrak{m}_x^2$, and hence $v_1, \ldots, v_n$ are local parameters at x. But $v_i \in k[Y]$ and $v_i(y) = 0$. Let us prove that $\mathfrak{m}_y = (v_1, \ldots, v_n)$. Let $h \in \mathfrak{m}_y \cap k[Y]$. Then $f^*(h) \in \mathfrak{m}_x$ and $f^*(h) = \sum h_i v_i$. Applying the operator S to this, in view of $S(f^*(h)) = f^*(h)$ and $S(v_i) = v_i$, we get that $f^*(h) = \sum S(h_i)v_i$. Thus $\dim \mathfrak{m}_y/\mathfrak{m}_y^2 \leq n$, and it follows that y is nonsingular.

It is important to note that nonsingularity of a point x is characterised by a purely algebraic property of the local ring $\mathcal{O}_x$. By definition $x \in X$ is nonsingular if and only if $\dim_k \mathfrak{m}_x/\mathfrak{m}_x^2 = \dim_x X$. The left-hand side of the equality is defined for any Noetherian local ring $\mathcal{O}$. The right-hand side can also be expressed as a property of the local ring $\mathcal{O}_x$. Namely, by Chap. I, 6.2, Theorem 5, Corollary 1, the dimension of X at x can be defined as the smallest r for which there exist r functions $u_1, \ldots, u_r \in \mathfrak{m}_x$ such that, in some neighbourhood of x, the set defined by $u_1 = \cdots = u_r = 0$ consists of x only. By the Nullstellensatz, this property is equivalent to $(u_1, \ldots, u_r) \supset \mathfrak{m}_x^k$ for some $k > 0$.

For an arbitrary Noetherian local ring $\mathcal{O}$ with maximal ideal $\mathfrak{m}$, the small-est number r for which there exist r functions $u_1, \ldots, u_r \in \mathfrak{m}_x$ such that $(u_1, \ldots, u_r) \supset \mathfrak{m}_x^k$ for some $k > 0$ is called the *dimension* of $\mathcal{O}$ and denoted by $\dim \mathcal{O}$. By Nakayama's lemma, the ideal $\mathfrak{m}$ itself is generated by n ele-ments, where $n = \dim_{\mathcal{O}/\mathfrak{m}}(\mathfrak{m}/\mathfrak{m}^2)$. Hence

$$\dim \mathcal{O} \leq \dim_{\mathcal{O}/\mathfrak{m}}(\mathfrak{m}/\mathfrak{m}^2).$$

If $\dim \mathcal{O} = \dim_{\mathcal{O}/\mathfrak{m}}(\mathfrak{m}/\mathfrak{m}^2)$ then the local ring $\mathcal{O}$ is said to be *regular*. We see that x is a nonsingular point if and only if the local ring $\mathcal{O}_x$ is regular. This is the algebraic meaning of nonsingularity of a point.

2.2. Power Series Expansions

The idea of associating power series with elements of the local ring $\mathcal{O}_x$ is based on the following arguments. For any function $f \in \mathcal{O}_x$, set $f(x) = \alpha_0$ and $f_1 = f - \alpha_0$; then $f_1 \in \mathfrak{m}_x$. Let $u_1, \ldots, u_n$ be a system of local parameters at x. By definition, $u_1, \ldots, u_n$ generate the whole of the vector space $\mathfrak{m}_x/\mathfrak{m}_x^2$. Thus there exist $\alpha_1, \ldots, \alpha_n \in k$ such that $f_1 - \sum_{i=1}^n \alpha_i u_i \in \mathfrak{m}_x^2$. Set $f_2 = f_1 - \sum_{i=1}^n \alpha_i u_i = f - \alpha_0 - \sum_{i=1}^n \alpha_i u_i$. Since $f_2 \in \mathfrak{m}_x^2$, we can write $f_2 = \sum g_j h_j$ with $g_j, h_j \in \mathfrak{m}_x$. As above, there exist $\beta_{ji}, \gamma_{ji} \in k$ such that

$$g_j - \sum_{i=1}^n \beta_{ji} u_i \in \mathfrak{m}_x^2 \quad \text{and} \quad h_j - \sum_{i=1}^n \gamma_{ji} u_i \in \mathfrak{m}_x^2.$$

Now set $\sum_j \left(\sum_i \beta_{ji} u_i \right) \left(\sum_i \gamma_{ji} u_i \right) = \sum_{l,k=1}^n \alpha_{lk} u_l u_k$. Then $f_2 - \sum \alpha_{lk} u_l u_k \in \mathfrak{m}_x^3$, and therefore $f_3 = f - \alpha_0 - \sum_{i=1}^n \alpha_i u_i - \sum \alpha_{lk} u_l u_k \in \mathfrak{m}_x^3$. Continuing in the same way, we can obviously find forms $F_i \in k[T_1, \ldots, T_n]$ of degree $\deg F_i = i$ such that $f - \sum_{i=0}^k F_i(u_1, \ldots, u_n) \in \mathfrak{m}_x^{k+1}$.

Definition 1. The *formal power series ring* in variables $(T_1, \ldots, T_n) = T$ is the ring whose elements are infinite expressions of the form

$$\Phi = F_0 + F_1 + F_2 + \cdots, \tag{1}$$

where $F_i \in k[T]$ is a form of degree i, and the ring operations are defined by the rules: if $\Psi = G_0 + G_1 + G_2 + \cdots$ then

$$\Phi + \Psi = (F_0 + G_0) + (F_1 + G_1) + (F_2 + G_2) + \cdots, \quad \text{and}$$

$$\Phi\Psi = H_0 + H_1 + H_2 + \cdots, \quad \text{where } H_i = \sum_{j+l=i} G_j F_l.$$

The formal power series ring is denoted by $k[[T]]$. It contains k as the power series with $F_i = 0$ for $i > 0$. If i is the first index for which $F_i \neq 0$ then F_i is called the *leading term* of (1). The leading term of a product is equal to the product of the leading terms, so that $k[[T]]$ has no zerodivisors.

The arguments discussed above allow us to associate a power series $\Phi = F_0 + F_1 + F_2 + \cdots$ with a function $f \in \mathcal{O}_x$.

We arrive at the following definition.

Definition 2. A formal power series Φ is called the *Taylor series* of a function $f \in \mathcal{O}_x$ if for every $k \geq 0$ we have

$$f - S_k(u_1, \ldots, u_n) \in \mathfrak{m}_x^{k+1}, \quad \text{with } S_k = \sum_{i=0}^k F_i. \tag{2}$$

Example. Let $X = \mathbf{A}^1$ with coordinate t, and let x be the point $t = 0$. Then $\mathfrak{m}_x = (t)$, and one can associate a power series $\sum_{m=0}^{\infty} \alpha_m t^m$ with any rational function $f(t) = P(t)/Q(t)$ with $Q(0) \neq 0$ such that

$$\frac{P(t)}{Q(t)} - \sum_{m=0}^{k} \alpha_m t^m \equiv 0 \quad \mathrm{mod}\, t^{k+1},$$

that is,

$$P(t) - Q(t) \sum_{m=0}^{k} \alpha_m t^m \equiv 0 \quad \mathrm{mod}\, t^{k+1}.$$

This is the usual procedure for finding the coefficients of a power series of a rational function by the method of unknown coefficients. For example,

$$\frac{1}{1-t} = \sum_{m=0}^{\infty} t^m, \quad \text{because} \quad \frac{1}{1-t} - \sum_{m=0}^{k} t^m = \frac{t^{k+1}}{1-t} \equiv 0 \quad \mathrm{mod}\, t^{k+1}.$$

The correspondence $f \mapsto \Phi$ depends in an essential way on the choice of the system of local parameters $u_1, \ldots, u_n$.

The arguments we have just given prove the following assertion.

Theorem 3. *Every function f has at least one Taylor series.*

Up to now we have used in essence not that x is nonsingular, but only that $u_1, \ldots, u_n$ generate $\mathfrak{m}_x/\mathfrak{m}_x^2$. Now we make use of the nonsingularity of x.

Theorem 4. *If x is nonsingular, then a function has a unique Taylor series.*

Proof. It is obviously enough to prove that any Taylor series of the function $f = 0$ is equal to 0. By (2), this is equivalent to the assertion that if $u_1, \ldots, u_n$ are local parameters of a nonsingular point x, then for a form $F_k(T_1, \ldots, T_n)$ of degree k,

$$F_k(u_1, \ldots, u_n) \in \mathfrak{m}_x^{k+1} \implies F_k(T_1, \ldots, T_n) = 0. \tag{3}$$

Suppose that this is not the case. By means of a nondegenerate linear transformation, we can arrange that the coefficient of T_n^k in F_k is nonzero. Indeed, this coefficient equals $F_k(0, \ldots, 0, 1)$, and if $F_k(\alpha_1, \ldots, \alpha_n) \neq 0$ (and such $\alpha_1, \ldots, \alpha_n$ certainly exist, given that $F_k \neq 0$), then we just have to carry out a linear transformation taking the vector $(\alpha_1, \ldots, \alpha_n)$ to $(0, \ldots, 0, 1)$. Thus we can assume that

$$F_k(T_1, \ldots, T_n) = \alpha T_n^k + G_1(T_1, \ldots, T_{n-1})T_n^{k-1} + \cdots + G_k(T_1, \ldots, T_{n-1}),$$

where $\alpha \neq 0$ and G_i is a form of degree i in $T_1, \ldots, T_{n-1}$. By 2.1, Theorem 2, it follows easily that any element of m_x^{k+1} can be written as a form of degree k in $u_1, \ldots, u_n$ with coefficients in m_x. Hence the left-hand side of (3) can be expressed in the form

$$
\begin{aligned}
&\alpha u_n^k + G_1(u_1, \ldots, u_{n-1})u_n^{k-1} + \cdots + G_k(u_1, \ldots, u_{n-1}) \\
&= \mu u_n^k + H_1(u_1, \ldots, u_{n-1})u_n^{k-1} + \cdots + H_k(u_1, \ldots, u_{n-1}),
\end{aligned}
\tag{4}
$$

where $\mu \in m_x$ and H_i are forms of degree i. It follows that $(\alpha - \mu)u_n^k \in (u_1, \ldots, u_{n-1})$. Since $\alpha \neq 0$, it follows that $\alpha - \mu \notin m_x$ and $(\alpha - \mu)^{-1} \in \mathcal{O}_x$, and hence $u_n^k \in (u_1, \ldots, u_{n-1})$. We see that $V(u_n) \supset V(u_1) \cap \cdots \cap V(u_{n-1})$. It follows that $\Theta_n \supset \Theta_1 \cap \cdots \cap \Theta_{n-1}$, where Θ_i is the tangent space to $V(u_i)$ at x, and hence $\Theta_1 \cap \cdots \cap \Theta_n = \Theta_1 \cap \cdots \cap \Theta_{n-1}$. Therefore $\dim \Theta_1 \cap \cdots \cap \Theta_n \geq 1$, and this contradicts 2.1, Theorem 1. The theorem is proved.

Thus we have a uniquely determined map $\tau \colon \mathcal{O}_x \to k[[T]]$ that takes each function to its Taylor series. A simple verification based on the definition (2) of τ shows that it is a homomorphism. We leave this verification to the reader.

What is the kernel of τ? If $\tau(f) = 0$ for a function $f \in \mathcal{O}_x$, then by (2) this means that $f \in m_x^{k+1}$ for all k. In other words, $f \in \bigcap_{k=0}^\infty m_x^k$. Thus we are talking about functions that are analogues of the functions in analysis with every derivative at some point equal to 0. In our case such a function must be equal to 0. This follows from Appendix, §6, Proposition 4 and 1.1, Lemma.

As a corollary we get the following result.

Theorem 5. *A function $f \in \mathcal{O}_x$ is uniquely determined by any of its Taylor series. In other words, τ is an isomorphic inclusion of the local ring $\mathcal{O}_x$ into the formal power series ring $k[[T]]$.* $\square$

Recall that in this section we have nowhere used that the variety X is irreducible. Conversely, Theorem 5 allows us to make certain deductions concerning irreducibility.

Theorem 6. *If x is a nonsingular point of X then there is a unique component of X passing through x.*

Proof. We replace X by an affine neighbourhood U of x contained in $X' = X \setminus \bigcup Z_i$, where Z_i are the components of X not passing through x. Then $k[U] \subset \mathcal{O}_x$. By Theorem 5, $\mathcal{O}_x$ is isomorphic to a subring of the formal power series ring $k[[T]]$. Since $k[[T]]$ has no zerodivisors, the same holds for $k[U]$, which is isomorphic to a subring of $k[[T]]$. Hence U is irreducible, as asserted in the theorem.

Corollary. *The set of singular points of an algebraic variety X is closed.*

Proof. Let $X = \bigcup X_i$ be a decomposition into irreducible components. It follows from Theorem 6 that the set of singular points of X is the union of the sets $X_i \cap X_j$ for $i \neq j$ and the sets of singular points of X_i. As a union of a finite number of closed sets, it is closed.

If x is a singular point, the best we can do is to send an element $f \in \mathcal{O}_x$ into the sequence of residue classes $\xi_n = f + \mathfrak{m}_x^n \in \mathcal{O}_x/\mathfrak{m}_x^n$. This sequence has the following compatibility property: if $\theta_{n+1} : \mathcal{O}_x/\mathfrak{m}_x^{n+1} \to \mathcal{O}_x/\mathfrak{m}_x^n$ is the quotient map then $\theta_{n+1}(\xi_{n+1}) = \xi_n$. The set of all such compatible sequences $\{\xi_n\}$ under componentwise addition and multiplication forms a ring $\widehat{\mathcal{O}}_x$, called the *completion* of $\mathcal{O}_x$. We have just defined a homomorphism $\tau : \mathcal{O}_x \to \widehat{\mathcal{O}}_x$ by $\tau(f) = \{\xi_n\}$, where $\xi_n = f + \mathfrak{m}_x^n \in \mathcal{O}_x/\mathfrak{m}_x^n$. The same argument as in the case of a nonsingular point shows that τ is an inclusion. The ring $\widehat{\mathcal{O}}_x$ is local, with maximal ideal $\mathfrak{M}$ consisting of all compatible sequences $\{\xi_n\}$ with $\xi_n \in \mathfrak{m}_x$. It can be shown that applying the same construction again to $\widehat{\mathcal{O}}_x$ gives nothing new, that is, $(\widehat{\mathcal{O}}_x)\widehat{} = \widehat{\mathcal{O}}_x$, and τ in this case is an isomorphism. If x is nonsingular, then $\widehat{\mathcal{O}}_x$ is just the formal power series ring. In the general case, $\widehat{\mathcal{O}}_x$ is an important characteristic of a singular point. If for $x \in X$ and $y \in Y$ the completed local rings $\widehat{\mathcal{O}}_x$ and $\widehat{\mathcal{O}}_y$ are isomorphic, we say that the varieties X and Y are *formally analytically equivalent* in neighbourhoods of these points. Since for a nonsingular point x of an n-dimensional variety the local ring $\widehat{\mathcal{O}}_x$ is isomorphic to that of a point $x' \in \mathbb{A}^n$, all nonsingular points of all varieties of the same dimension have formally analytically equivalent neighbourhoods. Compare §3, Ex. 8–16.

2.3. Varieties over the Reals and the Complexes

Suppose that $k = \mathbb{R}$ or $\mathbb{C}$. We prove that in this case, the formal Taylor series of a function $f \in \mathcal{O}_x$ converges for small values of $T_1, \ldots, T_n$.

Let $X \subset \mathbb{A}^N$ be a variety, with $\mathfrak{A}_X = (F_1, \ldots, F_m)$, and suppose that $\dim_x X = n$. If $x \in X$ is a nonsingular point then the matrix

$$\left(\frac{\partial F_i}{\partial T_j}(x) \right)_{\substack{i=1\ldots m \\ j=1\ldots N}}$$

has rank $N - n$. Suppose that the minor

$$\det \left| \frac{\partial F_i}{\partial T_j}(x) \right|_{\substack{i=1\ldots N-n \\ j=n+1\ldots N}} \neq 0, \tag{1}$$

and that x is the origin. Then the restrictions $t_1, \ldots, t_n$ to X of the first n coordinates form a system of local parameters on X at x. Write X' for the union of all components of the variety defined by

$$F_1 = \cdots = F_{N-n} = 0 \tag{2}$$

that pass through x. By (1), the dimension of the tangent space Θ' to X' at x equals n, and by the theorem on dimension of intersections, $\dim_x X' \geq n$. Since $\dim \Theta' \geq \dim_x X'$, then $\dim_x X' = n$ and x is a nonsingular point of X'. From this it follows by Theorem 6 that X' is irreducible. Obviously $X' \supset X$, so that $\dim X' = \dim X$ implies that $X' = X$.

We see that X can be defined in some neighbourhood of x by the $N - n$ equations (2), and that these satisfy (1). By the implicit function theorem (see for example Goursat [30], Vol. 1, Chap. IX, §§187–190 or Fleming [25], 4.6), there exists a system of power series $\Phi_1, \ldots, \Phi_{N-n}$ in n variables $T_1, \ldots, T_n$ and an $\varepsilon > 0$ such that $\Phi_j(T_1, \ldots, T_n)$ converges for all T_i with $|T_i| < \varepsilon$, and

$$F_i(T_1, \ldots, T_n, \Phi_1(T), \ldots, \Phi_{N-n}(T)) = 0; \tag{3}$$

moreover, the coefficients of the power series $\Phi_1, \ldots, \Phi_{N-n}$ are uniquely determined by the relation (3).

However, assuming that $t_1, \ldots, t_n$ are chosen as local parameters, the formal power series $\tau(T_{n+1}), \ldots, \tau(T_N)$ also satisfy (3), and hence must coincide with $\Phi_1, \ldots, \Phi_{N-n}$, and it therefore follows that $\tau(T_{n+1}), \ldots, \tau(T_N)$ converge if $|T_j| < \varepsilon$ for $j = 1, \ldots, n$.

Any function $f \in \mathcal{O}_x$ can be written in the form $f = P/Q$, where $P = P(T_1, \ldots, T_n)$ and $Q = Q(T_1, \ldots, T_n)$ are polynomials and $Q(x) \neq 0$; and then

$$\tau(f) = \frac{P(\tau(T_1), \ldots, \tau(T_n))}{Q(\tau(T_1), \ldots, \tau(T_n))}.$$

The convergence of $\tau(f)$ then follows from standard theorems on convergence of power series.

In the same way, one can show that if $u_1, \ldots, u_n$ is another system of local parameters then

$$\det \left| \frac{\partial \tau(u_i)}{\partial T_j}(0, \ldots, 0) \right|_{\substack{i=1\ldots n \\ j=1\ldots n}}$$

is nonzero, the Taylor series of $t_1, \ldots, t_n$ with respect to the system of local parameters $u_1, \ldots, u_n$ are obtained by inverting the series $\tau(u_i) = \Phi_i(T_1, \ldots, T_n)$ for $i = 1, \ldots, n$, and hence they also have positive radius of convergence. Therefore for $f \in \mathcal{O}_x$, with respect to any choice of the system of local parameters, the series $\tau(f)$ has positive radius of convergence.

The implicit function theorem asserts not only that the convergent series $\Phi_1, \ldots, \Phi_{N-n}$ exist, but also that there exists $\eta > 0$ such that any point $(t_1, \ldots, t_N) \in X$ with $|t_i| < \eta$ for $i = 1, \ldots, N$ is given by the form $t_{n+i} = \Phi_i(t_1, \ldots, t_n)$ for $i = 1, \ldots, N - n$. It follows that $(t_1, \ldots, t_N) \mapsto (t_1, \ldots, t_n)$ is a homeomorphism of the set $\{(t_1, \ldots, t_N) \in X \mid |t_i| < \eta\}$ to a domain of n-dimensional space.

Since in our case $k = \mathbb{R}$ or $\mathbb{C}$, projective space $\mathbb{P}^N$ over k is a topological space. An algebraic variety $X \subset \mathbb{P}^N$ is also a topological space. In the respective cases $k = \mathbb{R}$ or $\mathbb{C}$, this topology on X is called the *real* or *complex topology* of X. This topology and notions deduced from it should not be confused with the terms of topological nature such as closed set, neighbourhood, open set, closure, etc. used up to now.

The preceding arguments show that for an n-dimensional variety X, any nonsingular point has a neighbourhood in the real topology that is homeomorphic to a domain of $\mathbb{R}^n$. Hence if every point of X is nonsingular then X is an n-dimensional manifold in the sense of topology. If $k = \mathbb{C}$ then a nonsingular point $x \in X$ has a neighbourhood in the complex topology that is homeomorphic to a domain in n-dimensional complex space $\mathbb{C}^n$, and hence to a domain in $\mathbb{R}^{2n}$. Therefore if all points of X are nonsingular, X is a $2n$-dimensional manifold.

It is easy to prove that $\mathbb{P}^N$ over $k = \mathbb{R}$ or $\mathbb{C}$ is compact in the real or complex topology. Thus if X is projective, it is compact. If $k = \mathbb{C}$, the converse also holds: a quasiprojective variety X that is compact in its complex topology is projective. See Chap. VII, §2, Ex. 2.

In conclusion we note that everything we have said in this section (excluding the preceding paragraph) carries over word-for-word to the case that k is a p-adic number field.

Exercises to §2

1. Prove that for an n-dimensional variety X, the set of points where n given functions fail to form a system of local parameters is closed.

2. Prove that a polynomial $f \in k[T] = k[\mathbb{A}^1]$ is a local parameter at the point $T = \alpha$ if and only if α is a simple root of f.

3. Prove that a formal power series $\Phi = F_0 + F_1 + \cdots$ has an inverse in $k[[T]]$ if and only if $F_0 \neq 0$.

4. Let T be an indeterminate. Consider the ring $k((T))$ of expressions of the form $\alpha_{-n}T^{-n} + \alpha_{-n+1}T^{-n+1} + \cdots + \alpha_0 + \alpha_1 T + \cdots$. Prove that $k((T))$ is a field, and is isomorphic to the field of fractions of $k[[T]]$. (It is called the *field of formal Laurent series*.)

5. Let $S \subset \mathbb{A}^2$ be the circle given by $X^2 + Y^2 = 1$, over a field k of characteristic 0. Prove that X is a local parameter at $x = (0, 1)$, and that the Taylor series expansion of Y is given by

$$\tau(Y) = \sum_{n=0}^{\infty} (-1)^n \frac{(1/2)(1/2 - 1) \cdots (1/2 - n + 1)}{n!} X^{2n}.$$

6. Prove that if x is a singular point then any function $f \in \mathcal{O}_x$ has an infinite number of different Taylor series.

7. Let $X = \mathbb{A}^1$ and $x \in X$. Prove that $\tau(\mathcal{O}_x)$ is not the whole of $k[[T]]$.

3. Properties of Nonsingular Points

3.1. Codimension 1 Subvarieties

The theory of local rings allows us to prove an important property of non-singular varieties analogous to Chap. I, 6.1, Theorem 3. The question under discussion is that of defining a codimension 1 subvariety $Y \subset X$ by means of a single equation. For a singular variety, this property fails in general; (compare Chap. I, 6.2, Corollary 5, Remark 2). We prove, however, that it holds locally on a nonsingular variety. To state the result, we introduce the following definition.

Definition. Functions $f_1, \ldots, f_m \in \mathcal{O}_x$ are *local equations* of a subvariety $Y \subset X$ in a neighbourhood of x if there exists an affine neighbourhood X' of x such that $f_1, \ldots, f_m \in k[X']$ and $\mathfrak{a}_{Y'} = (f_1, \ldots, f_m)$ in $k[X']$, where $Y' = Y \cap X'$.

It is convenient to restate this condition in terms of the local ring $\mathcal{O}_x$ of x. For this, consider the ideal $\mathfrak{a}_{Y,x} \subset \mathcal{O}_x$ consisting of functions $f \in \mathcal{O}_x$ that are equal to 0 on Y in some neighbourhood of x. For an affine variety X, we obviously have

$$\mathfrak{a}_{Y,x} = \{ f = u/v \mid u, v \in k[X] \text{ with } u \in \mathfrak{a}_Y \text{ and } v(x) \neq 0 \},$$

and if all components of Y pass through x, then $\mathfrak{a}_Y = \mathfrak{a}_{Y,x} \cap k[X]$.

Lemma. *Functions $f_1, \ldots, f_m \in \mathcal{O}_x$ are local equations of Y in a neighbourhood of x if and only if $\mathfrak{a}_{Y,x} = (f_1, \ldots, f_m)$.*

Proof. If $\mathfrak{a}_Y = (f_1, \ldots, f_m)$ in $k[X]$ then obviously also $\mathfrak{a}_{Y,x} = (f_1, \ldots, f_m)$ in $\mathcal{O}_x$.

Conversely, suppose that $\mathfrak{a}_{Y,x} = (f_1, \ldots, f_m)$ with $f_i \in \mathcal{O}_x$, and let $\mathfrak{a}_Y = (g_1, \ldots, g_s)$ with $g_i \in k[X]$. For $i = 1, \ldots, s$, since $g_i \in \mathfrak{a}_{Y,x}$, we can write

$$g_i = \sum_{j=1}^{m} h_{ij} f_j \quad \text{with } h_{ij} \in \mathcal{O}_x. \tag{1}$$

The functions f_i and h_{ij} are all regular in some principal open neighbourhood U of x. Suppose that $U = X \setminus V(g)$ with $g \in k[X]$. Then $k[U]$ consists of elements of the form u/g^l with $u \in k[X]$ and $l \geq 0$. Then by (1), inside $k[U]$ we have

$$(g_1, \ldots, g_s) = \mathfrak{a}_Y k[U] \subset (f_1, \ldots, f_m).$$

We prove that $\mathfrak{a}_Y k[U] = \mathfrak{a}_{Y'}$, where now $Y' = Y \cap U$. From this, it then follows that $\mathfrak{a}_{Y'} \subset (f_1, \ldots, f_m)$, and since $f_i \in \mathfrak{a}_{Y'}$, this implies the assertion of the lemma.

It remains to prove that $\mathfrak{a}_Y k[U] = \mathfrak{a}_{Y'}$. The inclusion $\mathfrak{a}_Y k[U] \subset \mathfrak{a}_{Y'}$ is obvious. Let $v \in \mathfrak{a}_{Y'}$. Then $v = u/g^l$ with $u \in k[X]$, and hence $u = vg^l$; hence $u \in \mathfrak{a}_Y$, and since $1/g^l \in k[U]$, we get $v = u/g^l \in \mathfrak{a}_Y k[U]$. The lemma is proved.

Our aim is to prove the following result.

Theorem 1. *An irreducible subvariety $Y \subset X$ of codimension 1 has a local equation in a neighbourhood of any nonsingular point $x \in X$.*

The proof follows exactly the steps of the proof of Chap. I, 6.1, Theorem 3. There, however, we used the fact that $k[T]$ is a UFD. Here the part of $k[T]$ is played by the local ring $\mathcal{O}_x$. It has the analogous property.

Theorem 2. *The local ring $\mathcal{O}_x$ of a nonsingular point is a UFD.*

The proof of Theorem 2 is based on first establishing that the power series ring $k[[T]]$ is a UFD. This is a fairly elementary fact, similar to the corresponding result for polynomial rings. We indicate only the main steps of the proof. An entirely elementary proof (not depending on the remainder of the book) can be found in Zariski and Samuel [78], Vol. 2, Chap. VII, §1, Theorem 6.

We say that a power series $\Phi(T_1, \ldots, T_n)$ is regular with respect to the variable T_n if its initial form is of degree m, say, and contains the term $c_m T^m$ with $c_m \neq 0$. A linear transformation of the variables $T_1, \ldots, T_n$ obviously induces an automorphism of $k[[T]]$. We can, in particular, carry out a linear transformation so that any given nonzero power series Φ becomes regular with respect to T_n.

Lemma 1 (Weierstrass preparation theorem). *Suppose that a power series $\Phi \in k[[T]]$ is regular with respect to T_n and has initial form of degree m; then there exists a power series $U \in k[[T]]$ with nonzero constant term such that the series ΦU is a polynomial in T_n over $k[[T_1, \ldots, T_{n-1}]]$, that is,*

$$\Phi U = T_n^m + R_1 T_n^{m-1} + \cdots + R_m,$$

with $R_i = R_i(T_1, \cdots, T_{n-1}) \in k[[T_1, \cdots, T_{n-1}]]$ for $i = 1, \ldots, m$.

Proof. See Zariski and Samuel [78], Vol. 2, Chap. VII, §1, Theorem 5. $\square$

Lemma 2. *The formal power series ring $k[[T]]$ is a UFD.*

Lemma 1 allows us to prove this assertion by induction on the number of variables $T_1, \ldots, T_n$ by reducing it to the analogous statement for polynomials in T_n with coefficients in $k[[T_1, \ldots, T_{n-1}]]$. The proof is carried out in detail in Zariski and Samuel [78], Vol. 2, Chap. VII, §1, Theorem 6.

Proof of Theorem 2. We write $\widehat{\mathcal{O}}_x$ for the ring of formal power series, and view $\mathcal{O}_x$ as a subring $\mathcal{O}_x \subset \widehat{\mathcal{O}}_x$ (this is possible by 2.2, Theorem 5). Write $\widehat{\mathfrak{m}}_x$ for the ideal of $\widehat{\mathcal{O}}_x$ consisting of formal power series with constant term 0. Then $\widehat{\mathfrak{m}}_x^k$ is the ideal of formal power series having no terms of degree $< k$. By definition of the inclusion $\mathcal{O}_x \hookrightarrow \widehat{\mathcal{O}}_x$ (see 2.2, (2)), it follows that $\widehat{\mathfrak{m}}_x^k \cap \mathcal{O}_x = \mathfrak{m}^k$. Thus the assumptions of Appendix, §7, Proposition 1 are satisfied, and this guarantees that $\widehat{\mathcal{O}}_x$ a UFD (Lemma 2) implies $\mathcal{O}_x$ a UFD. The theorem is proved.

Proof of Theorem 1. As we have already said, the proof of Theorem 1 is exactly the same as that of Chap. I, 6.1, Theorem 3. Since the assertion is local in nature, we can assume that X is affine. Let $f \in \mathcal{O}_x$ be any function that vanishes on Y. Factorise f into prime factors in $\mathcal{O}_x$. By the irreducibility of Y, one factor must also vanish on Y. We denote this by g, and prove that it is a local equation of Y. Replacing X by a smaller affine neighbourhood of x, we can assume that g is regular on X.

Since $V(g) \supset Y$, and both are codimension 1 subvarieties, we have $V(g) = Y \cup Y'$. If $Y' \ni x$ there exist functions h and h' such that $hh' = 0$ on $V(g)$, but neither h nor h' are 0 on $V(g)$. Therefore, g divides $(hh')^r$ in $k[X]$ for some r, and so a fortiori $g \mid (hh')^r$ in $\mathcal{O}_x$. Since $\mathcal{O}_x$ is a UFD it then follows that g divides either h or h' in $\mathcal{O}_x$. Then either h or h' vanishes on $V(g)$ in some neighbourhood of x, and hence, after passing to a smaller neighbourhood, on the whole of $V(g)$. This contradicts the assumption $Y' \ni x$. Thus $Y' \not\ni x$, and again replacing X by a sufficiently small affine neighbourhood of x, we can assume that $V(g) = Y$. If now u vanishes on Y then g divides u^s in $k[X]$ for some $s > 0$, and hence a fortiori in $\mathcal{O}_x$. It follows that g divides u in $\mathcal{O}_x$. Thus $\mathfrak{a}_{Y,x} = (g)$ and the theorem is proved.

Theorem 1 has many applications. Here is the first of these (compare Chap. I, 1.6, Theorem 2).

Theorem 3. *If X is a nonsingular variety and $\varphi\colon X \to \mathbb{P}^N$ a rational map to projective space, then the set of points at which φ is not regular has codimension ≥ 2.*

Proof. Recall that the set of points at which a rational map φ is not regular (the *locus of indeterminacy* of φ) is a closed set. The assertion of the theorem is local in nature, and it is enough to prove it in a neighbourhood of a nonsingular point $x \in X$. We can write φ in the form $\varphi = (f_0 : \cdots : f_n)$ with $f_i \in k(X)$, and without changing φ, we can multiply the f_i through by a common factor in such a way that all the $f_i \in \mathcal{O}_x$, and they have no common factor in $\mathcal{O}_x$. Now φ fails to be regular only at points at which $f_0 = \cdots = f_n = 0$. But no codimension 1 subvariety Y can be contained in the locus defined by these equations; indeed, by Theorem 1, $\mathfrak{a}_{Y,x} = (g)$

and all the f_i would have g as a common factor in $\mathcal{O}_x$, which contradicts the assumption. The theorem is proved.

Corollary 1. *Any rational map of a nonsingular curve to projective space is regular.* □

Corollary 2. *If two nonsingular projective curves are birational then they are isomorphic.* □

Let $k = \mathbb{C}$ be the complex number field. It follows from Corollary 2 that if two curves X' and X'' are birational then the set of points of X' and X'' with their complex topologies are homeomorphic. Indeed, in this case, regular functions, hence also regular maps, are defined by convergent power series, and are hence continuous.

The same holds for the set of real points of curves X' and X'' defined by equations with real coefficients, and such that there exists a birational map $\varphi \colon X' \to X''$ defined over $\mathbb{R}$, that is, defined by rational functions with real coefficients. This sometimes allows us to deduce easily that two curves are not birational over $\mathbb{R}$. For example the curve X defined by $y^2 = x^3 - x$ consists of two connected components (Figure 8). Therefore X is not rational over $\mathbb{R}$, since $\mathbb{P}^1$ is homeomorphic to the circle, and has only one connected component.

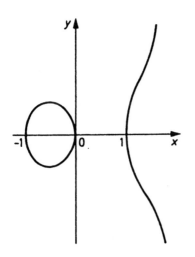

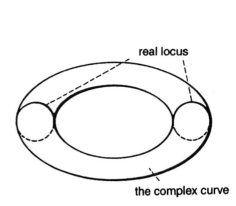
real locus

the complex curve

Figure 8. $y^2 = x^3 - x$ over $\mathbb{R}$ **Figure 9.** $y^2 = x^3 - x$ over $\mathbb{C}$

Using similar ideas, we can prove that the curve X given by $y^2 = x^3 - x$ is also irrational over $\mathbb{C}$. For this, we need to compare the topological spaces of complex points of X and of $\mathbb{P}^1$ with their complex topologies, and prove that they are not homeomorphic. Indeed, the first is homeomorphic to the torus and the second to the sphere. This is a particular case of results proved in Chap VII, 3.3. Figure 9 shows what the real points of X look like as a subset of its complex points.

3.2. Nonsingular Subvarieties

Theorem 1 does not generalise to subvarieties $Y \subset X$ of codimension greater than 1; compare, for example, Chap. I, §6, Ex. 2. But a similar statement does hold for a subvariety that is nonsingular at x. We prove a slightly more precise fact. We start with an auxiliary assertion.

Theorem 4. *Let X be an affine variety, $x \in X$ a nonsingular point, and suppose that $u_1, \ldots, u_n$ are regular functions on X that form a system of local parameters at x. Then for $m \leq n$, the subvariety Y defined by $u_1 = \cdots = u_m = 0$ is nonsingular at x, we have $\mathfrak{a}_Y = (u_1, \ldots, u_m)$ in some affine neighbourhood of x, and $u_{m+1}, \ldots, u_n$ form a system of local parameters on Y at x.*

Proof. The proof is by induction on m. For $m = 1$, Theorem 1 shows that $\mathfrak{a}_Y = (f)$ in some affine neighbourhood of x. Suppose that $u_1 = fv$; then $d_x u_1 = v(x) d_x f$. Now $d_x u_1 \neq 0$, since u_1 is an element of a system of local parameters at x. Thus $v(x) \neq 0$, so that $\mathfrak{a}_Y = (u_1)$ in some smaller open set. Since $d_x u_1 \neq 0$ it follows that x is a nonsingular point of Y.

The tangent space $\Theta_{Y,x}$ to Y at x is obviously obtained from $\Theta_{X,x}$ by imposing the condition $d_x u_1 = 0$. Therefore $d_x u_2, \ldots, d_x u_n$ is a basis of $\Theta^*_{Y,x}$, that is, $u_2, \ldots, u_n$ is a system of local parameters on Y at x.

In the general case, we let $X' \subset X$ be the subvariety defined by $u_1 = 0$. Then $Y \subset X'$ is defined in X' by the equations $u_2 = \cdots = u_m = 0$, and we can use induction. The theorem is proved.

Now we prove that any subvariety $Y \subset X$ that is nonsingular at x is given by the procedure described in Theorem 4 in some neighbourhood of x.

Theorem 5. *Let X be a variety, $Y \subset X$ a subvariety, and suppose that $x \in Y$ is a nonsingular point of both X and Y. Then we can choose a system of local parameters $u_1, \ldots, u_n$ on X at x and an affine neighbourhood U of x such that $\mathfrak{a}_Y = (u_1, \ldots, u_m)$ in U.*

In the special case $X = \mathbb{A}^n$ and $k = \mathbb{R}$ or $\mathbb{C}$, a similar fact has already been proved in 2.3.

Proof. The inclusion of the tangent spaces $\Theta_{Y,x} \hookrightarrow \Theta_{X,x}$ corresponds to a surjective map of the dual spaces $\varphi: \mathfrak{m}_{X,x}/\mathfrak{m}^2_{X,x} \to \mathfrak{m}_{Y,x}/\mathfrak{m}^2_{Y,x}$, defined by restricting functions from X to Y. We can choose a basis $u_1, \ldots, u_n$ of

$m_{X,x}/m_{X,x}^2$ such that $u_1, \ldots, u_m \in a_Y$ and such that the restrictions to Y of $u_{m+1}, \ldots, u_n$ form a basis of $m_{Y,x}/m_{Y,x}^2$. Consider an affine neighbourhood of x in which all the u_i are regular, and in this consider the subvariety Y' defined by $u_1 = \cdots = u_m = 0$. By construction, $Y' \supset Y$. We prove that $Y = Y'$, so that the assertion will follow by Theorem 4.

By Theorem 4, Y' is nonsingular at x, and hence is irreducible in a neighbourhood of x by 2.2, Theorem 6. It follows by Theorem 4 that $\dim Y' = n - m$. It is clear by construction that $\dim \Theta_{Y,x} = n - m$. Hence $Y = Y'$, and since $a_{Y'} = (u_1, \ldots, u_m)$ by Theorem 4, also $a_Y = (u_1, \ldots, u_m)$ in some neighbourhood of x. The theorem is proved.

Exercises to §3

1. Prove that if t is a local parameter of a nonsingular point of an algebraic curve then any function $f \in \mathcal{O}_x$ can be uniquely written in the form $f = t^n u$ with $n \geq 0$ and u an invertible element of $\mathcal{O}_x$. Use this to deduce 3.1, Theorem 2 for curves.

2. Prove the converse of 2.1, Theorem 1: if codimension 1 subvarieties $D_1, \ldots, D_n$ intersect transversally at x and $u_1, \ldots, u_n$ are their local equations in a neighbourhood of x then $u_1, \ldots, u_n$ form a system of local parameters at x.

3. Is 3.1, Theorem 3, Corollary 2 true without the nonsingularity assumption? What about Theorem 3 itself?

4. Prove that a point x of an algebraic curve X is nonsingular if and only if x has a local equation on X.

5. $X \subset \mathbb{A}^3$ is the cone given by $x^2 + y^2 - z^2$. Prove that the generator L defined by the equations $x = 0$, $y = z$ does not have a local equation in any neighbourhood of $(0, 0, 0)$.

6. Let $\varphi \colon \mathbb{P}^2 \to \mathbb{P}^2$ be the rational map defined by

$$\varphi(x_0 : x_1 : x_2) = (x_1 x_2 : x_0 x_2 : x_0 x_1).$$

Consider the point $x = (1 : 0 : 0)$ and a curve $C \subset \mathbb{P}^2$ that is nonsingular at x. By 3.1, Theorem 3, the map φ restricted to C is regular at x, and therefore maps x to some point that we denote by $\varphi_C(x)$. Prove that $\varphi_{C_1}(x) = \varphi_{C_2}(x)$ if and only if C_1 and C_2 touch at x, that is, $\Theta_{C_1,x} = \Theta_{C_2,x}$.

7. Prove that if $\varphi = f/g$ is a rational function, f and g are regular at a nonsingular point x and the power series $\tau(f)$ is divisible by $\tau(g)$ then φ is regular at x. [Hint: Use the arguments of Appendix, §7, Proposition 1.]

8. We use the following assertion in subsequent exercises. Let $X \subset \mathbf{A}^n$ be an affine variety and $x \in X$. Suppose that $\mathfrak{a}_X = (f_1, \ldots, f_m)$. Prove that

$$\hat{\mathcal{O}}_x = k[[T_1, \ldots, T_n]]/\bar{\mathfrak{a}}_X, \qquad \text{where } \bar{\mathfrak{a}}_X = \left(\tau(f_1), \ldots, \tau(f_m)\right),$$

and $\tau(f_i)$ is the Taylor series of f_i as in 2.2. [Hint: Use the results of Atiyah and Macdonald [7], Chap. 10.]

9. Prove that a formal analytic equivalence of $\mathbf{A}^n$ with itself (that is, a *formal analytic automorphism*) in a neighbourhood of 0 is given by power series $\Phi_1, \ldots, \Phi_n$ with no constant terms such that the determinant formed by the linear terms is nonzero.

10. Prove that two plane curves with equations $F = 0$ and $G = 0$ passing through the origin $0 \in \mathbf{A}^2$ are formally analytically equivalent in a neighbourhood of 0 if and only if there exists a formal analytic automorphism of $\mathbf{A}^2$ given by power series Φ_1, Φ_2 such that $F(\Phi_1, \Phi_2) = GU$, where U is a power series with nonzero constant term.

11. Prove that any nonsingular algebraic curve having the origin 0 as a double point with distinct tangents is formally analytically equivalent in a neighbourhood of 0 to the curve $xy = 0$. [Hint: Use Ex. 10. Look for Φ_1, Φ_2 modulo higher and higher powers of the ideal (x, y).]

12. Classify double points of algebraic plane curves up to formal analytic equivalence over a field k of characteristic 0.

13. Let X be the hypersurface in $\mathbf{A}^n$ with equation $F = F_2(T) + F_3(T) + \cdots + F_k(T) = 0$, where $F_2(T)$ is a quadratic form of maximal rank n. Prove that X is formally analytically equivalent in a neighbourhood of 0 to the quadratic cone $T_1^2 + \cdots + T_n^2 = 0$.

14. Construct an infinite number of nonsingular projective curves, with no two isomorphic over $\mathbf{R}$.

15. Suppose that a nonsingular irreducible affine n-dimensional variety $X \subset \mathbf{A}^n$ is given by equations $F_1 = \cdots = F_m = 0$, and that for every $x = (x_1, \ldots, x_n) \in X$ the space defined by $\sum\left(\partial F_i/\partial T_j\right)(x)(T_j - x_j) = 0$ is n-dimensional. Prove that then $\mathfrak{a}_X = (F_1, \ldots, F_m)$.

16. Deduce from Ex. 15 that if the Plücker equations $x \wedge x = 0$ of the Grassmannian $\mathrm{Grass}(2, r)$ are written out as $F_1 = \cdots = F_m = 0$ then $F_1, \ldots, F_m$ generate the ideal of $\mathrm{Grass}(2, r)$. (Compare Chap. I, 4.1, Example 1 and the remark after Chap. II, 1.3, Example 1.)

4. The Structure of Birational Maps

4.1. Blowup in Projective Space

We proved in the preceding section that a birational map between nonsingular projective curves is an isomorphism (3.2, Theorem 3, Corollary 2). This is no longer true for higher dimensional varieties. For example, stereographic projection establishes a birational equivalence between a nonsingular quadric surface $Q \subset \mathbb{P}^3$ and the projective plane $\mathbb{P}^2$, but this is not a regular map (see Chap. I, §4, Ex. 7 and compare Chap. I, 6.2, Proposition, Corollary 5). This section defines and studies the simplest and most typical case of a birational map that is not an isomorphism, the blowup.

We consider the two projective spaces $\mathbb{P}^n$ with homogeneous coordinates $x_0, \ldots, x_n$ and $\mathbb{P}^{n-1}$ with homogeneous coordinates $y_1, \ldots, y_n$. For points $x = (x_0 : \cdots : x_n) \in \mathbb{P}^n$ and $y = (y_1 : \cdots : y_n) \in \mathbb{P}^{n-1}$, we denote the point $(x, y) \in \mathbb{P}^n \times \mathbb{P}^{n-1}$ of the product also by $(x_0 : \cdots : x_n; y_1 : \cdots : y_n)$. Consider the closed subvariety $\Pi \subset \mathbb{P}^n \times \mathbb{P}^{n-1}$ defined by the equations

$$x_i y_j = x_j y_i \qquad \text{for } i, j = 1, \ldots, n. \tag{1}$$

Definition 1. The map $\sigma \colon \Pi \to \mathbb{P}^n$ defined by restricting the first projection $\mathbb{P}^n \times \mathbb{P}^{n-1} \to \mathbb{P}^n$ is called the *blowup*[6] of $\mathbb{P}^n$ centred at ξ.

Write ξ for the point $\xi = (1 : 0 : \cdots : 0) \in \mathbb{P}^n$. If $(x_0 : \cdots : x_n) \neq \xi$ then equations (1) imply that $(y_1 : \cdots : y_n) = (x_1 : \cdots : x_n)$, so that the map $\sigma^{-1} \colon \mathbb{P}^n \setminus \xi \to \Pi$ defined by

$$(x_0 : \cdots : x_n) \mapsto \big((x_0 : \cdots : x_n), (x_1 : \cdots : x_n)\big) \tag{2}$$

is the inverse of σ. However, if $(x_0 : \cdots : x_n) = \xi$ then equations (1) are satisfied by any values of the y_i. Thus $\sigma^{-1}(\xi) = \xi \times \mathbb{P}^{n-1}$, and σ defines an isomorphism between $\mathbb{P}^n \setminus \xi$ and $\Pi \setminus (\xi \times \mathbb{P}^{n-1})$. The point ξ is called the *centre* of the blowup σ.

Let us describe the structure of Π in a neighbourhood of points of the form $(\xi; y_1, \ldots, y_n)$. We have $y_i \neq 0$ for some i, so that the chosen point is contained in the open set U_i of Π defined by $x_0 \neq 0$, $y_i \neq 0$; in U_i we can even assume that $x_0 = 1$, $y_i = 1$. Then equations (1) take the form $x_j = y_j x_i$ for $j = 1, \ldots, n$ with $j \neq i$. It follows that U_i is isomorphic to affine space $\mathbf{A}^n$ with coordinates $y_1, \ldots, \widehat{y_i}, x_i, \ldots, y_n$.

We see in particular that Π is nonsingular, and thus by 2.2, Theorem 6 is irreducible in a neighbourhood of every point. We will see presently that Π is actually irreducible.

---

[6] This notion appears in the literature under many other names: *σ-process, monoidal transformation, dilation, quadratic transformation,* etc.

For this, to get a clearer idea of the effect of the blowup, we consider σ over some line L through ξ. Suppose that L is given by $x_j = \alpha_j x_i$ for some i and $j = 1, \ldots, n$ with $j \neq i$. On L the map (2) takes the form

$$\sigma^{-1}(x_0 : \cdots : x_n) = (x_0 : \cdots : x_n; \alpha_1 : \cdots : 1 : \cdots : \alpha_n),$$

with $\alpha_i = 1$ in the ith place. We see thus that σ^{-1} is regular on L and maps it to a curve $\sigma^{-1}(L) \subset \Pi$ that intersects $\xi \times \mathbb{P}^{n-1}$ in the point $(\xi; \alpha_1 : \cdots : 1 : \cdots : \alpha_n)$. We can interpret this result as follows: σ^{-1} is not regular at ξ, but considering it on L we get a regular map $\sigma^{-1} \colon L \to \Pi$. We can use this to extend the definition of σ^{-1} on L over the point ξ; over $\mathbb{R}$ or $\mathbb{C}$, this means that we define $\sigma^{-1}(x)$ for $x \in L \backslash \xi$ and define $\sigma^{-1}(\xi)$ by letting x tend to ξ along the direction of L. However, the result depends on the choice of L, since passing to the limit $x \to \xi$ depends on the direction along which x approaches ξ. Choosing different lines L we get all possible points of $\xi \times \mathbb{P}^{n-1}$. Thus, although σ^{-1} is not regular at ξ, on resolving the indeterminacy arising from this we don't get arbitrary points of Π, but only points of $\xi \times \mathbb{P}^{n-1}$. Bearing this picture in mind one says that σ^{-1} blows up ξ to $\xi \times \mathbb{P}^{n-1}$.

Note that at the same time we have proved the irreducibility of Π. Indeed,

$$\Pi = \left(\xi \times \mathbb{P}^{n-1}\right) \cup \left(\Pi \backslash (\xi \times \mathbb{P}^{n-1})\right).$$

Since $\Pi \backslash (\xi \times \mathbb{P}^{n-1})$ is isomorphic to $\mathbb{P}^n \backslash \xi$ it is irreducible, hence so is its closure $\overline{\Pi \backslash (\xi \times \mathbb{P}^{n-1})}$. Thus we need only show that

$$\xi \times \mathbb{P}^{n-1} \subset \overline{\Pi \backslash (\xi \times \mathbb{P}^{n-1})}.$$

But obviously $\sigma^{-1}(L) \subset \overline{\Pi \backslash (\xi \times \mathbb{P}^{n-1})}$, so that also

$$\sigma^{-1}(L) \cap \left(\xi \times \mathbb{P}^{n-1}\right) \subset \overline{\Pi \backslash (\xi \times \mathbb{P}^{n-1})}.$$

But we have just seen that for suitable choice of L the left-hand side here is an arbitrary point of $\xi \times \mathbb{P}^{n-1}$.

For $n = 2$ we have an intuitive picture of the map $\sigma \colon \Pi \to \mathbb{P}^2$ and its effect on the lines $L \colon \sigma^{-1}(L)$ intersects the line $\xi \times \mathbb{P}^1$ in a point that moves as L rotates around ξ. Thus Π looks like one twist of a helix (Figure 10).

4.2. Local Blowup

For an arbitrary quasiprojective variety X and a nonsingular point $\xi \in X$, we now construct a variety Y and a map $\sigma \colon Y \to X$ analogous to that constructed in 4.1 for $X = \mathbb{P}^n$ and $\xi = (1 : 0 : \cdots : 0)$.

We begin with an auxiliary construction. Let X be a quasiprojective variety and $\xi \in X$ a nonsingular point, and suppose that $u_1, \ldots, u_n$ are functions that are regular everywhere on X and such that (a) the equations $u_1 = \cdots = u_n = 0$ have the single solution ξ in X; and (b) $u_1, \ldots, u_n$ form a local system of parameters on X at ξ.

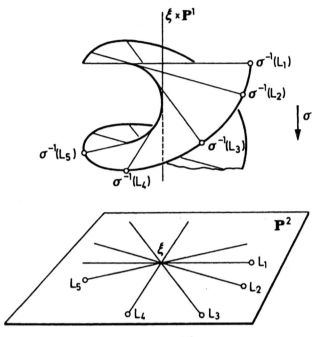

Figure 10. The Blowup $\sigma\colon \Pi \to \mathbb{P}^2$

Consider the product $X \times \mathbb{P}^{n-1}$ and the subvariety $Y \subset X \times \mathbb{P}^{n-1}$ consisting of points $(x; t_1 : \cdots : t_n)$ with $x \in X$ and $(t_1 : \cdots : t_n) \in \mathbb{P}^{n-1}$, such that

$$u_i(x)t_j = u_j(x)t_i \qquad \text{for } i, j = 1, \ldots, n.$$

The regular map $\sigma\colon Y \to X$ obtained as the restriction to Y of the first projection $X \times \mathbb{P}^{n-1} \to X$ is called the *local blowup* of X with centre in ξ.

Note that in general this construction does not apply to the case that X is projective, since we require the existence of nonconstant everywhere regular functions $u_1, \ldots, u_n$ on X. Thus the new notion does not include the previous notion of blowup in the case $X = \mathbb{P}^n$. The two are related as follows: write $X \subset \mathbb{P}^n$ for the affine subset defined by $x_0 \neq 0$, and set $Y = \sigma^{-1}(X)$. Then the map $\sigma\colon Y \to X$ induced on Y by the blowup $\Pi \to \mathbb{P}^n$ is a local blowup.

The following properties proved in 4.1 for the blowup of $\mathbb{P}^n$ are proved in exactly the same way for a local blowup. The map $\sigma\colon Y \to X$ is regular and defines an isomorphism

$$Y \setminus (\xi \times \mathbb{P}^{n-1}) \xrightarrow{\sim} X \setminus \xi.$$

At a point $y \in \sigma^{-1}(\xi)$, we have $t_i \neq 0$ for some i, and we can set $s_j = t_j/t_i$ for $j \neq i$. Then the equations of Y take the form $u_j = u_i s_j$ for $j = 1, \ldots, n$ with $j \neq i$. We see from this that the maximal ideal of y is given by

$$\mathfrak{m}_y = \big(u_1 - u_1(y), \ldots, u_n - u_n(y), s_1 - s_1(y), \ldots, s_n - s_n(y)\big)$$
$$= \big(s_1 - s_1(y), \ldots, u_i - u_i(y), \ldots, s_n - s_n(y)\big).$$

Hence $\dim \Theta_{Y,y} \leq n$, and since $\dim \sigma^{-1}(X \setminus \xi) = n$, the variety Y is nonsingular at every point $y \in \sigma^{-1}(X \setminus \xi)$. Since

$$Y = \big(\xi \times \mathbb{P}^{n-1}\big) \cup \sigma^{-1}(X \setminus \xi),$$

Y is either irreducible, equal to the closure $\overline{\sigma^{-1}(X \setminus \xi)}$ of the set $\sigma^{-1}(X \setminus \xi)$, or has a second component isomorphic to $\mathbb{P}^{n-1}$. In the second case, the two components would have to intersect: for otherwise $\sigma^{-1}(X \setminus \xi)$ would be closed in $X \times \mathbb{P}^{n-1}$, but then by Chap. I, 5.2, Theorem 3 also its image $X \setminus \xi \subset X$ would be closed. But a point of intersection of the two components is nonsingular, and this contradicts 2.2, Theorem 6. Thus Y is irreducible and nonsingular and $s_1 - s_1(y), \ldots, u_i - u_i(y), \ldots, s_n - s_n(y)$ are local parameters at a point $y \in \sigma^{-1}(\xi)$ at which $t_i \neq 0$.

A local blowup is obviously a proper map (see Chap. I, 5.3, Remark after the proof of Theorem 3).

We now prove a property that can reasonably be called the independence of the local blowup of the choice of the system of local parameters $u_1, \ldots, u_n$.

Lemma. *Let $v_1, \ldots, v_n$ be another system of functions on X satisfying the above conditions (a) and (b) and $\sigma' : Y' \to X$ the local blowup constructed as above in terms of $v_1, \ldots, v_n$. Then there exist an isomorphism $\varphi : Y \to Y'$ such that the diagram*

$$
\begin{array}{ccc}
Y & \xrightarrow{\varphi} & Y' \\
& \searrow{\scriptstyle \sigma} \quad \swarrow{\scriptstyle \sigma'} & \\
& X &
\end{array}
$$

is commutative.

Proof. We have $Y' \subset X \times \mathbb{P}^{n-1}$, where the homogeneous coordinates in $\mathbb{P}^{n-1}$ are $t_1', \ldots, t_n'$. In the open sets $Y \setminus \sigma^{-1}(\xi)$ and $Y' \setminus \sigma'^{-1}(\xi)$, we set

$$
\begin{aligned}
\varphi(x; t_1 : \cdots : t_n) &= \big(x; v_1(x) : \cdots : v_n(x)\big), \\
\psi(x; t_1' : \cdots : t_n') &= \big(x; u_1(x) : \cdots : u_n(x)\big).
\end{aligned}
\tag{1}
$$

It follows from property (a) of the u_i that φ and ψ are regular maps

$$\varphi : Y \setminus \sigma^{-1}(\xi) \to Y' \qquad \text{and} \qquad \psi : Y' \setminus \sigma'^{-1}(\xi) \to Y.$$

We now consider the open set of Y in which $t_i \neq 0$ and set $s_j = t_j / t_i$. Since $v_k(\xi) = 0$, and $u_1, \ldots, u_n$ is a basis of the ideal $\mathfrak{m}_\xi$, we have

$$v_k = \sum_{j=1}^{n} h_{kj} u_j \qquad \text{with } h_{kj} \in \mathcal{O}_\xi. \tag{2}$$

Since in our open set $u_j = u_i s_j$, it follows that

$$v_k = u_i \sum_{j=1}^n \sigma^*(h_{kj})s_j = u_i g_k, \qquad \text{where} \qquad g_k = \sum_{j=1}^n \sigma^*(h_{kj})s_j. \qquad (3)$$

We set $\varphi(x; t_1 : \cdots : t_n) = (x; g_1 : \cdots : g_n)$. Obviously this map coincides with (1) wherever both are defined, since there $g_k = v_k/u_i$. Let us check that φ is regular. For this, we must prove that $g_1, \ldots, g_n$ are not simultaneously 0 at any point $\eta \in \sigma^{-1}(\xi)$. Suppose that $g_1(\eta) = \cdots = g_n(\eta) = 0$. Since not all the $s_j(\eta) = 0$ (because $s_i = 1$), it follows from (3) that $\det |h_{kj}(\xi)| = 0$. But $v_k \equiv \sum h_{kj}(\xi)u_j$ modulo $\mathfrak{m}_\xi^2$ and it would follow from this that the v_k are linearly dependent in $\mathfrak{m}_\xi/\mathfrak{m}_\xi^2$, whereas they form a system of local coordinates at ξ. Thus we have defined a global map $\varphi \colon Y \to Y'$, and similarly a map $\psi \colon Y' \to Y$. It is enough to prove that these are mutually inverse on the open sets where the formulas (1) hold; and there it is obvious. The lemma is proved.

4.3. Behaviour of a Subvariety under a Blowup

Let $X \subset \mathbb{P}^N$ be a quasiprojective subvariety, and $\sigma \colon \Pi \to \mathbb{P}^N$ the blowup defined in 4.1. We investigate the inverse image $\sigma^{-1}(X)$ of the subvariety X, which is, of course, a quasiprojective subvariety of Π.

Theorem 1. *Suppose that $X \subset \mathbb{P}^N$ is an irreducible quasiprojective variety, with $\overline{X} \neq \mathbb{P}^N$, and that X is nonsingular at ξ. Then the inverse image $\sigma^{-1}(X)$ of X under the blowup of $\mathbb{P}^N$ centred at ξ is reducible, consisting of two components*

$$\sigma^{-1}(X) = \left(\xi \times \mathbb{P}^{N-1} \right) \cup Y. \qquad (1)$$

The restriction of σ to the component Y defines a regular map $\sigma \colon Y \to X$, which is an isomorphism of some neighbourhood U of x if $x \neq \xi$ and a local blowup $\sigma^{-1}(U) \to U$ with centre ξ if $x = \xi$.

Proof. Let Y denote the closure $\overline{\sigma^{-1}(X \setminus \xi)}$ of $\sigma^{-1}(X \setminus \xi)$. Since σ^{-1} is an isomorphism on $\mathbb{P}^N \setminus \xi$, it follows that $\sigma^{-1}(X \setminus \xi)$ is isomorphic to $X \setminus \xi$, and is hence irreducible. Hence also Y is irreducible. (1) is obvious by definition: if $x \in X \setminus \xi$ then $\sigma^{-1}(x) \in Y$, and $\sigma^{-1}(\xi) = \xi \times \mathbb{P}^{N-1}$.

The fact that $\sigma \colon Y \to X$ is an isomorphism in a neighbourhood of any points $x \in X$ other than $x = \xi$ has already been noted. It remains to study $\sigma \colon Y \to X$ over a neighbourhood of ξ.

Now we use the fact that the blowup can be described as a local blowup for an affine space $\mathbb{A}^N$ containing ξ, together with the independence of the local blowup of the choice of local parameters. Namely, by 3.2, Theorem 5, we can choose a system of local parameters $u_1, \ldots, u_N$ at $\xi \in \mathbb{P}^N$ such that in some neighbourhood of ξ, the subvariety X has local equations

$$u_{n+1} = \cdots = u_N = 0, \tag{2}$$

and the functions $u_1, \ldots, u_n$ define a system of local parameters on X at ξ. We can choose a neighbourhood $U \subset \mathbb{P}^N$ of ξ such that $u_1, \ldots, u_N$ satisfy conditions (a) and (b) of 4.2, Lemma, so that the proof of the theorem reduces to the special case when X is given by equations (2).

From conditions (a) and (b) and $u_i t_j = u_j t_i$ we get that $t_{n+1}(x) = \cdots = t_N(x) = 0$ for $x \neq \xi$. Hence Y is contained in the subspace $Y' \subset X \times \mathbb{P}^{N-1}$ defined by the equations

$$t_{n+1} = \cdots = t_N = 0 \tag{3}$$
$$u_i t_j = u_j t_i \qquad \text{for } i, j = 1, \ldots, n. \tag{4}$$

If we write $\mathbb{P}^{n-1}$ for the subspace of projective space $\mathbb{P}^{N-1}$ of points satisfying (3) then we see that $Y' \subset X \times \mathbb{P}^{n-1}$ and is defined there by equations (4). Thus Y' is the same thing as the variety obtained as a result of the local blowup. We have proved that $Y' = \sigma^{-1}(X \setminus \xi)$. Hence $Y = Y'$, which proves the theorem.

Now we can give the most general definition of blowup centred at a point. If $X \subset \mathbb{P}^N$ is a quasiprojective variety, ξ a nonsingular point of X and Y the variety introduced in the statement of Theorem 1 then $\sigma \colon Y \to X$ is called the *blowup* of X with *centre* ξ. From what we proved concerning the local blowup, it follows that Y is irreducible if X is, and $\sigma^{-1}(\xi) \cong \xi \times \mathbb{P}^{n-1}$, with all points of $\sigma^{-1}(\xi)$ nonsingular points of Y.

Notice that a blowup is an isomorphism if X is a curve. Thus nontrivial blowups are a typical phenomenon of higher dimensional algebraic geometry.

4.4. Exceptional Subvarieties

The example of a blowup shows a difference of principle between algebraic curves and varieties of dimension $n > 1$: whereas for nonsingular projective curves a birational map is an isomorphism, a blowup shows that this is not always the case in higher dimensions.

Notice one peculiarity of a blowup: it is a regular map, and only fails to be an isomorphism because the rational map σ^{-1} is not regular at a point ξ.

In this section we study a map $f \colon X \to Y$ where f is a regular map and is birational, that is, $g = f^{-1} \colon Y \to X$ is a rational, but not regular, map. In the example of a blowup we saw that a codimension 1 subvariety in Y is contracted to the point ξ. We prove that the same property always holds in this situation.

Theorem 2. *Let $f \colon X \to Y$ be a regular birational map. For $x \in X$, assume that $y = f(x)$ is a nonsingular point of Y and that the inverse map $g = f^{-1}$*

is not regular at y. *Then there exists a subvariety* $Z \subset X$ *with* $Z \ni x$ *such that* $\operatorname{codim} Z = 1$, *but* $\operatorname{codim} f(Z) \geq 2$.

Proof. We can if necessary replace X by an affine neighbourhood of x, and thus assume that X is affine. Suppose that $X \subset \mathbf{A}^N$, with coordinates $t_1, \ldots, t_N$, and that $g = f^{-1}$ is the map given by $t_i = g_i$ for $i = 1, \ldots, N$, with $g_i \in k(Y)$.

Obviously $g_i = g^*(t_i)$; since g is not regular at y, at least one of the functions g_i is not regular at y. Suppose this is g_1, so that $g_1 \notin \mathcal{O}_y$. We can write g_1 in the form $g_1 = u/v$ with $u, v \in \mathcal{O}_y$ and $v(y) = 0$, and since $\mathcal{O}_y$ is a UFD (because we assume that y is nonsingular), we can suppose that u and v have no common factors. Since $g = f^{-1}$, we have $t_1 = f^*(g_1) = f^*(u/v) = f^*(u)/f^*(v)$, and hence

$$f^*(v)t_1 = f^*(u). \tag{1}$$

Now $f^*(v)(x) = v(y) = 0$, so that $x \in V(f^*(v))$. Set $Z = V(f^*(v))$. By the theorem on dimension of intersection, $\operatorname{codim} Z = 1$, and since $x \in Z$ it is nonempty. It follows from (1) that $f^*(u) = 0$ on Z, so that t_1 is a regular function. Hence also $u = 0$ on $f(Z)$, and thus $f(Z) \subset V(u) \cap V(v)$.

It remains to check that $\operatorname{codim}(V(u) \cap V(v)) \geq 2$. But if $V(u) \cap V(v)$ contained a component Y' with $y \in Y'$ and $\operatorname{codim} Y' = 1$ then by 3.1, Theorem 1, Y' would have a local equation h. This means that $u, v \in (h)$, which contradicts the assumption that u and v have no common factor in $\mathcal{O}_y$. The theorem is proved.

Definition. Let $f \colon X \to Y$ be a regular birational map. A subvariety $Z \subset X$ is *exceptional* for f if $\operatorname{codim} Z = 1$, but $\operatorname{codim} f(Z) \geq 2$.

Corollary 1. *If* $f \colon X \to Y$ *is a regular birational map between nonsingular varieties, not an isomorphism, then* f *has an exceptional subvariety.*

Corollary 2. *Let* $f \colon X \to Y$ *be a regular birational map between curves* X *and* Y, *and suppose that* Y *is nonsingular; then* $f(X)$ *is open in* Y *and* f *defines an isomorphism from* X *to* $f(X)$.

Proof. $f(X)$ open in Y follows from the fact that X and Y have isomorphic open subsets U and V (4.3, Proposition); indeed, since $f(U) = V$ is obtained by discarding a finite number of points, a fortiori so is $f(X)$, and therefore it is open in Y. If $f \colon X \to f(X)$ were not an isomorphism, we would get a contradiction to Theorem 2, since in our case only the empty set has codimension ≥ 2.

4.5. Isomorphism and Birational Equivalence

Consider a birational equivalence class of quasiprojective varieties, that is, a class consisting of all quasiprojective varieties birationally equivalent to one another. A representative of this class is called a *model*.

In 5.3 below, we prove that there exists a nonsingular projective model X_0 in every birational equivalence class of algebraic curves. 3.1, Theorem 3, Corollary 2 asserts that there is at most one such model up to isomorphism in every birational equivalence class. Therefore, sending a birational equivalence class of curves to the unique nonsingular projective model contained in it reduces the question of the classification of curves up to birational equivalence to that of the classification of nonsingular projective curves up to isomorphism.

The function field $k(X)$ of an algebraic curve X is an extension field of k generated over k by finitely many elements and of transcendence degree 1. Hence we can establish a one-to-one correspondence between such fields K and nonsingular projective curves X. This correspondence is given by $K = k(X)$. We will also say that X is a *model* of K.

One can attempt to find the nonsingular projective model X directly from algebraic properties of K. We make this question more precise by asking how the local rings $\mathcal{O}_x$ of points $x \in X$ are characterised within K. It is easy to see that every local ring $\mathcal{O}_x$ of a point $x \in X$ has the following properties:
(1) $\mathcal{O}$ is a subring of K with $k \subsetneqq \mathcal{O} \subsetneqq K$;
(2) $\mathcal{O}$ is a local ring, and its maximal ideal m is principal, that is $m = (u)$;
(3) K equals the field of fractions of $\mathcal{O}$.
It can be proved (see Ex. 7–9) that any subring $\mathcal{O} \subset K$ satisfying (1–3) is the local ring $\mathcal{O}_x$ of some point $x \in X$. Thus the nonsingular projective model X is universal: it contains all the local rings of K satisfying the natural conditions (1–3).

What about these questions in dimension $n > 1$? Things turn out reasonably well as far as the existence of a nonsingular projective model goes: it is proved for $n = 2$ or 3 (Walker, Zariski over fields of characteristic 0, Abhyankar over fields of characteristic $p > 5$), and for arbitrary n in characteristic 0 (Hironaka). For arbitrary fields and arbitrary n the existence of a nonsingular projective model seems extremely plausible. The uniqueness of the nonsingular projective model, on the contrary, is a wholly exceptional feature of the case $n = 1$. This can be seen already in the example of the projective plane $\mathbb{P}^2$ and the nonsingular quadric $Q \subset \mathbb{P}^3$, which are birational, but not isomorphic.

One might ask for the existence, in each birational equivalence class, of a model X that would be universal in the sense that the local rings $\mathcal{O}_x$ of points $x \in X$ exhaust all the local subrings of the field $K = k(X)$ that satisfy conditions (1), (2) and (3), as in the case $n = 1$, where $m = (u)$ in condition (2) is replaced by the appropriate n-dimensional version $m = (u_1, \ldots, u_n)$. However, no such model can exist, for the same reasons. Namely if $\sigma \colon X' \to X$

is the blowup of X with centre in ξ, then the local rings of points $y \in \sigma^{-1}(\xi)$ are not equal to any of the local rings $\mathcal{O}_x$ with $x \in X$. The reader can easily prove this as an exercise. Admittedly, putting together all the nonsingular points of all models of a birational equivalence class one does obtain a certain object, the so-called *Zariski Riemann surface* with this universal property, but this object is not a finite dimensional algebraic variety. Some information about this "infinite model" can be found in Zariski and Samuel [78], Vol. 2, Chap. VI, §17.

Given that there does not exist a distinguished model, the problem arises of studying the relations between the nonsingular projective models in each birational equivalence class. We describe here without proof the main results in this area known to date. From now on, all varieties considered will be assumed to be irreducible, nonsingular and projective.

We start with two definitions. We say that a model X' *dominates* X if there exists a regular birational map $f \colon X' \to X$. A variety X is a *relatively minimal model* if it does not dominate any variety not isomorphic to itself. For example, a nonsingular projective curve is always a relatively minimal model. By Theorem 2, a variety is a relative minimal model if it does not contain any exceptional subvarieties.

It can be proved that every variety dominates at least one relatively minimal model. Thus every birational equivalence class contains at least one relatively minimal model. The question thus arises of the uniqueness of a relatively minimal model. If every birational equivalence class had such a unique relatively minimal model then this would again reduce the birational classification of varieties to the classification up to isomorphism.

However, for $n > 1$ this does not work. An example is provided by the projective plane $\mathbb{P}^2$ and the nonsingular quadric $Q \subset \mathbb{P}^3$, which, as we know, are birational, that is, they both belong to the same birational equivalence class. We prove that $\mathbb{P}^2$ and Q are both relatively minimal models, that is, they do not contain any exceptional curves. Since $\mathbb{P}^2$ and Q are not isomorphic (Chap. I, 6.2, Remark 1), this gives the required example.

In our case, an irreducible exceptional curve $C \subset X$ must be contracted to a point by a regular birational map $f \colon X \to Y$, that is, $f(C) = y \in Y$. Here X and Y are projective surfaces. Curves of this type have a series of very special properties (hence the name "exceptional"). We discuss just one of these.

By 3.1, Theorem 3, the map f^{-1} is not regular at only finitely many points $y_i \in Y$. Suppose that U is a affine neighbourhood of y, sufficiently small that f^{-1} is regular at all points of U other than y. Set $V = f^{-1}(U)$ and $C = f^{-1}(y)$. Obviously V is an open subset of X and $V \supset C$. We prove that V does not contain any irreducible curve C' that is closed in Y and not contained in C. Indeed, C' is a projective curve and its image $f(C')$ is again projective. But $f(C') \subset U$, which is affine. According to Chap. I, 5.2, Theorem 3, Corollary 2, this is only possible if $f(C') = y'$ is a point. If $y' \neq y$

then, since f^{-1} is an isomorphism, C' would also have to be a point. If $y' = y$ then $C' \subset f^{-1}(y) = C$.

Thus C is isolated in X, in the sense that there exists a neighbourhood V of C that does not contain any irreducible projective curve except for those contained in C. In other words, it is impossible to "wiggle C slightly". We can deduce from this that many surfaces do not contain any exceptional curves.

For example, let $X = \mathbb{P}^2$ and let C be an exceptional curve with $C \subset V = \mathbb{P}^2 \setminus D$. Then $\dim D = 0$, since otherwise, by the theorem on the dimension of intersection, C and D would intersect. But if $\dim D = 0$, that is, D is a finite set of points, then V contains any number of curves C not intersecting D, for example lines.

Now let $X = Q$ be the nonsingular quadric of $\mathbb{P}^3$. Here we make use of the existence of a group G of projective transformations taking Q to itself. Recall that transformations of G are given by 4×4 matrixes satisfying the relation $A^* F A = F$, where F is the matrix of the quadratic form defining Q. It follows that G is an algebraic variety in the space of all 4×4 matrixes. Hence we will from now on assume that G is an algebraic affine variety. It is easy to see that G acts transitively on Q.

If C is a curve and $C \subset Q \setminus D$, then we construct a transformation $\varphi \in G$ such that $\varphi(C) \not\subset C$ and $\varphi(C) \subset Q \setminus D$, which contradicts the property of exceptional curves obtained above. For this, it is enough to prove that the set of $\varphi \in G$ such that $\varphi(C) \cap D \neq \emptyset$ is closed. Then we have at our disposal a whole neighbourhood of the identity transformation $e \in G$ consisting of elements with the required property. In order to describe the set S of elements $\varphi \in G$ such that $\varphi(C) \cap D \neq \emptyset$ we consider the direct product $G \times Q$ and the subset $\Gamma \subset G \times Q$ of pairs (φ, x) such that $x \in C$ and $\varphi(x) \in D$. Obviously Γ is closed. If $f \colon G \times Q \to G$ is the projection then $S = f(\Gamma)$, and $f(\Gamma)$ is closed by Chap. I, 5.2, Theorem 2. This complete the proof the Q is a relatively minimal model, and hence the existence of two different relatively minimal model.

Thus it is all the more surprising that uniqueness of minimal models does hold for algebraic surfaces, provided only that we exclude some special types of surfaces. Namely, as proved by Enriques, a birational equivalence class of surfaces contains a unique relatively minimal model provided that it does not contain a surface of the form $C \times \mathbb{P}^1$, with C an algebraic curve. (A surface birational to $C \times \mathbb{P}^1$ is called a *ruled surface*.) The proof of Enriques' theorem is treated in Shafarevich [67], Chap. II or Barth, Peters and van de Ven [8].

There has recently been significant progress in the direction of constructing a theory of minimal models in dimension ≥ 3. In this case, minimal models cannot exist in the class of nonsingular varieties, but there is reason to hope that the theory can be generalised if we allow a certain class of rather well-controlled singularities. For this, see for example the surveys Kawamata, Matsuda and Matsuki [44] and Shokurov [68].

Exercises to §4

1. Suppose that $\dim X = 2$ and that $\xi \in X$ is a nonsingular point. Let $C_1, C_2 \subset X$ be two curves passing through ξ and nonsingular there, $\sigma: Y \to X$ the blowup centred at ξ, and set $C_i' = \sigma^{-1}(C_i \setminus \xi)$ and $Z = \sigma^{-1}(\xi)$. Prove that $C_1' \cap Z = C_2' \cap Z$ if and only if C_1 and C_2 touch at ξ.

2. Suppose that $\dim X = 2$ and that $\xi \in X$ is a nonsingular point. Let C be a curve passing through ξ and f the local equation of C in a neighbourhood of ξ. In local parameters u, v at ξ, suppose that $f \equiv \Pi_{i=1}^r (\alpha_i u + \beta_i v)^{l_i}$ modulo m_ξ^{k+1}, where $k = \sum l_i$ and the forms $\alpha_i u + \beta_i v$ are not proportional.

As in Ex. 1, $\sigma: Y \to X$ and $C' = \overline{\sigma^{-1}(C \setminus \xi)}$. Prove that $C' \cap Z$ consists of r points.

3. Use the notation of Ex. 2, and suppose also that $f \equiv (\alpha_1 u + \beta_1 v)(\alpha_2 u + \beta_2 v)$ modulo m_ξ^3, where the linear forms $\alpha_1 u + \beta_1 v$ and $\alpha_2 u + \beta_2 v$ are not proportional. Prove that both the points of $C' \cap Z$ are nonsingular on C'.

4. Consider the rational map $\varphi: \mathbb{P}^2 \to \mathbb{P}^4$ given by

$$\varphi(x_0 : x_1 : x_2) = (x_0 x_1 : x_0 x_2 : x_1^2 : x_1 x_2 : x_2^2).$$

Prove that φ is a birational map to a surface $\overline{\varphi(\mathbb{P}^2)}$, and that the inverse map $\overline{\varphi(\mathbb{P}^2)} \to \mathbb{P}^2$ coincides with the blowup.

5. In the spirit of Ex. 4, study the map $\varphi: \mathbb{P}^2 \to \mathbb{P}^6$ defined by all the monomials of degree 3 except for x_0^3, x_1^3 and x_2^3.

6. For any $n \geq 2$, construct an example of a regular birational map $f: X \to Y$ between n-dimensional nonsingular varieties having an exceptional codimension 1 subvariety Z whose image $f(Z) \subset Y$ has codimension 2.

7. Let X be a nonsingular projective curve and $\mathcal{O} \subset k(X)$ a local subring of the function field $k(X)$ satisfying the conditions 4.5, (1–3). Prove that for any $u \in k(X)$ either $u \in \mathcal{O}$ or $u^{-1} \in \mathcal{O}$. Suppose that $X \subset \mathbb{P}^n$ with $x_0, \ldots, x_n$ homogeneous coordinates of $\mathbb{P}^n$. Prove that there exists an i such that $x_j/x_i \in \mathcal{O}$ for $j = 0, \ldots, n$.

8. Use the notation of Ex. 7. Let X' be the affine curve $X' = X \cap \mathbb{A}_i^n$. Prove that $k[X'] \subset \mathcal{O}$, and that $k[X] \cap m$ is the ideal of some point $x \in X'$ with $\mathcal{O}_x \subset \mathcal{O}$.

9. Prove that if $\mathcal{O}_1$ and $\mathcal{O}_2$ are two subrings satisfying the conditions 4.5, (1–3). and $\mathcal{O}_1 \subset \mathcal{O}_2$ then $\mathcal{O}_1 = \mathcal{O}_2$. Deduce from this, using the results of Ex. 7–8, that $\mathcal{O} = \mathcal{O}_x$ (in the notation of Ex. 8).

10. Let $V \subset \mathbb{A}^3$ be the quadratic cone defined by $xy = z^2$; let $X' \to \mathbb{A}^3$ be the blowup of $\mathbb{A}^3$ with centre in the origin, and V' the closure of $\sigma^{-1}(V \setminus 0)$ in X'. Prove that V' is a nonsingular variety and that the inverse image of the origin under $\sigma: V' \to V$ is a nonsingular rational curve.

5. Normal Varieties

5.1. Normal Varieties

We start by recalling a notion of algebra: a ring A with no zerodivisors is *integrally closed* if every element of its field of fractions K that is integral over A (Chap. I, 5.3) is in A.

Definition. An irreducible affine variety X is *normal* if $k[X]$ is integrally closed. An irreducible quasiprojective variety X is normal if every point has a normal affine neighbourhood.

We will prove presently (Theorem 1) that a nonsingular variety is normal. Here is an example of a nonnormal variety: on the curve X defined by $y^2 = x^2 + x^3$, the rational function $t = y/x \in k(X)$ is integral over $k[X]$, since $t^2 = 1 + x$, but $t \notin k[X]$. (See Chap. I, §3, Ex. 7.)

This example shows that the condition that a variety is normal is somehow related to singular points of a variety. We now give an example of a variety that is normal although it has a singular point. This is the cone $X \subset \mathbb{A}^3$ given by $x^2 + y^2 = z^2$ (we assume that the ground field k has characteristic $\neq 2$).

Let us prove that $k[X]$ is integrally closed in $k(X)$. For this we use the simplest properties of integral elements (see Chap. I, 5.3 and Atiyah and Macdonald [7], Chap. 5). The field $k(X)$ consists of elements of the form $u + vz$ with $u, v \in k(x, y)$, where x and y are independent variables. Similarly, $k[X]$ consists of elements $u + vz = k(X)$ with $u, v \in k[x, y]$; hence $k[X]$ is a finite module over $k[x, y]$, and hence all elements of $k[X]$ are integral over $k[x, y]$. If $\alpha = u + vz \in k(X)$ is integral over $k[X]$ then it must be also integral over $k[x, y]$. Its minimal polynomial is

$$T^2 - 2uT + u^2 - (x^2 + y^2)v^2;$$

hence $2u \in k[x, y]$, so that $u \in k[x, y]$. Similarly, $u^2 - (x^2 + y^2)v^2 \in k[x, y]$, and hence also $(x^2 + y^2)v^2 \in k[x, y]$. Now since $x^2 + y^2 = (x + iy)(x - iy)$ is the product of two coprime irreducibles, it follows that $v \in k[x, y]$, and thus $\alpha \in k[X]$.

We prove some simple properties of normal varieties.

Lemma. *If X is a normal variety then its local ring $\mathcal{O}_Y$ at any irreducible subvariety $Y \subset X$ (see the end of 1.1 for the definition) is integrally closed. Conversely, if X is irreducible and the local ring $\mathcal{O}_x$ at each point $x \in X$ is integrally closed then X is normal.*

Proof. Since the definition of normal is local in nature, we can restrict to the case that X is affine. Suppose that X is normal, and let $Y \subset X$ be an irreducible subvariety. We prove that $\mathcal{O}_Y$ is integrally closed. Suppose that $\alpha \in k(X)$ is integral over $\mathcal{O}_Y$, that is,

$$\alpha^n + a_1\alpha^{n-1} + \cdots + a_n = 0 \qquad \text{with } a_i \in \mathcal{O}_Y. \tag{1}$$

Since $a_i \in \mathcal{O}_Y$ we have $a_i = b_i/c_i$ with $b_i, c_i \in k[X]$ and $c_i \notin \mathfrak{a}_Y$. Set $d_0 = c_1 \cdots c_n$, and multiply (1) by d_0. We get that

$$d_0\alpha^n + d_1\alpha^{n-1} + \cdots + d_n = 0 \qquad \text{with } d_i \in k[X] \text{ and } d_0 \notin \mathfrak{a}_Y. \tag{2}$$

Multiplying (2) through again by d_0^{n-1} and setting $d_0\alpha = \beta$, we get that β is integral over $k[X]$. By assumption $k[X]$ is integrally closed, and hence $d_0\alpha = \beta \in k[X]$. Then $\alpha = \beta/d_0 \in \mathcal{O}_Y$, because $d_0 \notin \mathfrak{a}_Y$. This proves that $\mathcal{O}_Y$ is integrally closed.

Conversely, suppose that all the local rings $\mathcal{O}_x$ are integrally closed. We prove that $k[X]$ is also. If $\alpha \in k(X)$ is integral over $k[X]$ then $\alpha^n + a_1\alpha^{n-1} + \cdots + a_n = 0$ with $a_i \in k[X]$. But then a fortiori $a_i \in \mathcal{O}_x$ for every $x \in X$, and since $\mathcal{O}_x$ is integrally closed by assumption, it follows that $\alpha \in \mathcal{O}_x$. Therefore $\alpha \in \bigcap_{x \in X} \mathcal{O}_x$. Now by Chap. I, 3.2, Theorem 4, $\bigcap_{x \in X} \mathcal{O}_x = k[X]$, and hence $\alpha \in k[X]$. The lemma is proved.

Theorem 1. *A nonsingular variety is normal.*

Proof. By the lemma, it is enough to show that if x is a nonsingular point then $\mathcal{O}_x$ is integrally closed. We know by 3.1, Theorem 2 that $\mathcal{O}_x$ is a UFD. Any element $\alpha \in k(X)$ can be represented in the form $\alpha = u/v$, were $u, v \in \mathcal{O}_x$ and have no common factors. If α is integral over $\mathcal{O}_x$ then $\alpha^n + a_1\alpha^{n-1} + \cdots + a_n = 0$ with $a_i \in \mathcal{O}_x$. Hence $u^n + a_1u^{n-1}v + \cdots + a_nv^n = 0$, and we see that $v \mid u^n$. But now, since u, v have no common factors and $\mathcal{O}_x$ is a UFD, it follows that $\alpha \in \mathcal{O}_x$. The theorem is proved.

Theorem 1 shows that the definition of normal is a certain weakening of the notion of nonsingularity. This is also reflected in the properties of normal varieties. In particular, we show that one of the basic properties of nonsingular varieties (3.1, Theorem 1) extends in a weak form to normal varieties.

Theorem 2. *If X is a normal variety and $Y \subset X$ a codimension 1 subvariety then there exists an affine open set $X' \subset X$ with $X' \cap Y \neq \emptyset$ such that the ideal of $Y' = X' \cap Y$ in $k[X']$ is principal.*

Proof. We can of course assume that X is affine. Moreover, it is enough to prove that the maximal ideal $\mathfrak{m}_Y$ is principal in the local ring $\mathcal{O}_Y$. Indeed, if $\mathfrak{m}_Y = (u)$ with $u \in \mathcal{O}_Y$ then $u = a/b$ with $a, b \in k[X]$ and $b \notin \mathfrak{a}_Y$. Suppose that $\mathfrak{a}_Y = (v_1, \ldots, v_m)$. Since $\mathfrak{a}_Y \subset \mathfrak{m}_Y$, we can write $v_i = uw_i$, where $w_i = c_i/d_i$, with $c_i, d_i \in k[X]$ and $d_i \notin \mathfrak{a}_Y$. Then the ideal $\mathfrak{a}_{Y'}$ of the variety $Y' = X' \cap Y$ is the principal ideal (u) in $k[X']$, where we set

$$X' = X \setminus \left(V(b) \bigcup_{i=1}^{m} V(d_i) \right).$$

Suppose that $0 \neq f \in k[X]$ and $f \in a_Y \subset \mathcal{O}_Y$. Then $Y \subset V(f)$, and since both of these are codimension 1 subvarieties (by assumption and by the theorem on dimension of intersection), Y consists of components of $V(f)$. Suppose that $V(f) = Y \cup Y_1$ with $Y \not\subset Y_1$. Setting $X_1 = X \setminus Y_1$, we get that $Y \cap X_1 \neq \emptyset$ and $Y \cap X_1 = V(f) \cap X_1$. Thus we can assume from the start that $Y = V(f)$.

By the Nullstellensatz, $Y = V(f)$ in X implies that $a_Y^k \subset (f)$ for some $k > 0$, and hence $m_Y^k \subset (f)$ in $\mathcal{O}_Y$. Suppose that k is the minimal number having this property. Then there exist $\alpha_1, \ldots, \alpha_{k-1} \in m_Y$ such that $\alpha_1 \cdots \alpha_{k-1} \notin (f)$ but $\alpha_1 \cdots \alpha_{k-1} m_Y \subset (f)$. That is, setting $g = \alpha_1 \cdots \alpha_{k-1}$ we have $g \notin (f)$ but $g m_Y \subset (f)$, or in other words, $u = f/g$ satisfies

$$u^{-1} \notin \mathcal{O}_Y, \qquad \text{but} \qquad u^{-1} m_Y \subset \mathcal{O}_Y.$$

Now we use the fact that, by the lemma, $\mathcal{O}_Y$ is integrally closed. It follows from this that $u^{-1} m_Y \not\subset m_Y$; for otherwise, by the basic relation between finite modules and integral elements, the "determinant trick" (Atiyah and Macdonald [7], Proposition 2.4), u^{-1} would be integral over $\mathcal{O}_Y$ and therefore contained in it, which is not the case. But m_Y is the maximal ideal of $\mathcal{O}_Y$, so that $u^{-1} m_Y \subset \mathcal{O}_Y$ but $u^{-1} m_Y \not\subset m_Y$ implies that $u^{-1} m_Y = \mathcal{O}_Y$. This means that $m_Y = (u)$. The theorem is proved.

Theorem 3. *The set of singular points of a normal variety has codimension ≥ 2.*

Proof. Suppose that X is normal, with $\dim X = n$, and let S be the set of singular points of X. We have seen in 1.4 that S is closed in X. Suppose that S contains an irreducible component Y of dimension $n - 1$. Let X' be the open subset whose existence we established in Theorem 2 and $Y' = Y \cap X'$. There is at least one point $y \in Y'$ that is a nonsingular point of the variety Y' (but not of X', by assumption). Let $\mathcal{O}_{Y',y}$ be the local ring of Y' at y, and $u_1, \ldots, u_{n-1}$ local parameters.

By Theorem 2, $a_{Y'} = (u)$ is a principal ideal of $k[X']$, and hence $k[Y'] = k[X']/(u)$. Similarly $\mathcal{O}_{Y',y} = \mathcal{O}_{X',y}/(u)$, and obviously $m_{X',y}$ is equal to the inverse image of $m_{Y',y}$ under the natural map $\mathcal{O}_{X',y} \to \mathcal{O}_{Y',y}$. Choose arbitrary inverse images $v_1, \ldots, v_{n-1} \in \mathcal{O}_{X',y}$ of the local parameters $u_1, \ldots, u_{n-1} \in \mathcal{O}_{Y',y}$. Then $m_{X',y} = (v_1, \ldots, v_{n-1}, u)$. This proves that $\dim m_{X',y}/m_{X',y}^2 \leq n$, and hence that y is a nonsingular point of X, which contradicts the assumption $y \in Y \subset S$. The theorem is proved.

Corollary. *For algebraic curves, normal and nonsingular are equivalent conditions.*

Example. Let X be a normal affine variety and G a finite group of automorphisms of X. We prove that the quotient variety $Y = X/G$ (see Chap. I, 2.3, Example 11) is normal. Suppose that $h \in k(Y)$ is integral over $k[Y]$. Then h is a fortiori integral over $k[X]$, and hence $h \in k[X]$. But $h \in k(Y)$, so that $g^*(h) = h$ for any $g \in G$, and hence $h \in k[X]^G = k[Y]$.

In particular, suppose that $X = \mathbf{A}^2$ and $G = \{1, g\}$, where $g(x, y) = (-x, -y)$. It is easy to check that $k[X]^G = k[x, y]^G$ is generated by $w = xy$, $u = x^2$ and $v = y^2$. In other words, Y is the quadratic cone defined by $uv = w^2$ constructed at the start of this section. Since X is normal by Theorem 1, we get another proof that Y is normal.

We now compare the nonsingular and normal properties of varieties we have introduced. We first note that the proof of Theorem 1 did not make full use of the nonsingularity of X; we only used that $\mathcal{O}_x$ is a UFD. In this connection, it is natural to distinguish the class of varieties with the property that each local ring $\mathcal{O}_x$ is a UFD; these are called *factorial varieties*. Thus nonsingular varieties are factorial, and factorial varieties normal; in essence, that is what is proved in Theorem 1. One can show that these three classes of varieties are really different. For example, it is known that for $n \geq 5$, a hypersurface $X \subset \mathbf{A}^n$ with just one singular point is factorial (Grothendieck [32] (SGA2), Chap. XI, 3.14). A beautiful example of a surface that is singular, but factorial, is the surface given by $x^2 + y^3 + z^5 = 0$. An example of a variety that is normal, but not factorial, is given by the quadratic cone considered above: $z^2 = (x + iy)(x - iy)$ are two different factorisations into irreducibles of the same element.

Theorem 3 focuses attention on a new property of varieties: the set of singular points has codimension ≥ 2. A variety with this property is said to be *nonsingular in codimension* 1. Theorem 3 asserts that this class includes, in particular, normal varieties. These two classes of varieties are also distinct. Constructing a counterexample is a bit more complicated here; the point is that normal is equivalent to nonsingular in codimension 1 for a hypersurface. Hence the simplest possible example would be a surface in $\mathbf{A}^4$. An irreducible variety X is not normal if there exists an affine variety Y and a surjective regular map $f\colon Y \to X$, not an isomorphism, that restricts to an isomorphism of open subsets $V \subset Y$ and $U \subset X$, and such that $k[Y]$ is a finite module over $f^*(k[X])$. A first approximation to the counterexample is thus given by $X = L_1 \cup L_2$, where L_1 and L_2 are two planes of $\mathbf{A}^4$ meeting in a point (defined by $xz = xt = yz = yt = 0$ in coordinates x, y, z, t of $\mathbf{A}^4$) and $Y = L_1 \sqcup L_2$ the disjoint union of L_1 and L_2 (for example, in $\mathbf{A}^5$). But this is a reducible variety, and our definition of normal assumes irreducibility. We therefore construct an example that imitates this situation near the singular point. For this, it is enough to construct a finite regular map $f\colon \mathbf{A}^2 \to \mathbf{A}^4$ birational onto its image $X = f(\mathbf{A}^2)$, with $X \subset \mathbf{A}^4$ closed in $\mathbf{A}^4$ such that two points, y_1, y_2, say, have the same image $z \in X$ and $f\colon \mathbf{A}^2 \setminus \{y_1, y_2\} \to X \setminus \{z\}$

is an isomorphism. Thus f is very similar to the parametrisation Chap I, 1.2, (2) of the curve (1). The existence of the map f means that X is not normal, and z will be the unique singular point of X.

Writing ξ, η for coordinates in $\mathbf{A}^2$ and x, y, z, t for coordinates in $\mathbf{A}^4$, we define f by

$$x = \xi(1 - \eta), \quad y = \eta(\eta - 1)^2, \quad z = \xi\eta, \quad t = \eta^2(\eta - 1).$$

One sees easily that the ideal $\mathfrak{A}_X$ is generated by the four equations

$$xz = -(t - y)(x + z)^2, \quad xt = -yz = (t - y)^2(x + z), \quad yt = (t - y)^3.$$

The relations $\xi = x + z$ and $\eta^2 - \eta = t - y$ prove that ξ and η are integral over $f^*(k[X])$, so that f is finite. The remaining properties of f we need are very easy to check. It is easy to see that the tangent cone to X at origin (see 1.5) is $L_1 \cup L_2$.

5.2. Normalisation of an Affine Variety

Consider the simplest possible example of a nonnormal variety, the curve X defined by $y^2 = x^2 + x^3$. Its parametrisation, using the parameter $t = y/x$, defines a map $f \colon \mathbf{A}^1 \to X$, or equivalently, an inclusion $k[X] \hookrightarrow k[t]$. Since f is birational, we have $k[X] \subset k[t] \subset k(X) = k(t)$. The line $\mathbf{A}^1$ is normal, of course, corresponding to that fact that $k[t]$ is integrally closed. Moreover, the ring $k[t]$ can be characterised as the set of all elements $u \in k(X)$ that are integral over $k[X]$. Indeed, $t^2 = 1 + x$, hence t and all elements of $k[t]$ are integral over $k[X]$; moreover, if u is integral over $k[X]$ then it is also integral over $k[t]$, and hence $u \in k[t]$ since $k[t]$ is integrally closed. Finally, in geometric terminology, $k[t]$ integral over $k[X]$ says that f is a finite map. We show that for any irreducible affine variety X, there exists a variety X' and a map $X' \to X$ having the same properties. We start with a definition that relates to arbitrary irreducible varieties.

Definition. A *normalisation* of an irreducible variety X is an irreducible normal variety X^ν, together with a regular map $\nu \colon X^\nu \to X$, such that ν is finite and birational. (If $X = \bigcup X_i$ is a reducible variety then one can define $X^\nu = \bigsqcup X_i^\nu$.)

Theorem 4. *An affine irreducible variety has a normalisation that is also affine.*

Proof. Let A be the integral closure of $k[X]$ in $k(X)$, that is, the set of elements $u \in k(X)$ that are integral over $k[X]$. It follows from elementary properties of integral elements that A is a ring, and is integrally closed. Suppose that we can find an affine variety X' such that $A = k[X']$. Then X'

is normal and the inclusion $k[X] \hookrightarrow k[X']$ defines a regular birational map $f: X' \to X$. Obviously X' is a normalisation of X.

By Chap. I, 2.3, Theorem 1, the required affine variety X' exists if and only if A is finitely generated over k and has no zerodivisors. We will prove more, that A is a finite module over $k[X]$. If $A = k[X]w_1 + \cdots + k[X]w_m$ then $w_1, \ldots, w_m$, together with the generators of the algebra $k[X]$ over k, provide a finite system of generators of A as a k-algebra.

To prove that A is a finite $k[X]$-module, we use Noether normalisation, Chap. I, 5.4, Theorem 10. By this theorem, there exists a subring $B \subset k[X]$ such that B is isomorphic to a polynomial ring $B \cong k[T_1, \ldots, T_r]$ and $k[X]$ is integral over B. We write out all the current rings and fields:

$$k(T_1, \ldots, T_r) \quad \subset \quad k(X) = K$$
$$\cup \qquad\qquad\qquad \cup$$
$$B \quad \subset k[X] \subset A$$

From this diagram and from basic properties of integral elements, one sees that A is equal to the integral closure of B in $k(X)$. Moreover, $K = k(X)$ is a finite field extension of $k(T_1, \ldots, T_r)$, since $T_1, \ldots, T_r$ is a transcendence basis of $k(X)$. Finally, B is integrally closed, since $\mathbf{A}^r$ is normal, indeed, nonsingular. Thus the final result we are aiming for, that A is a finite $k[X]$-module, follows from the fact that if $B = k[T_1, \ldots, T_r]$, $L = k(T_1, \ldots, T_r)$, and K is any finite extension field of L, then the integral closure of B in K is a finite B-module. For the proof of this assertion, see Appendix, §8, Proposition 1. The theorem is proved.

Theorem 5. *(i) If $g: Y \to X$ is a finite regular birational map, then there exists a regular map $h: X^\nu \to Y$ such that the diagram*

$$
\begin{array}{ccc}
 & X^\nu & \\
{}^h\swarrow & & \searrow^\nu \\
Y & \xrightarrow[g]{} & X
\end{array}
$$

is commutative.

(ii) If $g: Y \to X$ is a regular map, $g(Y)$ is dense in X and Y is normal then there exists a regular map $h: Y \to X^\nu$ such that the diagram

$$
\begin{array}{ccc}
 & Y & \\
{}^h\swarrow & & \searrow^g \\
X^\nu & \xrightarrow[\nu]{} & X
\end{array}
$$

is commutative.

Proof of (i). By assumption we have inclusions $k[X] \subset k[Y] \subset k(X)$, with $k[Y]$ integral over $k[X]$. Now by definition of integral closure, $k[Y] \subset k[X^\nu]$, which provides the required regular map $h: X^\nu \to Y$.

Proof of (ii). An element $u \in k[X^\nu]$ is integral over $k[X]$ and contained in $k(X) \subset k(Y)$; since $k[X] \subset k[Y]$, it is a fortiori integral over $k[Y]$, and thus, since $k[Y]$ is integrally closed, $u \in k[Y]$. Hence $k[X^\nu] \subset k[Y]$, which provides the regular map $h: Y \to X^\nu$ with the required properties.

The theorem is proved.

Corollary. *The normalisation of an affine variety X is unique. More precisely, if $\nu: X^\nu \to X$ and $\nu': X^{\nu\prime} \to X$ are two normalisations of X then there exists an isomorphism $g: X^\nu \xrightarrow{\sim} X^{\nu\prime}$ such that the diagram*

$$X^\nu \xrightarrow{g} X^{\nu\prime}$$
$$\nu \searrow \qquad \swarrow \nu'$$
$$X$$

is commutative.

This follows from either of the assertions of Theorem 5.   □

We do not prove the existence of the normalisation for arbitrary quasiprojective varieties; the proof is discussed in Chap. VI, 1.1 in a more general context. Note that for those varieties for which the normalisation is known to exist, it has the properties established in Theorem 5, as follows at once from considering affine covers.

5.3. Normalisation of a Curve

Theorem 6. *An irreducible quasiprojective curve X has a normalisation X^ν, and X^ν is again quasiprojective.*

Proof. Let $X = \bigcup U_i$ be a cover of X by affine open sets. Write U_i^ν for the normalisation of U_i, which exists by Theorem 4, and $f_i: U_i^\nu \to U_i$ for the natural regular map, which is birational and finite.

We embed the affine space containing U_i^ν into projective space, and write V_i for its closure in projective space. Note that all the varieties appearing so far are birational to X: for $U_i \subset X$ is open, $f: U_i^\nu \to U_i$ is a birational map, and $U_i^\nu \subset V_i$ is also open. Therefore U_i^ν and V_j are birational; write $\varphi_{ij}: U_i^\nu \to V_j$ for the corresponding birational map.

By 5.1, Theorem 3, Corollary, U_i^ν is a nonsingular curve, and, since V_j is projective, $\varphi_{ij}: U_i^\nu \to V_j$ is a regular map by 3.1, Theorem 3, Corollary 1. Set $W = \prod_j V_j$ and $\varphi_i = \prod_j \varphi_{ij}: U_i^\nu \to W$, that is, $\varphi_i(u) = (\varphi_{i1}(u), \varphi_{i2}(u), \dots)$. Write $X' = \bigcup \varphi_i(U_i^\nu) \subset W$ for the union of all the $\varphi_i(U_i^\nu)$. We claim that $X' = X^\nu$. For this we have to show that X' is (a) quasiprojective, (b) irreducible and (c) normal, and (d) that it has a finite birational map $\nu: X' \to X$.

To prove these statements, set $U_0 = \bigcap U_i$; this is an open subset of X. By construction of φ_i it follows easily that $U_0^\nu \subset U_i^\nu$, and all the φ_i coincide on U_0^ν. Write φ for their common restriction to U_0^ν. Then

$$\varphi(U_0^\nu) \subset \varphi_i(U_i^\nu) \subset \overline{\varphi(U_0^\nu)},$$

where $\overline{\varphi(U_0^\nu)}$ is the closure of $\varphi(U_0^\nu)$ in W. Obviously $\varphi(U_0^\nu)$ is an irreducible quasiprojective curve and $\overline{\varphi(U_0^\nu)} \setminus \varphi(U_0^\nu)$ consists of a finite number of points. By construction, $\varphi(U_0^\nu) \subset X' \subset \overline{\varphi(U_0^\nu)}$ and hence $\overline{\varphi(U_0^\nu)} \setminus X'$ also consists of a finite number of points. This proves (a) and (b).

Let $x \in X'$; then $x \in \varphi_i(U_i^\nu)$ for some i, and $\varphi_i(U_i^\nu)$ is a neighbourhood of x. We prove that $\varphi_i \colon U_i^\nu \to \varphi_i(U_i^\nu) \subset W$ is an isomorphism; since U_i^ν is normal, it follows that X' is normal, proving (c). For this, note that φ_{ii} is an embedding of U_i^ν to its projective closure V_i. Hence $(u_1, u_2, \ldots) \mapsto \varphi_{ii}^{-1}(u_i)$ is an inverse to φ_i, which proves that φ_i is an isomorphic embedding.

Finally for the proof of (d) we construct the map

$$g_i = f_i \circ \varphi_i^{-1} \colon \varphi_i(U_i^\nu) \to U_i \subset X.$$

By what we have said above, all the g_i are finite maps. We prove that all the g_i define on X' a single finite map $f \colon X' \to X$. For this, note that all the g_i coincide on U_0^ν. If $g \colon U_0^\nu \to U_0$ is the normalisation map then $g_i = g$ on U_0^ν. Hence the maps g_i and g_j coincide on the open set of $\varphi(U_0^\nu)$ contained in $\varphi_i(U_i^\nu) \cap \varphi_j(U_j^\nu)$. But two regular maps that coincide on an nonempty open set coincide everywhere. This follows from the same statement for functions. Thus g_i and g_j coincide at all points at which they are both defined, so that they all define a regular map $\nu \colon X' \to X$. Obviously ν is birational. The theorem is proved.

Theorem 7. *The normalisation of a projective curve is projective.*

Proof. Let X be a projective curve and X^ν its normalisation, with $\nu \colon X^\nu \to X$ the normalisation map. Suppose that X^ν is not projective, and write Y for its closure in projective space. Choose a point $x \in Y \setminus X^\nu$; let U be an affine neighbourhood of x in Y and U^ν its normalisation, with $\nu' \colon U^\nu \to U$ the normalisation map. We have a diagram

$$\begin{array}{ccc}
U^\nu & \xrightarrow{h} X^\nu \xrightarrow{\nu} X \\
\nu' \downarrow & \quad \downarrow \varphi \\
U & \xrightarrow[\psi]{} Y
\end{array}$$

where $\varphi \colon X^\nu \hookrightarrow Y$ and $\psi \colon U \hookrightarrow Y$ are the inclusions of open sets. The composite map $\nu \circ \varphi^{-1} \circ \psi \circ \nu' \colon U^\nu \to X$ is birational, and since U^ν is nonsingular, it is regular by 3.1, Theorem 3, Corollary 1. By Theorem 5, there exists a regular map $h \colon U^\nu \to X^\nu$ as in the diagram, making the square commute, $\varphi \circ h = \psi \circ \nu'$. However, the existence of h leads to a contradiction: $\varphi(h(U^\nu)) \subset X^\nu$, and $\psi(\nu'(U^\nu)) \ni x$, since the normalisation map is finite, and hence surjective by Chap. I, 5.3, Theorem 4. This contradiction proves the theorem.

Corollary. *An irreducible algebraic curve is birational to a nonsingular projective curve.*

This is a combination of 5.1, Theorem 3, Corollary and Theorem 7. Normalisation allows us to study properties of curves in more detail.

Theorem 8. *A regular map $\varphi \colon X \to Y$ from an irreducible nonsingular projective curve X is finite (Chap. I, 5.3) if $Y = \varphi(X)$ is a variety with $\dim Y > 0$.*

Proof. Write V for an affine neighbourhood of a point $y \in Y$, and $B = k[V]$. We view $k(Y)$ as a subfield of $k(X)$ under the inclusion φ^*. In particular, $B \subset k(X)$; let A be the integral closure of B in $k(X)$. In the proof of the existence of the normalisation of an algebraic variety, we proved that A is a finite B-module, and hence $A = k[U]$, where U is an affine normal curve. Since U is birational to X, by 4.4, Theorem 2, Corollary 2, we can assume that U is an open subset of X. Let us prove that $U = \varphi^{-1}(V)$. This will guarantee the finiteness of φ.

Suppose that for some point $y_0 \in V$ there is a point $x_0 \notin U$ with $\varphi(x_0) = y_0$. Consider a function f such that $f \notin \mathcal{O}_{x_0}$ but $f \in \mathcal{O}_{x_i}$ for all $x_i \in U$ with $\varphi(x_i) = y_0$ and $x_i \neq x_0$. Such a function can easily be constructed by putting x_0 and x_i in one affine open set. If f has poles at points $x' \in U$, then $\varphi(x') = y' \neq y_0$, and hence we can find a function $h \in B$ such that $h(y_0) \neq 0$ and $fh \in \mathcal{O}_{x'}$, that is, $fh \in A$; for this, we need only take a sufficiently high power of a function that vanishes at y'. Then $f_1 = fh$ is integral over B, that is

$$f_1^n + b_1 f_1^{n-1} + \cdots + b_n = 0 \qquad \text{with } b_i \in B,$$

so that $f_1 = -b_1 - b_2/f_1 - \cdots - b_n/f_1^{n-1}$. Since $f_1 \notin \mathcal{O}_{x_0}$, we get $f_1^{-1} \in \mathfrak{m}_{x_0}$. Hence the final equality is a contradiction: the right-hand side is regular at x_0, but the left-hand side is not. The theorem is proved.

Another application concerns curve singularities: the existence of the normalisation allows us to introduce some useful invariants of singular points of curves.

Let X be a curve and $p \in X$ a point, possibly singular; write $\nu \colon X^\nu \to X$ for the normalisation and $q_1, \ldots, q_k$ for the inverse images of p in X^ν. The points q_i are called *branches* of X at p. The terminology is explained in that if $k = \mathbb{C}$ (or $\mathbb{R}$) and U_i are sufficiently small complex (or real) neighbourhoods of the q_i, then some neighbourhood of X is the union of the branches $\nu(U_i)$.

Write Θ_i for the tangent line to X^ν at q_i. The differential $d_{q_i}\nu$ of ν maps Θ_i onto a linear subspace of the tangent space to X at p. Obviously $(d_{q_i}\nu)(\Theta_i)$ is either the point p or a line; in the second case, we say that q_i is a *linear branch*, and $(d_{q_i}\nu)(\Theta_i)$ the *tangent line* to the branch.

A branch q_i is linear if and only if ν^* takes m_p/m_p^2 onto the whole of $m_{q_i}/m_{q_i}^2$. Suppose that p is the origin in $\mathbf{A}^n$ with coordinates $t_1,\ldots,t_n$. Then $\nu^*(m_p/m_p^2)$ is generated by $\nu^*(t_1)+m_{q_i}^2,\ldots,\nu^*(t_n)+m_{q_i}^2$. Since q_i is nonsingular we have $\dim(m_{q_i}/m_{q_i}^2)=1$, and therefore a branch is linear if and only if $\nu^*(t_s) \notin m_{q_i}^2$ for at least one $s=1,\ldots,n$. In other words, $\nu^*(t_s)$ should be a local parameter at q_i. Since $m_p = (t_1,\ldots,t_n)$, in invariant form the condition for q_i to be a linear branch takes the form $\nu^*(m_p) \not\subset m_{q_i}^2$. As a measure of how far q_i fails to be a linear branch, we can take the number k such that $\nu^*(m_p) \subset m_{q_i}^k$ but $\nu^*(m_p) \not\subset m_{q_i}^{k+1}$. This is called the *multiplicity* of the branch q_i.

The point $(0,0)$ of $y^2 = x^2 + x^3$ gives an example of two linear branches with the tangent lines $y = x$ and $y = -x$, and the point $(0,0)$ of the cusp $y^2 = x^3$ an example of a nonlinear branch of multiplicity 2. If x is the centre of a single branch, and this is linear, then x is a nonsingular point. This is a corollary of a lemma that we prove at the end of this section. Thus the simplest invariants measuring how singular a point is are its number of branches, and their multiplicities. We say that a point of an algebraic curve in the plane is an *ordinary singularity*, or a *singular point with distinct tangent lines* if it has only linear branches and all its branches have distinct tangent lines.

Suppose that X is given by $F(x,y) = 0$, and that char $k = 0$. Let $(0,0) = p \in X$ and let $q \in X^\nu$ be one of the branches corresponding to p. If t is a local parameter at q then there are formal power series expansions

$$x = a_n t^n + a_{n+1} t^{n+1} + \cdots, \qquad \text{with } n, m > 0 \text{ and } a_n, b_m \neq 0. \quad (1)$$
$$y = b_m t^m + a_{m+1} t^{m+1} + \cdots,$$

There exists a formal power series $\tau = r_1 t + r_2 t^2 + \cdots$ with $r_1 \neq 0$ such that $\tau^n = x$. This is easy to check: we have first to set $r_1 = a_n^{1/n}$, and from then on, for each $i > 1$ we get a linear equation for r_i, to solve which we must divide by n, which is possible under the assumption char $k = 0$. On the other hand, t can also be expressed in terms of τ as a formal power series, $t = r_1^{-1}\tau + s_2\tau^2 + \cdots$, as can also be checked at once by equating coefficients. Finally, substituting this expression in (1), we get a parametrisation $x = \tau^n$, $y = c_m\tau^m + c_{m+1}\tau^{m+1} + \cdots$, that can be rewritten

$$y = c_m x^{m/n} + c_{m+1} x^{m+1/n} + \cdots \quad (2)$$

A parametrisation of a branch of this type is called a *Puiseux expansion* of y. This is particularly useful in problems of analysis, where y is viewed as a function of x.

To find explicitly the Puiseux expansions corresponding to different branches, there is an extremely convenient method using the *Newton polygon* of a polynomial F. Suppose that $F(x, y) = \sum A_{ij} x^i y^j$. In the plane, we draw the points with coordinates (i, j) for which $A_{ij} \neq 0$ (Figure 11).

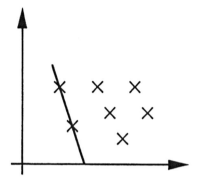

Figure 11. The Newton Polygon

A necessary condition for the expansion (2) to satisfy $F(x, y) = 0$ is that after substituting (2) in F, the lowest powers of x arising from the various monomials $A_{ij} x^i y^j$ must cancel out. In order for this to be possible, at least two monomials $A_{i'j'} x^{i'} y^{j'}$ and $A_{i''j''} x^{i''} y^{j''}$ must give terms of the same degree d in x, and other monomials terms of degree $\geq d$. In other words, the exponent $\alpha = m/n$ should satisfy the condition $i' + j'\alpha = i'' + j''\alpha \leq i + j\alpha$ for all (i, j) with $A_{ij} \neq 0$. In Figure 11 this is expressed by saying that α is minus the slope of the line through points (i', j') and (i'', j''), with all other points drawn in the picture either on or above the line. In other words, the only exponents α that can appear are minus the slopes of the lower convex boundary of the convex hull of the set of points drawn in the picture.

We rewrite the expansion (2) in the form $y = \sum c_{\nu_i} x^{\nu_i}$, where ν_i are increasing rational exponents, and $c_{\nu_i} \neq 0$. Certain of these exponents play an especially important role as invariants of the singularity. Suppose that the first nonintegral exponent is m_1/n_1. Obviously, $n_1 \mid n$, and if $n_1 \neq n$ then there must be exponents with denominators strictly divisible by n_1. Suppose that the first of these beyond m_1/n_1 is $m_2/(n_1 n_2)$; then suppose that $m_3/(n_1 n_2 n_3)$ is the first exponent with denominator strictly divisible by $n_1 n_2$, and so on, up to $m_k/(n_1 \cdots n_k)$, where $n_1 \cdots n_k = n$. The pairs (m_1, n_1), (m_2, n_2), $\ldots$, (m_k, n_k) are called the *characteristic pairs* of the branch. We state in its simplest form a result that illustrates the significance of characteristic pairs. Consider only singularities having a single branch. For any sequence of characteristic pairs there exists a natural number l such that the singularities with given characteristic pairs are uniquely determined up to formal analytic equivalence (see 2.2 for the definition) by the first l terms of the expansion (2). Thus singularities with given sequence of characteristic pairs form a finite dimensional space. For a simple proof and various generalisations, see Hironaka [37] or Walker [74], Chap. IV, §§2–3.

5.4. Projective Embedding of Nonsingular Varieties

The nonsingular projective model of an algebraic curve constructed in the preceding section is contained in some projective space $\mathbb{P}^n$. The natural question arises as to how small n we can take to be. We answer this by proving a general result on varieties of arbitrary dimension.

Theorem 9. *A nonsingular projective n-dimensional variety is isomorphic to a subvariety of* $\mathbb{P}^{2n+1}$.

Let $X \subset \mathbb{P}^N$ be a nonsingular projective variety. Theorem 9 will be proved if for $N > 2n + 1$ we can choose a point $\xi \in \mathbb{P}^N \setminus X$ such that the projection from ξ is an isomorphic embedding of X into $\mathbb{P}^{N-1}$. We therefore start by elucidating when a regular map is an isomorphic embedding.

Lemma. *A finite map f from a variety X is an isomorphic embedding if and only if it is one-to-one and $d_x f$ is an isomorphic embedding of the tangent space Θ_x for every $x \in X$.*

Proof. Set $f(X) = Y$ and $\varphi = f^{-1}$. The lemma will be proved if we show that φ is a regular map. The assertion is local in nature. For $y \in Y$, let $x \in X$ be such that $f(x) = y$. Write U and V for affine neighbourhoods of x and y with $f(U) = V$ and such that $k[U]$ is integral over $k[V]$, and continue to write $f: U \to V$ for the restriction of f. It is enough to prove that f is an isomorphism for suitable choice of U and V, since then $\varphi = f^{-1}$ is a regular map at y.

Recall that the tangent space Θ_x is the dual vector space to $\mathfrak{m}_x/\mathfrak{m}_x^2$, where $\mathfrak{m}_x$ is the maximal ideal of the local ring $\mathcal{O}_x$. The second hypothesis of the lemma is that $f^*: \mathfrak{m}_y/\mathfrak{m}_y^2 \to \mathfrak{m}_x/\mathfrak{m}_x^2$ is surjective. In other words, if $\mathfrak{m}_y = (u_1, \ldots, u_k)$, then the elements $f^*(u_i) + \mathfrak{m}_x^2$ generate $\mathfrak{m}_x/\mathfrak{m}_x^2$. We apply Nakayama's lemma (Appendix, §6, Proposition 3) to $\mathfrak{m}_x$ as an $\mathcal{O}_x$-module. It then follows from this that $\mathfrak{m}_x = (f^*(u_1), \ldots, f^*(u_k))$, or in other words

$$\mathfrak{m}_x = f^*(\mathfrak{m}_y)\mathcal{O}_x. \tag{1}$$

We check that $\mathcal{O}_x$ is a finite module over $f^*(\mathcal{O}_y)$. Since $k[U]$ is a finite $k[V]$-module, it is enough to prove that each element of $\mathcal{O}_x$ can be expressed in the form $\xi/f^*(a)$ with $\xi \in k[U]$ and $a \notin \mathfrak{m}_y$. For this, it is enough to check that for $\alpha \in k[U]$ with $\alpha \notin \mathfrak{m}_x$ there exists an element $a \in k[V]$ with $a \notin \mathfrak{m}_y$ such that $f^*(a) = \alpha\beta$ with $\beta \in k[U]$. By Chap. I, 5.3, Theorem 4 the set $f(V(\alpha))$ is closed, and since f is one-to-one, $y \notin f(V(\alpha))$. Hence there exists a function $c \in k[V]$ such that $c = 0$ on $f(V(\alpha))$ and $c(y) \neq 0$. Then

$f^*(c) = 0$ on $V(\alpha)$ and $f^*(c)(x) \neq 0$. By the Nullstellensatz, $f^*(c)^n = \alpha\beta$ for some $n > 0$ and $\beta \in k[U]$. We can set $a = c^n$.

Now we can apply Nakayama's lemma to $\mathcal{O}_x$ as a $f^*(\mathcal{O}_y)$-module. The equality (1) shows that $\mathcal{O}_x / f^*(\mathfrak{m}_y)\mathcal{O}_x = \mathcal{O}_x/\mathfrak{m}_x = k$, and hence is generated by the single element 1. It now follows by Nakayama's lemma that $\mathcal{O}_x = f^*(\mathcal{O}_y)$.

Let $u_1, \ldots, u_l$ be a basis of $k[U]$ as a module over $k[V]$. By what we have proved, $u_i \in \mathcal{O}_x = f^*(\mathcal{O}_y)$. Write $V' = V \setminus V(h)$ for a principal affine neighbourhood of y such that all $(f^*)^{-1}(u_i)$ are regular in $U' = U \setminus V(f^*(h))$. Then $k[U'] = \sum f^* k[V']u_i$. By assumption $u_i \in f^*(k[V'])$, and it follows that $k[U'] = k[V']$, which means that $f: U' \to V'$ is an isomorphism. The lemma is proved.

Corollary 1. *Let $X \subset \mathbb{P}^N$ be a variety and $\xi \in \mathbb{P}^N \setminus X$. Suppose that every line through ξ intersects X in at most one point, and ξ is not contained in the tangent space to X at any point then the projection from ξ is an isomorphic embedding $X \hookrightarrow \mathbb{P}^{N-1}$.*

It it enough to apply the lemma, together with Chap. I, 5.3, Theorem 7. $\square$

Proof of Theorem 9. It is enough to prove that if $X \subset \mathbb{P}^N$ is a nonsingular n-dimensional variety and $N > 2n + 1$ then there exists ξ as in Corollary 1; this is a standard dimension count. Let U_1 and $U_2 \subset \mathbb{P}^N$ be the sets of points $\xi \in \mathbb{P}^N$ not satisfying the two assumptions of Corollary 1.

In $\mathbb{P}^N \times X \times X$ consider the set Γ of triples (a, b, c) with $a \in \mathbb{P}^N$, $b, c \in X$ such that a, b, c are collinear. Γ is obviously a closed subset of $\mathbb{P}^N \times X \times X$. The projections of $\mathbb{P}^N \times X \times X$ to $\mathbb{P}^N$ and to $X \times X$ define regular maps $\varphi: \Gamma \to \mathbb{P}^N$ and $\psi: \Gamma \to X \times X$. Obviously if $y \in X \times X$ with $y = (b, c)$, and $b \neq c$ then $\psi^{-1}(y)$ consists of points (a, b, c) where a is any point of the line through b and c. Hence $\dim \psi^{-1}(y) = 1$ and it follows by Chap. I, 6.3, Theorem 7 that $\dim \Gamma = 2n + 1$. By definition $U_1 = \varphi(\Gamma)$, and the same theorem gives $\dim U_1 \leq \dim \Gamma = 2n + 1$.

In a similar way, to study the set U_2 we consider in $\mathbb{P}^N \times X$ the set Γ' consisting of points (a, b) such that $a \in \Theta_b$. In exactly the same way we have projections $\psi: \Gamma' \to X$ and $\varphi: \Gamma' \to \mathbb{P}^N$. For $b \in X$ we have $\dim \psi^{-1}(b) = n$ since X is nonsingular, and hence $\dim \Gamma' = 2n$, and since $U_2 = \varphi(\Gamma')$, also $\dim U_2 \leq 2n$.

We see that $\dim U_1 \leq \dim \Gamma = 2n + 1$ and $\dim U_2 \leq 2n$; therefore if $N > 2n + 1$ then $U_1 \cup U_2 \neq \mathbb{P}^N$, which was what we wanted. The theorem is proved.

Corollary 2. *Any nonsingular quasiprojective curve is isomorphic to a curve in $\mathbb{P}^3$.*

We will see later that not every curve is isomorphic to a curve in the projective plane. That is, not every algebraic curve has a nonsingular plane projective model.

However, it can be proved that, continuing the process of projection used in the proof of Theorem 9, we can obtain a plane curve all of whose singular points are ordinary double points (we assume here that char $k = 0$). By Theorem 9, every nonsingular projective surface is isomorphic to a surface in $\mathbb{P}^5$; in general, it cannot be projected isomorphically into $\mathbb{P}^4$. However, one can choose a projection to $\mathbb{P}^4$ so that it is an isomorphism outside finitely many points. In this way we easily arrive at isolated surface singularities that are not normal, one example of which was constructed at the end of 5.1.

Exercises to §5

1. Let X be an affine variety and K a finite extension of $k(X)$. Prove that there exists an affine variety Y and a map $f : Y \to X$ with the properties (1) f is finite; (2) Y is normal; (3) $k(Y) = K$ with $f^* : k(X) \hookrightarrow k(Y) = K$ the given inclusion. Prove that Y is uniquely determined by these properties. It is called the *normalisation* of X in K.

2. Let X be the cone $z^2 = xy$. Prove that the normalisation of X in the field $k(X)(\sqrt{x})$ equals the affine plane, and the normalisation map is of the form $x = u^2$, $y = v^2$, $z = uv$.

3. Prove the assertions analogous to those of Ex. 1 for an arbitrary quasiprojective curve X. Prove that if X is projective then so is Y.

4. How is the normalisation of $X \times Y$ related to those of X and Y?

5. Prove that x is a normal point of X if the completed local ring $\widehat{O}_x$ of 2.2 has no zerodivisors and is integrally closed. [Hint: Extend §3, Ex. 7 to singular points and apply.]

6. Prove that the cone $X \subset \mathbf{A}^n$ given by $x_1^2 + \cdots + x_n^2 = 0$ is normal.

7. In §3, Ex. 13, prove that the origin is a normal point of the hypersurface X.

8. Is the Steiner surface of §1, Ex. 15 normal?

9. Prove that any algebraic curve has a plane projective model all of whose singularities have only linear branches.

6. Singularities of a Map

When studying a regular map $f: X \to Y$, the following question arises: to what extent do the fibres $f^{-1}(y)$ over points $y \in Y$ inherit properties of X. As a rule, there are relations that do not hold everywhere, but hold over "most" points $y \in Y$, that is, over points of some dense open set $U \subset Y$. Over other points $y \notin U$, the fibres $f^{-1}(y)$ may suffer some kind of degeneration, or acquire singularities not present on X. The situation should be compared with that of Chap. I, 6.3, Theorem 7 on the dimensions of fibres.

6.1. Irreducibility

Of course, even if X is irreducible, we cannot hope that almost all the fibres of $f: X \to Y$ are irreducible. For example, if f is finite, its fibres are finite collections of points. We now formulate a restriction that allows us to guarantee the irreducibility of "most" fibres.

Suppose that X and Y are irreducible, and that $f(X)$ is dense in Y. A variety X defined over k can also be viewed as a variety over the bigger field $k(Y) \supset k$. Since all our considerations so far related to algebraically closed fields, we have to view it over an even bigger field, the algebraic closure $\overline{k(Y)}$ of $k(Y)$. Now over $\overline{k(Y)}$, our variety X may no longer be irreducible. For example, let X be the pencil of conics defined by $\sum_{i,j=0}^{2} a_{ij}(t)\xi_i\xi_j = 0$ in $\mathbb{P}^2 \times \mathbb{A}^1$. Set $D(t) = \det|a_{ij}(t)|$. If $D(t)$ is not identically 0, the conic $\sum_{i,j=0}^{2} a_{ij}(t)\xi_i\xi_j = 0$ is nondegenerate, and X is irreducible over $\overline{k(t)}$. But if $D(t) \equiv 0$, then over $k(t)$, the equation of the conic can be reduced to $a(t)\xi_0^2 + b(t)\xi_1^2 = 0$. If $-b(t)/a(t)$ is not a square in $k(t)$ then $a(t)\xi_0^2 + b(t)\xi_1^2$ is irreducible over $k(t)$, but nevertheless reducible over $\overline{k(t)}$.

In the general case, it can be shown that a variety X is irreducible over $\overline{k(Y)}$ if and only if the map $f: X \to Y$ cannot be factored as a composite $X \to Y' \to Y$ where $k(Y')$ is a nontrivial finite extension field of $k(Y)$, or in other words, if and only if $f^*: k(Y) \hookrightarrow k(X)$ embeds $k(Y)$ as an algebraically closed subfield of $k(X)$. This is a purely algebraic fact, see Zariski and Samuel [78], Chap. VII, §11, Theorem 38.

Theorem 1 (The first Bertini theorem). *Let X and Y be irreducible varieties defined over a field of characteristic 0, and $f: X \to Y$ a regular map such that $f(X)$ is dense in Y. Suppose that X remains irreducible over the algebraic closure $\overline{k(Y)}$ of $k(Y)$. Then there exists an open dense set $U \subset Y$ such that all the fibres $f^{-1}(y)$ over $y \in U$ are irreducible.*

Remark 1. The theorem also holds over a field of characteristic p; the proof just becomes slightly more complicated.

Remark 2. By the remark just before the statement of the theorem, the only reason for "most" fibres of $f: X \to Y$ to be reducible is the existence of a factorisation $X \to Y' \to Y$ where $Y' \to Y$ is a generically finite maps.

Proof. We can replace Y by an affine open subset Y_1, and so by Chap. I, 6.3, Theorem 7, we can assume that for $y \in Y_1$ all the components of the fibres $f^{-1}(y)$ have the same dimension $r = \dim X - \dim Y$. In this situation, we can also replace X by any open subset X_1. Indeed, set $X \setminus X_1 = Z$ and let $Z = \bigcup Z_i$ be its decomposition into irreducible components. By passing to a smaller open set $Y_2 \subset Y_1$ if necessary, we can discard the components Z_i for which $\overline{f(Z_i)} \neq Y_2$. If $f(Z_i)$ is dense in Y_2, possibly shrinking Y_2 still further, we can once more assume that all components of fibres of $f: Z_i \to Y_2$ have the same dimension equal to $\dim Z_i - \dim Y_2 < r$. Therefore they meet the fibres of $f: X \to Y_2$ in subsets of smaller dimension, and since all components of these fibres have equal dimension, discarding subsets of smaller dimension from them does not affect their irreducibility.

We now make use of the fact that our fields have characteristic 0. We can find $r + 1$ elements $u_1, \ldots, u_r, u_{r+1} \in k(X)$ such that $u_1, \ldots, u_r$ are algebraically independent over $k(Y)$, and such that u_{r+1} is a primitive element for the field extension $k(Y)(u_1, \ldots, u_r) \subset k(X)$, and is integral over $k[Y_2]$. Let X_2 be the affine variety for which $k[X_2] = k[Y_2][u_1, \ldots, u_r, u_{r+1}]$. By construction X_2 is birational to X, and hence they contain isomorphic open subsets, so that it is enough to prove the theorem for X_2 in place of X, with the map $f: X_2 \to Y_2$ defined by the inclusion $k[Y_2] \subset k[Y_2][u_1, \ldots, u_r, u_{r+1}]$.

Let $F = T^k + a_1(u_1, \ldots, u_r)T^{k-1} + \cdots + a_k(u_1, \ldots, u_r)$ be the irreducible polynomial with $a_i \in k[Y_2][u_1, \ldots, u_r]$ of which u_{r+1} is a root. The assumption that X is irreducible over the field $\overline{k(Y)}$ means that F is irreducible over the ring $\overline{k(Y)}[T, u_1, \ldots, u_r]$. Now the thing that we have to prove is that there exists an open subset $U \subset Y_2$ such that F remains irreducible on making the substitution $a_i \mapsto a_i(y)$ for each $y \in U$, that is, replacing each coefficient $a_i \in k[Y_2]$ of F by its value $a_i(y)$ at y. But this follows at once from Chap. I, 5.2, Proposition, Remark 2 according to which the reducibility of a polynomial is expressed by polynomial relations $R_j(a_1, \ldots, a_k) = 0$ between its coefficients. At least one of these relations fails for F, say $R_j(a_1, \ldots, a_k) = R \neq 0 \in k[Y_2]$; but then for any point $y \in Y_2$ with $R(y) \neq 0$, the polynomial obtained by substituting $a_i \mapsto a_i(y)$ for the coefficients of F is also irreducible. In other words, $U = Y_2 \setminus V(a)$. The theorem is proved.

6.2. Nonsingularity

In the theory of differentiable manifolds, one proves that for a smooth map $f: X \to Y$, the points $y \in Y$ over which the fibre $f^{-1}(y)$ is not a smooth manifold form a subset of measure 0 in Y (an analogue in differential topology of a subvariety of smaller dimension). This is called *Sard's theorem* (see Lang [54], Chap. VIII, §1 or Abraham and Robbin, [2], §15). Theorem 2 below is an algebraic geometric equivalent of this over a field of characteristic 0. We will see in 6.4 that the same assertion in characteristic p is false.

Theorem 2 (The second Bertini theorem). *Let $f: X \to Y$ be a regular map of varieties defined over a field of characteristic 0, with $f(X)$ dense in Y; assume that X is nonsingular. Then there exists a dense open set $U \subset Y$ such that the fibre $f^{-1}(y)$ is nonsingular for every $y \in U$.*

Theorems 1 and 2 and their various generalisations are called the *Bertini Theorems*.

Set $\dim X = n$ and $\dim Y = m$. By Chap I, 6.3, Theorem 7, there exists a dense open subset of Y over which all components of the fibres $f^{-1}(y)$ are of the same dimension $n - m$. We can assume that Y is the whole of this open set. In the same way, we can assume that Y is nonsingular. We prove first two lemmas.

Lemma 1. *The fibre $f^{-1}(y)$ is nonsingular if $d_x f: \Theta_{X,x} \to \Theta_{Y,y}$ is surjective for all points $x \in f^{-1}(y)$.*

Proof. Note that the tangent space $\Theta_{f^{-1}(y),x}$ to the fibre $f^{-1}(y)$ is contained in the kernel of $d_x f$. Indeed, the composite of $\Theta_{f^{-1}(y),x} \hookrightarrow \Theta_{X,x}$ with $d_x f$ is 0. To check this, by duality we must check that the composite of the dual $m_y/m_y^2 \to m_x/m_x^2$ of $d_x f$ with $m_x/m_x^2 \to \overline{m}_x/\overline{m}_x^2$ is 0, where $\overline{m}_x$ is the maximal ideal of x on the fibre $f^{-1}(y)$, and $m_x \to \overline{m}_x$ the restriction from X to the fibre. But this is obvious. Thus $d_x f$ surjective implies that

$$\dim \Theta_{f^{-1}(y),x} \le \dim \ker d_x f = \dim \mathcal{O}_{X,x} - \dim \Theta_{Y,y} \le n - m;$$

(here we use the fact that X is nonsingular, that is, $\dim \Theta_{X,x} = n$). Since all the components of the fibre $f^{-1}(y)$ have dimension $n - m$, it follows that it is nonsingular.

Lemma 2. *There exists a nonempty open subset $V \subset X$ such that $d_x f$ is surjective for $x \in V$.*

Proof. The surjectivity of $d_x f: \Theta_{X,x} \to \Theta_{Y,y}$ is dual to the injectivity of $m_y/m_y^2 \to m_x/m_x^2$, that is, if $u_1, \ldots, u_m$ are local parameters at Y, to the linearly independence of $d_x u_1, \ldots, d_x u_m$. Using the inclusion of $\mathcal{O}_y$ to the formal power series ring as in 2.2, it is easy to see that $u_1, \ldots, u_m$ are algebraically independent, and since $f(X)$ is dense in Y, it follows that they are also algebraically independent as functions on X. We complete them to a system $u_1, \ldots, u_n$ of $n = \dim X$ algebraically independent functions.

Lemma 2 will be proved if we check that for any system $u_1, \ldots, u_n$ of algebraically independent functions on X, the set of points at which $u_1, \ldots, u_n$ are local parameters is open and nonempty. We can assume that X is affine, $X \subset \mathbf{A}^N$, with coordinates $x_1, \ldots, x_N$. We prove that for points x of a nonempty open set $U \subset X$ all the $\mathrm{d}_x x_i$ can be expressed as linear combinations of $\mathrm{d}_x u_1, \ldots, \mathrm{d}_x u_n$. If these were linearly dependent it would then follow that $\dim \Theta_{X,x} < n$.

Each x_i is related to $u_1, \ldots, u_n$ by a relation $F_i(x_i, u_1, \ldots, u_n) = 0$, with F_i an irreducible polynomial, and hence (using that $\operatorname{char} k = 0$), $\partial F_i / \partial x_i$ is not identically 0. Suppose that $F_i = a_0 x_i^{n_i} + a_1 x_i^{n_i - 1} + \cdots + a_n$, with $a_j \in k[u_1, \ldots, u_n]$. Now $\mathrm{d}_x a_j$ are linear combinations of $\mathrm{d}_x u_1, \ldots, \mathrm{d}_x u_n$. Using the basic properties 1.3, (5) of the differential d_x, it follows from $F_i(x_i, u_1, \ldots, u_n) = 0$ that

$$\frac{\partial F_i}{\partial x_i}(x) \mathrm{d}_x x_i + x_i^{n_i} \mathrm{d}_x a_0 + \cdots + \mathrm{d}_x a_n = 0$$

at any point $x \in X$. The points at which all $\partial F_i / \partial x_i(x) \neq 0$ form a nonempty open set, and at such points $\mathrm{d}_x x_i$ can be written as linear combinations of $\mathrm{d}_x u_1, \ldots, \mathrm{d}_x u_n$. Lemma 2 is proved.

Proof of Theorem 2. It is now easy to complete the proof of Theorem 2. Let $Z \subset X$ denote the subset of points $x \in X$ at which $\mathrm{d}_x f$ is not surjective. It is easy to see that it is a closed subset, since it is defined by the vanishing of certain minors. We need to prove that $f(Z)$ is contained in a proper closed subset of Y. If not, then $f(Z)$ is dense in Y. Applying Lemma 2 to Z, we find a nonempty open set $V \subset Z$ such that $\Theta_{Z,x} \to \Theta_{Y,f(x)}$ is surjective at all points of V. But $\Theta_{Z,x} \subset \Theta_{X,x}$, and thus a fortiori the map $\Theta_{X,x} \to \Theta_{Y,x}$ must be surjective. This contradiction proves the theorem.

6.3. Ramification

We consider now an especially simple case of maps, those with 0-dimensional fibres. For a finite map $f \colon X \to Y$, as we saw in Chap. I, 5.3, Theorem 5, the inverse image $f^{-1}(y)$ of any point $y \in Y$ is a finite number of points. Let us study this number. By analogy with the theorem on dimension of fibres, it is natural to expect that it is constant for all y in some open set, deviating from this value only on some closed subset $Z \subset Y$. This is what happens in the simplest case

$$f \colon \mathbf{A}^1 \to \mathbf{A}^1 \quad \text{given by} \quad y = f(x) = x^2. \tag{1}$$

To state in a general form the specific property of this example, we introduce the following notion.

Definition. Let X and Y be irreducible varieties of the same dimension and $f: X \to Y$ a regular map such that $f(X) \subset Y$ is dense. The degree of the field extension $f^*(k(Y)) \subset k(X)$, which is finite under these assumptions, is called the *degree* of f:

$$\deg f = [k(X) : f^*(k(Y))].$$

The map (1) has $\deg f = 2$, and if char $k \neq 2$, every point $y \neq 0$ has two distinct inverse images, and the point $y = 0$ one only. Is it always true that the number of inverse images is $\leq$ the degree of a map? This is not so for the example of the parametrisation $f: \mathbf{A}^1 \to Y$ of the cubic curve with an ordinary double point (Chap. I, 1.2, (1–2)): here $\deg f = 1$, but the inverse image of the singular point consists of two points. It turns out that the reason here is that Y is not normal.

Theorem 3. *If $f: X \to Y$ is a finite map of irreducible varieties, and Y is normal, then the number of inverse images of any point $y \in Y$ is $\leq \deg f$.*

Proof. In view of the definition of a finite map, we can restrict to the case that X and Y are affine. Set

$$k[X] = A, \quad k(X) = K, \qquad \text{with } [K : L] = \deg f = n.$$
$$k[Y] = B, \quad k(Y) = L,$$

Since Y is normal, B is integrally closed, and since f is finite, A is a finite B-module. Hence for any $a \in A$, the coefficients of the minimal polynomial of a are in B. This is a simple property of integrally closed rings, whose proof can be found in Atiyah and Macdonald [7], Proposition 5.15.

Suppose that $f^{-1}(y) = \{x_1, \dots, x_m\}$. Consider an element $a \in A$ taking distinct values $a(x_i)$ at the points x_i for $i = 1, \dots, m$; if $X \subset \mathbf{A}^N$, the point is to find a polynomial on $\mathbf{A}^N$ with this property, which is entirely elementary. Let $F \in B[T]$ be the minimal polynomial of a. Obviously $\deg F \leq n$. We replace all the coefficients of F by their values at y, writing $\overline{F}(T)$ for the resulting polynomial. Then this has m distinct roots $a(x_i)$. Thus

$$m \leq \deg \overline{F} = \deg F \leq n,$$

so that $m \leq n$, as asserted. The theorem is proved.

In what follows, throughout this section, we consider a finite map $f: X \to Y$ between irreducible varieties, and assume that Y is normal.

Definition. f is *unramified* over $y \in Y$ if the number of inverse images of y equals the degree of the map. Otherwise, we say that f is *ramified* at y, or that y is a *ramification point* or a *branch point* of f.

Theorem 4. *The set of points at which a map is unramified is open, and is nonempty if $f^*(k(Y)) \subset k(X)$ is a separable field extension.*

Proof. We preserve the notation introduced in the proof of Theorem 3. If f is unramified at y then $\deg \overline{F} = \deg F = n$, and $\overline{F}$ has n distinct roots. Write $D(F)$ for the discriminant of F. As we have seen, a sufficient condition for f to be unramified at a point y can be written

$$D(\overline{F}) = D(F)(y) \neq 0. \tag{2}$$

But then $D(F)(y') \neq 0$ for points y' in some neighbourhood of y. This is what we had to prove. Thus the set of branch point is a closed set; it is called the *branch locus* or *ramification locus* of f.

The question remains as to whether it is a strict subset. Suppose that $f^*(k(Y)) \subset k(X)$ is separable. In this case we also say that f is *separable*. We can again assume that X and Y are affine, and use the previous notation. If $a \in A$ is a primitive element for the field extension $f^*(k(Y)) \subset k(X)$ and $F(T)$ its minimal polynomial, then $\deg F = n$ and $D(F) \neq 0$. Therefore, there exist points $y \in Y$ such that $D(F)(y) \neq 0$, so that f is unramified. This proves Theorem 4.

Remark. In the case of an inseparable map, every point is a ramification point; the standard example of this is the map $\mathbb{A}^1 \to \mathbb{A}^1$ defined by $x \mapsto x^p$.

We see that if $f : X \to Y$ is finite and separable, with X and Y irreducible and Y normal, then the picture is as in the example (1): points of some nonempty open subset $U \subset Y$ have $\deg f$ distinct inverse images, and points in the complement have fewer inverse images. See Chap. VII, 3.1, Theorem 1 for a more concrete local description of the ramification of a map $f : X \to Y$ between algebraic curves of over $\mathbb{C}$.

Now suppose that Y is nonsingular. The preceding considerations allow us to describe finite unramified maps $f : X \to Y$ in a very explicit form. Consider a function $a \in A = k[X]$ that takes distinct values at all the points of the inverse image $f^{-1}(y)$ of some point $y \in Y$. Then $k(X) = k(Y)(a)$. If $B = k[Y]$, and $F = F(T) \in B[T]$ is the minimal polynomial of a, then by (2), the discriminant $D(F)(y) \neq 0$, and hence $F'(a)(x) \neq 0$ for $x \in f^{-1}(y)$. From now on, we write Y for an affine neighbourhood of y on which $D(F)$ is nonzero, and X for its inverse image. Set $A' = B[a] = B[T]/(F(T))$. Then $A' = k[X']$, where $X' \subset Y \times \mathbb{A}^1$ is defined by the equation $F(T) = 0$. We prove that, in view of the nonsingularity of Y, also X' is nonsingular. But then X' is normal, and therefore A' is integrally closed; since $A' \subset A$ and the two rings have the same field of fractions, we have $A = A'$ and $X = X'$, that is, the explicit construction of X' actually describes X.

It remains to prove that X' is nonsingular. Suppose that

$$F(T) = T^n + b_1 T^{n-1} + \cdots + b_n \qquad \text{with } b_i \in B.$$

We prove that the map $d_x f \colon \Theta_{X',x} \to \Theta_{Y,z}$ is an inclusion for any point $x \in X'$, where $z = f(x)$. By duality, this is equivalent to $\mathfrak{m}_z/\mathfrak{m}_z^2 \to \mathfrak{m}_x/\mathfrak{m}_x^2$ surjective. Let $u_1, \ldots, u_m$ be local parameters at z. We need to prove that $d_x u_1, \ldots, d_x u_m$ generate $\mathfrak{m}_x/\mathfrak{m}_x^2$. By definition this space is generated by elements $d_x b$ for $b \in B$ (which are linear combinations of $d_x u_1, \ldots, d_x u_m$) together with $d_x a$. It remains to prove that $d_x a$ can be written as a linear combination of $d_x u_1, \ldots, d_x u_m$. For this, we use the fact $F(a) = 0$, and the properties 1.3, (5) of differentials. We get

$$F'(a)(x) d_x a + a^{n-1}(x) d_x b_1 + \cdots + d_x b_n = 0.$$

Since $F'(a)(x) \neq 0$, this expresses $d_x a$ in terms of $d_x u_1, \ldots, d_x u_m$.

Now recall that Y is nonsingular and $\dim X' = \dim Y = m$. Hence $\dim \Theta_{Y,z} = m$, and in view of the inclusion $d_x f \colon \Theta_{X',x} \hookrightarrow \Theta_{Y,z}$, also $\dim \Theta_{X',x} = m$. Hence X' is nonsingular and $X' = X$. But we have proved a little more. We summarise what we have proved.

Theorem 5. *An unramified finite map $f \colon X \to Y$ to a nonsingular variety Y is locally described as the projection to Y of a subvariety $X \subset Y \times \mathbf{A}^1$, where X is defined by an equation $F(T) = 0$ and $D(F) \neq 0$ on Y. The differential $d_x f$ defines an isomorphism $\Theta_{X,x} \xrightarrow{\sim} \Theta_{Y,f(x)}$ on the tangent spaces.* $\square$

In the case $k = \mathbb{C}$, this theorem shows that as a map of topological spaces, $f \colon X \to Y$ is an unramified cover, that is, any point $y \in Y$ has a neighbourhood U such that $f^{-1}(U)$ decomposes as a disjoint union of open sets, each of which is mapped homeomorphically to U by f. Indeed, suppose that $f^{-1}(y) = \{x_1, \ldots, x_n\}$, and that $u_1, \ldots, u_m$ are local parameters in a neighbourhood of y and $v_1^{(i)}, \ldots, v_m^{(i)}$ local parameters at x_i. The isomorphism $d_{x_i} \colon \Theta_{X,x_i} \xrightarrow{\sim} \Theta_{Y,y}$ shows that $\det |\partial v_k^{(i)} / \partial u_j|(x_i) \neq 0$ for all $i = 1, \ldots, n$. By the implicit function theorem it follows from this that there exist neighbourhoods V_i of x_i and U of y such that f define a homeomorphism from each V_i to U. We can choose these neighbourhoods sufficiently small that V_i and V_j do not intersect for $i \neq j$. We check that $f^{-1}(U) = \bigcup V_i$. If $y' \in U$ then, since f is unramified, $f^{-1}(y')$ consists of $n = \deg f$ points. But since y' already has n inverse images in $\bigcup V_i$, we have $f^{-1}(U) = \bigcup V_i$.

6.4. Examples

Example 1. Pencil of quadrics. Assume that $\operatorname{char} k \neq 2$, and consider the hypersurface $X \subset \mathbb{P}^n \times \mathbf{A}^1$ defined by the equation $\sum_{i,j=0}^n a_{ij}(t)\xi_i\xi_j = 0$, where $a_{ij}(t) \in k[\mathbf{A}^1] = k[t]$, and the projection $X \to \mathbf{A}^1$ induced by the projection $\mathbb{P}^n \times \mathbf{A}^1 \to \mathbf{A}^1$. This is called a *pencil of quadrics*, and the polynomial $D(t) = \det |a_{ij}(t)|$ the *discriminant* of the pencil. Pencils of conics have already appeared in Chap. I, 6.2, Example 1.

We determine first of all when X is nonsingular, and secondly, over what points $\alpha \in \mathbf{A}^1$ the fibre of $X \to \mathbf{A}^1$ is singular.

Set $F = \sum a_{ij}(t)\xi_i\xi_j$. If $D(\alpha) \neq 0$ at a point $t = \alpha$ then the equations $\partial F/\partial \xi_i = 0$ for $i = 0, \dots, n$ and $t = \alpha$ have only the solution 0, so that points of the fibre over $t = \alpha$ are nonsingular both as points of X and as points of the fibre. It remains to consider the values $t = \alpha$ for which $D(\alpha) = 0$. We will assume that $\alpha = 0$.

Write $\overline{F}$ for the quadratic form $\overline{F} = \sum a_{ij}(0)\xi_i\xi_j$, and r for its rank. We can make a nondegenerate linear transformation with coefficients in k to put $\overline{F}$ in the form $\xi_0^2 + \cdots + \xi_{r-1}^2$. Now we apply to F the standard method of completing the square; we can make a linear transformation with coefficients in the local ring $\mathcal{O}_0$ of the origin of $\mathbf{A}^1$ (that is, the coefficients are rational function with no t in the denominators), and with determinant invertible in $\mathcal{O}_0$, to put F in the form

$$F = a_0(t)\xi_0^2 + \cdots + a_{r-1}(t)\xi_{r-1}^2 + tG(\xi_r, \dots, \xi_n),$$

with $a_i(t) \in \mathcal{O}_0$ and $a_i(0) \neq 1$ for $i = 0, \dots, r-1$. Any points with $t = 0$, $\xi_0 = \cdots = \xi_{r-1}$ (and arbitrary $\xi_r, \dots, \xi_n$) lie on X, and there $\partial F/\partial \xi_i = 0$ for all i. Suppose that

$$G(\xi_r, \dots, \xi_n) = \overline{G}(\xi_r, \dots, \xi_n) + tG_1(\xi_r, \dots, \xi_n),$$

with $\overline{G} \in k[\xi_r, \dots, \xi_n]$. Then at our point, $\partial F/\partial t = \overline{G}(\xi_r, \dots, \xi_n)$. If $r < n$ then there exist $\xi_r, \dots, \xi_n$, not all 0, such that $\overline{G}(\xi_r, \dots, \xi_n) = 0$, and the point is singular on X. For $r = n$, the equation looks like

$$F = a_0(t)\xi_0^2 + \cdots + a_{n-1}(t)\xi_{n-1}^2 + t^k a_n(t)\xi_n^2,$$

with $a_i(0) \neq 0$ for $i = 0, \dots, n$ and some $k \geq 1$. If $k > 1$ then $\partial F/\partial t = 0$ at the point $t = 0$, $(\xi_0, \dots, \xi_n) = (0, \dots, 0, 1)$, and this is a singular point of X. There remains the case $k = 1$, when it is easy to seen that no point of the fibre over $t = 0$ is a singular point of X. Thus we have proved the following result.

Proposition 1. *The quadric bundle X is a nonsingular variety if and only if its discriminant has no repeated roots. The singular fibres are precisely the fibres over the roots of the discriminant. In particular, the number of singular fibres of $X \to \mathbf{A}^1$ equals the degree of the discriminant.* $\square$

Example 2. Pencil of elliptic curves. Assume that the characteristic of k is not 2 or 3, and consider the surface $X \subset \mathbb{P}^2 \times \mathbb{A}^1$ defined by the equation

$$\xi_2^2 \xi_0 = \xi_1^3 + a(t)\xi_1 \xi_0^2 + b(t)\xi_0^3 \qquad \text{with } a(t), b(t) \in k[\mathbb{A}^1] = k[t].$$

The projection $\mathbb{P}^2 \times \mathbb{A}^1 \to \mathbb{A}^1$ defines a map $f: X \to \mathbb{A}^1$. The fibre $f^{-1}(\alpha)$ over a point α is the cubic curve $\xi_2^2 \xi_0 = \xi_1^3 + a(\alpha)\xi_1\xi_0^2 + b(\alpha)\xi_0^3$. This cubic has a unique point on the "line at infinity" $\xi_0 = 0$, the flex $(0 : 1 : 0)$; in the chart $\mathbb{A}^2$ with $\xi_0 \neq 0$, it is given in affine coordinates $x = \xi_1/\xi_0$ and $y = \xi_2/\xi_0$ by $y^2 = x^3 + a(\alpha)x + b(\alpha)$. If the fibre $f^{-1}(\alpha)$ is nonsingular, then as in Example 1, X has no singular points on it.

Suppose that $f^{-1}(\alpha)$ is singular. It is easy to see that $(0 : 1 : 0)$ is nonsingular. Thus there must be a simultaneous solution of $y = 0$, $3x^2 + a(\alpha) = 0$ and $y^2 = x^3 + a(\alpha)x + b(\alpha)$, from which it follows that $4a(\alpha)^3 + 27b(\alpha)^2 = 0$. The polynomial $D(t) = 4a(t)^3 + 27b(t)^2$ is called the *discriminant* of the pencil $X \to \mathbb{A}^1$. We will assume that $D(t)$ is not identically 0. We have proved that if $D(\alpha) \neq 0$ then all points of the fibre $f^{-1}(\alpha)$ are nonsingular both on the fibre and on the surface X.

If $D(\alpha) = 0$ then the same argument shows that the fibre $f^{-1}(\alpha)$ has a singular point, and it follows from the equations $3x^2 + a(\alpha) = 0$ and $x^3 + a(\alpha)x + b(\alpha) = 0$ that the x-coordinate of this point is given by $2a(\alpha)x + 3b(\alpha) = 0$. In order for this to be a singular point of X, it must also satisfy $a'(\alpha)x + b'(\alpha) = 0$, whence $2ab' - 3b'a = 0$. Since moreover $4a(\alpha)^3 + 27b(\alpha)^2 = 0$, either $a(\alpha) = b(\alpha) = 0$ or $a(\alpha) \neq 0$ and $b(\alpha) \neq 0$. When $a(\alpha) = b(\alpha) = 0$ our relations are equivalent to $b'(\alpha) = 0$, and when $a(\alpha) \neq 0$ and $b(\alpha) \neq 0$, they can be expressed as $(a^3/b^2)'(\alpha) = 0$, or $D'(\alpha) = \big(b^2(4a^3/b^2 + 27)'\big)(\alpha) = 0$. This proves the following result.

Proposition 2. *The pencil of elliptic curves $X \to \mathbb{A}^1$ is a nonsingular surface if the discriminant has simple roots or are common roots of a and b that are simple roots of b. Singular fibres correspond to roots of the discriminant.* $\square$

Example 3. Pathologies in finite characteristic. We can construct examples in which the assertion of Theorem 2 fails in characteristic 2. For this, consider the finite part of the pencil of elliptic curves $\xi_2^2 \xi_0 = \xi_1^3 + a(t)\xi_1\xi_0^2 + b(t)\xi_0^3$, given by $y^2 = x^3 + a(t)x + b(t)$. In characteristic 2 every fibre $y^2 = x^3 + a(\alpha)x + b(\alpha)$ is singular at the point $x = a(\alpha)^{1/2}$, $y = b(\alpha)^{1/2}$, and at no other point. In order for this to be a singular point of the surface, we must have $a'(\alpha)x + b'(\alpha) = 0$, that is, $\big((a')^2 a + (b')^2\big)(\alpha) = 0$. Thus all the fibres of $X \to \mathbb{A}^1$ are singular, but the surface X itself only has singular points in the fibres $f^{-1}(\alpha)$, where α is a root of $(a')^2 a + (b')^2$. If S is the set of these roots, then the surface $X \setminus f^{-1}(S)$ is nonsingular, but all the fibres of $X \setminus f^{-1}(S) \to \mathbb{A}^1 \setminus S$ are singular.

There is a similar example in characteristic 3, the pencil with equation $y^2 = x^3 + a(t)$. It can be proved that such "pathological" pencils of cubic curves exist only in characteristic 2 or 3, although of course similar examples occur for all p for curves of higher degree, for example $y^2 = x^p + a(t)$.

An example of a finite map $f\colon X \to Y$ such that every point $y \in Y$ is a branch point is given by the Frobenius map, Chap. I, 2.3, Example 6 in characteristic $p > 0$. It has $\varphi(\alpha_1, \ldots, \alpha_n) = (\alpha_1^p, \ldots, \alpha_n^p)$, so that in characteristic p, every point x has a unique inverse image $\varphi^{-1}(x)$.

In the theory of curves, the Frobenius map particularly reflects the specific properties of finite characteristic. For this, we need to generalise somewhat. If C is the plane curve $f(x, y) = \sum a_{ij} x^i y^j = 0$, we let C' be the curve $g(x, y) = \sum a_{ij}^p x^i y^j = 0$. In characteristic p, the map $u = x^p$, $v = x^p$ obviously defines a rational map $\varphi\colon C \to C'$ (in fact, a regular map). This is also called the *Frobenius map*, and coincides with that introduced in Chap. I, 2.3, Example 6 if $a_{ij} \in \mathbb{F}_p$, when $a_{ij}^p = a_{ij}$, and therefore $C = C'$.

Theorem 6. *The Frobenius map of an algebraic curve has degree p. Every inseparable rational map of curves $f\colon X \to Y$ factors as a composite $f = g \circ \varphi$ where $g\colon X' \to Y$ is some map and $\varphi\colon X \to X'$ the Frobenius map.*

Proof. This follows from general properties of fields of characteristic p and transcendence degree 1; see Appendix, §5, Proposition 2. It is proved there that $[k(X) : \varphi^*(k(X'))] = p$, and this means that $\deg \varphi = p$. Moreover, $f^*(k(Y)) \supset k(X)^p$, but $k(X)^p = k(X')$. The inclusion of fields $f^*(k(Y)) \subset \varphi^*(k(X'))$ and the isomorphism $\varphi^*\colon k(X') \to \varphi^*(k(X'))$ define an inclusion $(\varphi^*)^{-1}(f^*(k(Y))) \subset k(X')$, that is, a rational map $g\colon X' \to Y$ such that $f = g \circ \varphi$. The theorem is proved.

Exercises to §6

1. Classify singular points of pencils of quadrics over the point $t = 0$ up to formal analytic equivalence, under the assumption that the rank of the quadric drops by 1 at $t = 0$.

2. Consider the net of conics X on $\mathbb{P}^2$ defined in $\mathbb{P}^2 \times \mathbb{A}^2$ by $\sum_{i,j=0}^2 a_{ij}(s, t) \xi_i \xi_j = 0$. Assume that the rank of a conic over every point $\alpha \in \mathbb{A}^2$ drops by at most 1. Prove that X is nonsingular if and only if the discriminant curve $\det |a_{ij}(s, t)| = 0$ is nonsingular.

3. Prove that if a pencil of elliptic curves (6.4, Example 2) is a nonsingular surface then its singular fibre is irreducible. Is this true for any family of cubics?

4. Determine the branch locus of the map $X \to \mathbb{P}^n$, where X is the normalisation of $\mathbb{P}^n$ in the quadratic extension of $k(\mathbb{P}^n) = k(x_1, \ldots, x_n)$ defined by the equation $y^2 = f(x_1, \ldots, x_n)$, where f is a polynomial of degree m. [Hint: The answer depends on the parity of m.]

5. Prove that if char $k = p$ then the curve $y^p + y = f(x)$ where f is a polynomial is an unramified cover of the line $\mathbb{A}^1$ with coordinate x.

6. Prove that for the surfaces $y^2 = x^3 + a(t)x + b(t)$ over a field of characteristic 2 and $y^2 = x^3 + a(t)$ over a field of characteristic 3, the singular points of fibres form a nonsingular curve having projection to the line $\mathbb{A}^1$ with coordinate t of degree $p = 2$ or 3 respectively.

Chapter III. Divisors and Differential Forms

1. Divisors

1.1. The Divisor of a Function

A polynomial in one variable is uniquely determined up to a constant factor by specifying its roots and their multiplicities; that is by specifying a set of points $x_1, \ldots, x_r \in \mathbf{A}^1$ with multiplicities $k_1, \ldots, k_r$. A rational function $\varphi(x) = f(x)/g(x)$ with $f, g \in k[\mathbf{A}^1]$ is determined by the zeros of f and g, that is, by the points at which it is 0 or is irregular. To distinguish the roots of g from those of f, we take their multiplicities with a minus sign. Thus the function φ is given by points $x_1, \ldots, x_r$ with arbitrary integer multiplicities $k_1, \ldots, k_r$.

The task we set ourselves here is to find a similar way of specifying a rational function on an arbitrary algebraic variety. The starting point is that, according to the theorem on the dimension of intersections, the set of points at which a regular function is 0 is a codimension 1 subvariety. Thus the object we associate with a function is a collection of irreducible codimension 1 subvarieties, together with assigned multiplicities; the multiplicities we assign are integers, both positive and negative.

Definition. Let X be an irreducible variety. A collection of irreducible closed subvarieties $C_1, \ldots, C_r$ of codimension 1 in X with assigned integer multiplicities $k_1, \ldots, k_r$ will be called a *divisor* on X. A divisor is written

$$D = k_1 C_1 + \cdots + k_r C_r. \tag{1}$$

If all the $k_i = 0$, we write $D = 0$. If all $k_i \geq 0$ and some $k_i > 0$ then we write $D > 0$; in this case D is said to be *effective*. An irreducible codimension 1 subvariety C_i taken with multiplicity 1 is called a *prime divisor*. If all the $k_i \neq 0$ in (1) then the variety $C_1 \cup \cdots \cup C_r$ is called the *support* of D and denoted by $\operatorname{Supp} D$.

We define an addition operation on divisors. For this, note that, provided we also allow the coefficients to take the value 0 in (1), any two divisors D and D' can be written

$$D = k_1 C_1 + \cdots + k_r C_r \quad \text{and} \quad D' = k'_1 C_1 + \cdots + k'_r C_r,$$

with the same collection of prime divisors $C_1, \ldots, C_r$. Then by definition,

$$D + D' = (k_1 + k'_1)C_1 + \cdots + (k_r + k'_r)C_r.$$

Thus divisors on X form a group, equal to the free $\mathbb{Z}$-module with the irreducible codimension 1 subvarieties C of X as generators. This group is denoted by Div X.[7]

We now describe the map taking a nonzero function $f \in k(X)$ into its divisor div f. Let C be a prime divisor; first of all, to each nonzero $f \in k(X)$, we assign an integer $v_C(f)$. If $X = \mathbf{A}^1$ then $v_C(f)$ is the order of zero or pole of a function at a point.

This can be done only under one restriction on X. Namely, we assume that X is nonsingular in codimension 1 (see Chap. II, 5.1); in other words, we assume that the set of singular points of X has codimension ≥ 2. Let $C \subset X$ be an irreducible codimension 1 subvariety, and U some affine open set intersecting C, consisting of nonsingular points, and such that C is defined in U by a local equation. Such an affine set U exists by the assumption on X and by Chap. II, 3.1, Theorem 1. Thus $\mathfrak{a}_C = (\pi)$ in $k[U]$. We prove that for any $0 \neq f \in k[U]$, there exists an integer $k \geq 0$ such that $f \in (\pi^k)$ and $f \notin (\pi^{k+1})$. If this were not the case, that is, if $f \in (\pi^k)$ for every k, then $f \in \bigcap(\pi^k)$; the same then holds in the local ring $\mathcal{O}_C$ at an irreducible subvariety C. Hence $f = 0$ by Chap. II, 2.2, Theorem 5 and Appendix, §6, Proposition 4.

The number k just determined is denoted by $v_C(f)$. It has the properties

$$v_C(f_1 f_2) = v_C(f_1) + v_C(f_2), \quad \text{and}$$
$$v_C(f_1 + f_2) \geq \min\{v_C(f_1), v_C(f_2)\}, \quad \text{if } f_1 + f_2 \neq 0, \tag{2}$$

as follows easily from the definition and the irreducibility of C. In the case of a nonsingular plane curve, we have already defined this function in Chap. I, 1.5, Theorem 1.

If X is irreducible, then any function $f \in k(X)$ can be written in the form $f = g/h$ with $g, h \in k[U]$. If $f \neq 0$ we set $v_C(f) = v_C(g) - v_C(h)$. It follows at once from (2) that $v_C(f)$ does not depend on the representation of f in the form g/h, and that (2) holds for all $f \in k(X)$ with $f \neq 0$.

Our definition of $v_C(f)$ depends at present on the choice of an open set U, and hence we temporarily write $v_C^U(f)$ in place of $v_C(f)$. Let us show that in fact $v_C^U(f)$ is independent of U. Suppose first that $V \subset U$ is an affine open set with $V \cap C \neq \emptyset$. Then π is a local equation for C also in V, and

---

[7] The current literature is inconsistent, some authors using Div X for the group of "ordinary" divisors $\sum k_i C_i$ (Weil divisors) described here, some for locally principal divisors (Cartier divisors) (see 1.2 below). In case of ambiguity, one can write WDiv or CDiv.

obviously then $v_C^U(f) = v_C^V(f)$. However, if V is any open set satisfying the same conditions as U then $U \cap C$ and $V \cap C$ are open in C and nonempty, and since C is irreducible they have nonempty intersection. Taking W to be an affine neighbourhood in $U \cap V$ of some point $x \in U \cap V \cap C$, by the preceding remark, we get that $v_C^U(f) = v_C^W(f)$ and $v_C^V(f) = v_C^W(f)$, and hence $v_C^U(f) = v_C^V(f)$. Thus we have justified that the notation $v_C(f)$ is well defined.

Notice that if $X = \mathbb{A}^1$ and $C = x$ is the point α then $v_x(f)$ equals the multiplicity of α as a root of f for any nonzero $f \in k[T]$; the general definition essentially copies this particular case.

If $v_C(f) = k > 0$ then we say that f has a *zero* of order k along C; if $v_C(f) = -k < 0$ that f has a *pole* of order k along C. Note that these notions are defined for codimension 1 subvarieties, rather than for points. For example, if f is the function $f = x/y$ on $\mathbb{A}^2$ then the point $(0,0)$ is contained both in the locus of zeros $(x = 0)$ and the locus of poles $(y = 0)$ of f.

We now prove that for a given function $f \in k(X)$, there are only a finite number of irreducible codimension 1 subvarieties C such that $v_C(f) \neq 0$. Consider first the case that X is an affine variety and $f \in k[X]$. Then it follows from the definition that if C is not a component of the subvariety $V(f)$ then $v_C(f) = 0$. If X is still affine, but $f \in k(X)$ then $f = g/h$ with $g, h \in k[X]$, and we see that $v_C(f) = 0$ if C is not a component of $V(g)$ or $V(h)$. Finally, in the general case, let $X = \bigcup U_i$ be a finite cover of X by affine open sets. Then any subvariety C intersects at least one U_i, so that $v_C(f) \neq 0$ only for C that is the closure of an irreducible codimension 1 subvariety $C' \subset U_i$ for some i, with $v_{C'}(f) \neq 0$ in U_i. Since there are only finitely many U_i and finitely many C' in each U_i, there are only finitely many C with $v_C(f) \neq 0$. Thus we can consider the divisor

$$\sum v_C(f)C, \tag{3}$$

where the sum takes place over all the irreducible codimension 1 subvarieties C for which $v_C(f) \neq 0$. This divisor is called the *divisor* of f, and denoted[8] by div f.

A divisor of the form $D = \operatorname{div} f$ for some $f \in k(X)$ is called a *principal divisor*. If div $f = \sum k_i C_i$ then the divisors

$$\operatorname{div}_0 f = \sum_{\{i|k_i>0\}} k_i C_i \quad \text{and} \quad \operatorname{div}_\infty f = \sum_{\{i|k_i<0\}} -k_i C_i$$

are called respectively the *divisor of zeros* and *divisor of poles* of f. Obviously $\operatorname{div}_0 f, \operatorname{div}_\infty f \geq 0$ and $\operatorname{div} v = \operatorname{div}_0 f - \operatorname{div}_\infty f$. Notice a number a simple properties: $\operatorname{div}(f_1 f_2) = \operatorname{div}(f_1) + \operatorname{div}(f_2)$, and $\operatorname{div} f = 0$ for $f \in k$; and $\operatorname{div} f \geq 0$ for $f \in k[X]$.

---

[8] The divisor div f is also traditionally denoted by (f) in the literature.

Let us prove that for a nonsingular irreducible variety the converse also holds, that is, if div $f \geq 0$ then f is regular on X. The same thing holds if X is only normal, but we omit the proof. Let $x \in X$ be a point at which f is not regular. Then $f = g/h$ with $g, h \in \mathcal{O}_x$ but $g/h \notin \mathcal{O}_x$. It follows from the fact that $\mathcal{O}_x$ is a UFD (Chap. II, 3.1, Theorem 2) that we can choose g and h without common factors. Let π be a prime element of $\mathcal{O}_x$ that divides h but not g. In some affine neighbourhood U of x, the variety $V(\pi)$ is irreducible and of codimension 1. Write C for its closure in X. Then obviously $v_C(f) < 0$. This proves that div $f \geq 0 \implies f$ regular.

Since an everywhere regular function on an irreducible projective variety X is a constant (Chap. I, 5.2, Theorem 3, Corollary 1), it follows from the result just proved that on a nonsingular projective variety X, if div $f \geq 0$ then $f = \alpha \in k$. In particular, on a nonsingular projective variety, a rational function is uniquely determined up to a constant factor by its divisor: if div $f =$ div g then div$(f/g) = 0$, so that $f = \alpha g$ with $\alpha \in k$.

Example 1. $X = \mathbf{A}^n$. By Chap. I, 6.1, Theorem 3, any irreducible codimension 1 subvariety C is defined by one equation, $\mathfrak{A}_C = (F)$ with $F \in k[X]$. It follows that $C =$ div F, that is, every prime divisor, hence every divisor, is principal.

Example 2. $X = \mathbb{P}^n$. Any irreducible codimension 1 subvariety C is defined by a single homogeneous equation F, and moreover, if F has degree k then in the affine chart U_i, we have $\mathfrak{a}_C = (F/T_i^{-k})$. From this we get a method of constructing the divisor of a function $f \in k(\mathbb{P}^n)$ as follows: represent f as $f = F/G$ with F and G forms of the same degree, and factor F and G into irreducibles: $F = \prod H_i^{k_i}$ and $G = \prod L_j^{m_j}$. Then

$$\operatorname{div} f = \sum k_i C_i - \sum m_j D_j, \tag{4}$$

where C_i and D_j are the irreducible hypersurfaces defined by $H_i = 0$ and $L_j = 0$.

Write deg F for the degree of the form F. Since $\deg F = \deg G$ it follows that $\sum k_i \deg H_i = \sum m_j \deg G_j$. Define the *degree* of a divisor $D = \sum k_i C_i$ as the integer $\deg D = \sum k_i \deg H_i$. We have proved that if D is a principal divisor then $\deg D = 0$. The converse is also easy to prove: if $\sum k_i \deg C_i = 0$ and C_i is defined by a form H_i then $f = \prod H_i^{k_i}$ is homogeneous of degree 0 and div $f = \sum k_i C_i$.

Example 3. $X = \mathbb{P}^{n_1} \times \cdots \times \mathbb{P}^{n_k}$. This case is treated similarly. A codimension 1 subvariety C is again given by one equation $H = 0$ by Chap. I, 6.1, Theorem 3′, although now H is homogeneous separately in each of the k sets of coordinates of $\mathbb{P}^{n_i}$, and correspondingly has k different degrees $\deg_i H$ for $i = 1, \ldots, k$. In the same way as in Example 2 one can introduce the degrees $\deg_i D$ of a divisor D on X and prove that a divisor D is principal if and only if $\deg_i D = 0$ for $i = 1, \ldots, k$.

Principal divisors form a subgroup $P(X)$ of the group $\operatorname{Div} X$ of all divisors. The quotient group $\operatorname{Div} X/P(X)$ is called the *divisor class group* of X, and is denoted by $\operatorname{Cl} X$. A coset of $\operatorname{Div} X/P(X)$ is called a *divisor class*. Divisors in the same coset of $\operatorname{Div} X/P(X)$ are said to be *linearly equivalent* $D_1 \sim D_2$ if $D_1 - D_2 = \operatorname{div} f$ for some nonzero $f \in k(X)$.

In the three examples just worked out, we have respectively

$$\operatorname{Cl}(\mathbb{A}^n) = 0, \quad \operatorname{Cl}(\mathbb{P}^n) = \mathbb{Z} \quad \text{and} \quad \operatorname{Cl}(\mathbb{P}^{n_1} \times \cdots \times \mathbb{P}^{n_k}) = \mathbb{Z}^k.$$

1.2. Locally Principal Divisors

Suppose that the variety X is nonsingular. In this case, for any prime divisor $C \subset X$ and any point $x \in X$ there exists an open set $U \ni x$ in which C is defined by a local equation π. If $D = \sum k_i C_i$ is any divisor, and each of the C_i is defined in U by a local equation π_i, then we have $D = \operatorname{div} f$ in U, where $f = \prod \pi_i^{k_i}$. Thus every point x has a neighbourhood in which D is principal. From among all such neighbourhoods we can choose a finite cover $X = \bigcup U_i$, and $D = \operatorname{div}(f_i)$ on U_i.

Obviously, the functions f_i cannot be chosen arbitrarily: f_i is not identically 0, and in $U_i \cap U_j$ the divisors $\operatorname{div}(f_i)$ and $\operatorname{div}(f_j)$ coincide. As we saw in 1.1, it follows from this that f_i/f_j is regular in $U_i \cap U_j$ and nowhere 0 there. We say that a system $\{f_i\}$ of functions corresponding to the open sets U_i of a cover $X = \bigcup U_i$ is *compatible* if the f_i satisfy these conditions, that is, f_i/f_j is regular in $U_i \cap U_j$ and nowhere 0 there.

Conversely, any compatible system of functions defines a divisor on X. Indeed, for any prime divisor C we set $k_C = v_C(f_i)$ if $U_i \cap C \neq \emptyset$, where f_i and C are considered as a function and a prime divisor for the variety U_i. From the compatibility of the system of functions it follows that this number is independent of the choice of U_i. Hence we can consider the divisor $D = \sum k_C C$. Obviously the given $\{f_i\}$ are then a compatible system corresponding to D.

Finally, it is easy to determine when a system of functions $\{f_i\}$ corresponding to the open sets of a cover $X = \bigcup U_i$ defines the same divisor as another system $\{g_j\}$ corresponding to the open sets of a cover $X = \bigcup V_j$. For this, a necessary and sufficient condition is that f_i/g_j should be regular in $U_i \cap V_j$ and nowhere 0 there. We leave the simple verification to the reader.

Specifying divisors in terms of compatible systems of functions allows us to study their behaviour under regular maps. Let $\varphi \colon X \to Y$ be a regular map of nonsingular irreducible varieties, and D a divisor on Y. Suppose that $\varphi(X) \not\subset \operatorname{Supp} D$. We prove that; under this restriction, one can define the pullback $\varphi^*(D)$ of a divisor D by analogy with the definition of the pullback of a regular function. First we determine when we can construct the pullback of a rational function f on Y, and when it will not be identically 0 on X. For this it is sufficient that there is at least one point $y \in \varphi(X)$ at which f is regular and $f(y) \neq 0$; these points then form a nonempty open set V, and f

is regular and nowhere 0 on V. Therefore $\varphi^*(f)$ defines a regular function on $\varphi^{-1}(V)$ that is not identically 0 (in fact, nowhere 0). Since $\varphi^{-1}(V)$ is open in X, $\varphi^*(f)$ defines a rational function on X. In terms of divisors, our condition on the map φ and the function f boil down to $\varphi(X) \not\subset \mathrm{Supp}(\mathrm{div}\, f)$.

Now suppose that D is given by a compatible system of functions $\{f_i\}$ with respect to a cover $X = \bigcup U_i$. We consider the U_i with $\varphi(X) \cap U_i \neq \emptyset$, and prove that $\varphi(X) \cap U_i \not\subset \mathrm{Supp}(\mathrm{div}\, f_i)$. Indeed, the irreducibility of X implies that $\overline{\varphi(X)}$ is irreducible in Y. If we assume that $\varphi(X) \cap U_i \subset \mathrm{Supp}(\mathrm{div}\, f_i)$ then since $\overline{\varphi(X)}$ is irreducible and $\varphi(X) \cap U_i \neq \emptyset$, it would follow that $\overline{\varphi(X)} \subset \mathrm{Supp}(\mathrm{div}\, f_i)$. Finally, from the fact that $\mathrm{Supp}(\mathrm{div}\, f_i) \cap U_i = \mathrm{Supp}\, D \cap U_i$, the irreducibility of $\overline{\varphi(X)}$ and the fact that $\varphi(X) \cap U_i$ is nonempty, it would follow that $\varphi(X) \subset \mathrm{Supp}\, D$, contradicting our assumption.

Therefore, for every i such that $\varphi(X) \cap U_i \neq \emptyset$, the rational function $\varphi^*(f_i)$ is defined on $V_i = \varphi^{-1}(U_i)$. Then $X = \bigcup V_i$ is an open cover of X, with respect to which $\{\varphi^*(f_i)\}$ is a compatible system of functions defining a divisor on X. This divisor is obviously unchanged if we define D using a different system of functions. The divisor obtained in this way is called the *pullback* or *inverse image* of D and denoted by $\varphi^*(D)$.

Example. Suppose that X and Y are two curves, and $f: X \to Y$ a map taking X to a point $a \in Y$. If $a \neq b \in Y$ and $D = b$ is the divisor consisting of b with multiplicity 1, then 1 is a local equation of D in a neighbourhood of a, so that $f^*(D) = 0$.

In particular if $\varphi(X)$ is dense in Y then the pullback of any divisor $D \in \mathrm{Div}\, Y$ is defined.

If D and D' are two divisors on Y defined by systems of functions $\{f_i\}$ and $\{g_j\}$ with respect to covers $X = \bigcup U_i$ and $X = \bigcup V_j$ then the divisor $D + D'$ is defined by the system of functions $\{f_i g_j\}$ with respect to the cover $X = \bigcup(U_i \cap V_j)$. It follows at once that $\varphi^*(D + D') = \varphi^*(D) + \varphi^*(D')$, so that if $\varphi(X)$ is dense in Y, the pullback φ^* defines a homomorphism

$$\varphi^*: \mathrm{Div}\, Y \to \mathrm{Div}\, X.$$

The principal divisor $\mathrm{div}\, f$ is given by the system of functions $f_i = f$, and hence $\varphi^*(\mathrm{div}\, f) = \mathrm{div}(\varphi^*(f))$. Therefore φ^* maps $P(Y)$ to $P(X)$, and so defines a homomorphism $\varphi^*: \mathrm{Cl}\, Y \to \mathrm{Cl}\, X$.

As an application of the idea of a divisor defined by a compatible system of functions, we show how one can associate a divisor not with a function, but with a form in the coordinates on a nonsingular projective variety. Suppose that $X \subset \mathbb{P}^N$ and let F be a form in the coordinates of $\mathbb{P}^N$ that is not identically 0 on X. For any $x \in X$, consider a form G of the same degree $d = \deg F$, but with $G(x) \neq 0$; such forms exist, of course: if, say, $x = (\alpha_0 : \cdots : \alpha_N)$ and $\alpha_i \neq 0$ then we can take $G = T_i^d$. Then $f = F/G$ is a rational function on X and is regular on the open set where $G \neq 0$.

It is easy to see that there exist forms G_i such that the open sets $U_i = X \setminus X_{G_i}$ form a cover of X. One checks just as easily that the functions $f_i = F/G_i$ form a compatible system of functions with respect to the open sets U_i, and therefore define a divisor on X. A different choice of the forms G_i does not change this divisor, which therefore depends only on F. It is called the *divisor* of F, and denoted by $\operatorname{div} F$. Since the functions f_i are regular in the U_i, it follows that $\operatorname{div} F \geq 0$. If F_1 is another form with $\deg F_1 = \deg F$ then $\operatorname{div} F - \operatorname{div} F_1 = \operatorname{div}(F/F_1)$ is the divisor of the rational function F/F_1. Therefore $\deg F_1 = \deg F$ implies that $\operatorname{div} F \sim \operatorname{div} F_1$.

In particular, all the divisors $\operatorname{div} L$, where L is a linear form, are linearly equivalent. Obviously $\operatorname{Supp}(\operatorname{div} L) = X_L$ is the section of X by the hyperplane $L = 0$. These divisors are thus called *hyperplane section divisors* of X.

Taking $F_1 = L^d$ as the form in the above argument, where $d = \deg F$, we get $\operatorname{div} F \sim d \operatorname{div} L$, where $\operatorname{div} L$ is a hyperplane section divisor.

All the arguments concerned with using compatible systems of functions to specify divisors generalise to arbitrary, possibly singular, varieties. However, for this, we must take the specification by compatible systems of functions as the definition of divisor. The object we get is called a *locally principal divisor*. More precisely, we have the following definition.

Definition. A *locally principal divisor* or *Cartier divisor* on an irreducible variety X is a system of rational functions $\{f_i\}$ corresponding to the open sets U_i of a cover $X = \bigcup U_i$ satisfying the conditions: (1) the f_i are not identically 0; (2) f_i/f_j and f_j/f_i are both regular on $U_i \cap U_j$. Here functions $\{f_i\}$ and open sets U_i define the same divisor as functions $\{g_j\}$ and open sets V_j if f_i/g_j and g_j/f_i are regular on $U_i \cap V_j$.

Every function $f \in k(X)$ defines a locally principal divisor $\operatorname{div} f$ if we set $f_i = f$. Divisors of this form are said to be *principal*.

The *product* of the two locally principal divisors defined by functions $\{f_i\}$ corresponding to open sets U_i and functions $\{g_j\}$ corresponding to open sets V_j is the divisor defined by functions $\{f_i g_j\}$ and open sets $U_i \cap V_j$. All locally principal divisors form a group, and principal divisors a subgroup. The quotient group is called the *Picard group* of X, and denoted by $\operatorname{Pic} X$.

Any locally principal divisor has a *support*. This is the closed subset which in U_i consists of points at which f_i is either not regular, or equal to 0. Just as for divisors on nonsingular varieties, one can define the pullback of a locally principal divisor D on Y under a regular map $\varphi \colon X \to Y$ if $\varphi(X)$ is not contained in $\operatorname{Supp} D$.

We note an important special case. If X is a nonsingular variety and Y a possibly singular subvariety of X, then any divisor D on X with $\operatorname{Supp} D \not\supset Y$ defines a locally principal divisor $\widetilde{D}$ on Y. For this, we need to consider the inclusion map $\varphi \colon Y \hookrightarrow X$ and set $\widetilde{D} = \varphi^*(D)$. We call $\widetilde{D}$ the *restriction* of D to Y, and denote it by $\rho_Y(D)$. From the definition it follows that for a principal divisor div f we have $\rho_Y(\mathrm{div}\, f) = \mathrm{div}(\tilde{f})$, where $\tilde{f}$ is the restriction of f to Y.

Of course, the distinction between divisors and locally principal divisors, and between the groups $\mathrm{Cl}\, X$ and $\mathrm{Pic}\, X$, occurs only for singular varieties.

1.3. Moving the Support of a Divisor away from a Point

Theorem 1. *For any divisor D on a nonsingular variety X, and any finite number of points $x_1, \ldots, x_m \in X$, there exists a divisor D' with $D' \sim D$ such that $x_i \notin \mathrm{Supp}(D')$ for $i = 1, \ldots, m$.*

Proof. We can assume that D is a prime divisor, since otherwise we need only apply the assertion to each component separately. Choose an open affine subset of X containing $x_1, \ldots, x_m$; it is enough to prove the theorem for this, so that we can assume that X is affine. By induction, we can assume that $x_1, \ldots, x_{m-1} \notin \mathrm{Supp}\, D$ but $x_m \in \mathrm{Supp}\, D$, and it is enough to find a divisor D' such that $D' \sim D$ and $x_1, \ldots, x_m \notin \mathrm{Supp}(D')$.

Consider some local equation π' of the prime divisor D in a neighbourhood of x_m. We prove that we can choose a local equation π for D with $\pi \in k[X]$ (by assumption X is affine). Indeed, π' is regular at x_m, so that, if π' has divisor of poles $\mathrm{div}_\infty(\pi') = \sum k_l F_l$, then $x_m \notin F_l$. Thus for each l there exists a function $f_l \in k[X]$ that vanishes along F_l and such that $f_l(x_m) \neq 0$. Then the function $\pi = \pi' \prod f_l^{k_l}$ is obviously regular on X and is a local equation of D in a neighbourhood of x_m.

For each $i = 1, \ldots, m-1$, since $x_i \notin D \cup \{x_1, \ldots, x_{i-1}, x_{i+1}, \ldots, x_m\}$ by assumption, there exists a function $g_i \in k[X]$ such that $g_i(x_i) \neq 0$, but $g_i = 0$ on that set. Now adjust the constants $\alpha_i \in k$ such that the function

$$f = \pi + \sum_{i=1}^{m-1} \alpha_i g_i^2, \qquad \text{satisfies} \qquad f(x_i) \neq 0 \text{ for } i = 1, \ldots, m-1. \qquad (1)$$

For this it is sufficient to take $\alpha_i \neq -\pi(x_i)/(g_i(x_i)^2)$. We claim that $D' = D - \mathrm{div}\, f$ satisfies the conclusions of the theorem. First, (1) shows that $x_i \notin \mathrm{div}\, f$, and hence $x_i \notin \mathrm{Supp}(D')$ for $i = 1, \ldots, m-1$.

Now since by construction the g_i vanish on D, we get $\pi \mid g_i$ in the local ring $\mathcal{O}_{x_m}$, so that $\sum \alpha_i g_i^2 = \pi^2 h$ with $h \in \mathcal{O}_{x_m}$, and therefore $f = \pi(1 + \pi h)$. Since $(1 + \pi h)(x_m) = 1$, it follows that f is a local equation of D in a neighbourhood of x_m. Therefore $\mathrm{div}\, f = D + \sum r_s D_s$, with no prime divisor D_s passing through x_m. This means that $x_m \notin \mathrm{Supp}(D')$. The theorem is proved.

The same holds for a locally principal divisor on a singular X (the proof is very similar).

Here is a first application of Theorem 1. In 1.2 we defined the pullback $f^*(D)$ of a divisor D on a variety X by a regular map $f : Y \to X$ under the assumption that $f(Y) \not\subset \operatorname{Supp} D$. Theorem 1 allows us to replace D by a linearly equivalent divisor D' so that $\operatorname{Supp}(D') \not\ni x$, where x is an arbitrarily chosen point of $f(Y)$. Then automatically $f(Y) \not\subset \operatorname{Supp}(D')$, so that the pullback $f^*(D')$ is defined. This shows that we can define the pullback of a divisor class $C \in \operatorname{Cl} X$ without any restriction on f. For this, we must choose a divisor D in the class C such that $f(Y) \not\subset \operatorname{Supp} D$ and consider the divisor class on Y containing the divisor $f^*(D)$. One checks easily that we thus obtain a homomorphism

$$f^*\colon \operatorname{Cl} X \to \operatorname{Cl} Y.$$

In other words, $\operatorname{Cl} X$ is a functor from the category of irreducible nonsingular algebraic varieties to the category of Abelian groups.

Example. Let $f\colon X \to Y$ be the constant map $f(X) = a \in Y$ (see 1.2, Example). Then by Theorem 1, the divisor a is linearly equivalent to $\sum r_i b_i$ with $b_i \neq a$, and if C_a is the divisor class containing a then again $f^*(C_a) = 0$.

1.4. Divisors and Rational Maps

Associating divisors with functions is useful for studying rational maps of varieties to projective space. Let X be a nonsingular variety and $\varphi\colon X \to \mathbb{P}^n$ a rational map. We determine the points of X at which φ is not regular.

A rational map is defined by the formulas

$$\varphi = (f_0 : \cdots : f_n), \qquad \text{with } f_i \in k(X), \tag{1}$$

and we can assume that none of the f_i is identically 0 on X. Suppose that

$$\operatorname{div}(f_i) = \sum_{j=1}^m k_{ij} C_j,$$

with the C_j prime divisors; here we allow some of the k_{ij} to be 0.

To determine whether φ is regular at a point $x \in X$, write π_j for a local equation of C_j at x. Then

$$f_i = \left(\prod \pi_j^{k_{ij}} \right) u_i \qquad \text{with } u_i \in \mathcal{O}_x \text{ and } u_i(x) \neq 0.$$

Since $\mathcal{O}_x$ is a UFD, there exists a highest common factor d of the elements $f_0, \ldots, f_n$, that is, an element $d \in k(X)$ such that $f_i/d \in \mathcal{O}_x$, and if $d_1 \in k(X)$ is some element for which $f_i/d_1 \in \mathcal{O}_x$ then $d_1 \mid d$, that is, $d/d_1 \in \mathcal{O}_x$. Since the local equations π_j of prime divisors are prime elements of $\mathcal{O}_x$, we have

$$d = \prod \pi_j^{l_j}, \qquad \text{where } l_j = \min_{0 \leq i \leq n} k_{ij}.$$

Now φ is regular at x if there exists a function $g \in k(X)$ such that $f_i/g \in \mathcal{O}_x$ for all $i = 0, \ldots, n$, and not all the $(f_i/g)(x)$ are 0 at x. By definition of the highest common factor d it follows that $g \mid d$. If $d = gh$ with $h \in \mathcal{O}_x$ and $h(x) = 0$ then $h \mid (f_i/g)$, and hence all the $(f_i/g)(x) = 0$. Thus the required conditions can only be satisfied if $d = gh$ with $h(x) \neq 0$. Then $f_i/g = (f_i/d)h$, that is

$$f_i/g = \left(\prod_j \pi_j^{k_{ij}-l_j}\right) u_i h,$$

and φ is regular at x if and only if not all the functions $\prod_j \pi_j^{k_{ij}-l_j}$ are zero there.

To translate this answer into the language of divisors, we define quite generally the *highest common divisor* of given divisors $D_i = \sum k_{ij} C_j$ for $i = 1, \ldots, n$ to be the divisor

$$\mathrm{hcd}\{D_1, \ldots, D_n\} = \sum l_j C_j, \qquad \text{where } l_j = \min_{1 \leq i \leq n} k_{ij}.$$

Obviously $D'_i = D_i - \mathrm{hcd}\{D_1, \ldots, D_n\} \geq 0$, and the D'_i have no common components. We set in particular $D = \mathrm{hcd}\{\mathrm{div}(f_0), \ldots, \mathrm{div}(f_n)\}$ and $D'_i = \mathrm{div}(f_i) - D$.

Then in some neighbourhood of x we have

$$\mathrm{div}\left(\prod_j \pi_j^{k_{ij}-l_j}\right) = D'_i,$$

and we can say that φ is regular at x if and only if not all the subvarieties $\mathrm{Supp}(D'_i)$ pass through x.

We have proved the following result.

Theorem 2. *The rational map (1) fails to be regular precisely at the points of $\bigcap \mathrm{Supp}(D'_i)$, where $D'_i = \mathrm{div}(f_i) - \mathrm{hcf}\{\mathrm{div}(f_0), \ldots, \mathrm{div}(f_n)\} \geq 0$ for $i = 0, \ldots, n$.* $\square$

Since the D'_i have no common irreducible components, $\bigcap \mathrm{Supp}(D'_i)$ is a subvariety of codimension ≥ 2. Thus Theorem 2 is a more precise version of Chap. II, 3.1, Theorem 3.

Remark. The divisors D'_i can be interpreted as the pullbacks of the hyperplanes $x_i = 0$ under the map $\varphi: X \to \mathbb{P}^n$. Indeed, if $x \notin \bigcap \mathrm{Supp}\, D'_i$ and $D = \mathrm{div}\, h$ in a neighbourhood U of x, then in U the regular map is defined by

$$\varphi = \left(\frac{f_0}{h} : \cdots : \frac{f_n}{h}\right).$$

The pullback of the hyperplane x_i has local equation f_i/h, that is, it is equal to D'_i.

More generally, if $\lambda = (\lambda_0 : \cdots : \lambda_n)$ and $E_\lambda \subset \mathbb{P}^n$ is the hyperplane $\sum \lambda_i x_i = 0$, then

$$\varphi^*(E_\lambda) = \mathrm{div}\left(\sum \lambda_i f_i\right) - D.$$

1.5. The Linear System of a Divisor

The fact that all the polynomials $f(t)$ of degree $\leq n$ form a finite dimensional vector space has the following interpretation in terms of divisors. Write x_∞ for the point at infinity on the projective line $\mathbb{P}^1$ with coordinate t. A polynomial in t of degree k has pole of order k at x_∞, and no other poles. Hence the condition $\deg f \leq n$ can be expressed as the statement that $\mathrm{div}\, f + nx_\infty$ is effective.

In the same way, for an arbitrary divisor D on a nonsingular variety X, we consider the set consisting of 0 together with the nonzero functions $f \in k(X)$ such that

$$\mathrm{div}\, f + D \geq 0. \tag{1}$$

This set is a vector space over k under the usual algebraic operations on functions. Indeed, if $D = \sum n_i C_i$ then (1) is equivalent to

$$v_{C_i}(f) \geq -n_i \quad \text{and} \quad v_C(f) \geq 0 \text{ for } C \neq C_i,$$

and because of this, our assertion follows at once from 1.1, (2).

The space of functions satisfying (1) is called the *associated vector space* of D, or the *Riemann–Roch space* of D, and denoted by $\mathcal{L}(D)$ or $\mathcal{L}(X, D)$.

The analogue of the finite dimensionality of the vector space of polynomials of degree $\leq n$ is the fact that $\mathcal{L}(D)$ is finite dimensional if X is a projective variety and D any divisor. We prove this theorem in 2.3, Theorem 5 for the case of algebraic curves. The proof in the general case can be deduced from this without special difficulty using induction of the dimension. However, the status of the theorem becomes clearer if it is obtained as a particular case of a much more general assertion on coherent sheaves; we prove it in this form in Chap. VI, 3.4, Corollary 1.

The dimension of $\mathcal{L}(D)$ is also called the *dimension* of D, and denoted by $\ell(D)$.

Theorem 3. *Linearly equivalent divisors have the same dimension.*

Proof. Suppose that $D_1 \sim D_2$. This means that $D_1 - D_2 = \mathrm{div}\, g$, with $g \in k(X)$. If $f \in \mathcal{L}(D_1)$ then $\mathrm{div}\, f + D_1 \geq 0$. It follows that $\mathrm{div}(fg) + D_2 = \mathrm{div}\, f + D_1 \geq 0$, that is, $fg \in \mathcal{L}(D_2)$, so that $g \cdot \mathcal{L}(D_1) = \mathcal{L}(D_2)$. Thus multiplying functions $f \in \mathcal{L}(D_1)$ by g defines an isomorphism of the vector spaces $\mathcal{L}(D_1)$ and $\mathcal{L}(D_2)$, and the theorem follows.

Thus we see that it makes sense to speak of the dimension $\ell(C)$ of a divisor class C, that is, the common dimension of all the divisors of this class. This number has the following meaning. If $D \in C$ and $f \in \mathcal{L}(D)$ then the divisor $D_f = \operatorname{div} f + D$ is effective. Obviously, since $D_f \sim D$ also $D_f \in C$. Conversely, any effective divisor $D' \in C$ is of the form D_f, for $f \in \mathcal{L}(D)$. Obviously, if X is projective, f is uniquely determined by D_f up to a constant factor. Thus we can set up a one-to-one correspondence between effective divisors in the class C and points of the $(\ell(C) - 1)$-dimensional projective space $\mathbb{P}(\mathcal{L}(D))$ corresponding to a divisor D (recall that the projective space $\mathbb{P}(L)$ of a vector space L consists of all the 1-dimensional vector subspaces of L).

The space $\mathcal{L}(D)$ is useful when specifying rational maps in terms of divisors, as described in 1.4. If

$$\varphi = (f_0 : \cdots : f_n) : X \to \mathbb{P}^n \tag{2}$$

is a rational map, and, in the notation of 1.4,

$$D = \operatorname{hcd}\{\operatorname{div}(f_0), \ldots, \operatorname{div}(f_n)\} \quad \text{with } D_i = \operatorname{div}(f_i) - D, \tag{3}$$

then $D_i \geq 0$ and hence all the $f_i \in \mathcal{L}(-D)$.

The choice of the functions f_i depended on the choice of the projective coordinate system in $\mathbb{P}^n$. Thus in an invariant way, φ corresponds to the set of all functions $\sum_{i=0}^{n} \lambda_i f_i$ that are linear combinations of the functions f_i. These functions form a vector subspace $M \subset \mathcal{L}(-D)$. From now on we assume that $\varphi(X)$ is not contained in any proper linear subspace of $\mathbb{P}^n$. Then $\sum \lambda_i f_i \neq 0$ on X, provided that not all the $\lambda_i = 0$. The set of effective divisors that correspond to these functions, that is, the divisors $\operatorname{div} g - D$ with $g \in M$, is called a *linear system* of divisors. If $M = \mathcal{L}(-D)$ then we have a *complete linear system*. The meaning of the divisors $\operatorname{div} f - D$ for $f \in M$ is very simple: they are the pullbacks of the hyperplane divisors of $\mathbb{P}^n$ under φ. In this way we can construct all rational maps of a given nonsingular variety X into different projective spaces. For this, we need to take an arbitrary divisor D, and a finite dimensional subspace $M \subset \mathcal{L}(-D)$. If $f_0, \ldots, f_n$ is a basis of M then (2) gives the required map. Note that the divisors $D_i \in \mathcal{L}(-D)$ have an additional property: they have no common components.

Since multiplying all the f_i through by a common factor $g \in k(X)$ does not change the map φ, and replaces the divisor D by the linearly equivalent divisor $\operatorname{div} g + D$, the class of the divisor D is an invariant of a rational map. Thus we have the following method of constructing all rational maps $\varphi \colon X \to \mathbb{P}^m$ such that $\varphi(X)$ is not contained in any proper subspace of $\mathbb{P}^m$: take an arbitrary divisor class on X, and for any divisor D in this class, a finite dimensional vector subspace $M \subset \mathcal{L}(-D)$ such that the effective divisors $\operatorname{div} f - D$ for $f \in M$ have no common components. If $f_0, \ldots, f_n$ is a basis of M then our map is given by (2). Of course, it can happen that $\mathcal{L}(-D) = 0$, or that all the divisors $\operatorname{div} f - D$ for $f \in \mathcal{L}(-D) = 0$ have common components, and then this divisor class does not lead to any map.

We observe one interesting feature of the picture we obtain. Among all the rational maps corresponding to a divisor class C, there is a maximal one: that obtained by taking M to be the whole space $M = \mathcal{L}(-D)$ with $D \in C$. (Here we take on trust the so far unproved theorem that $\mathcal{L}(-D)$ is finite dimensional.) All other maps corresponding to this class are obtained by composing this map $X \to \mathbb{P}^N$ with the various projection maps $\mathbb{P}^N \to \mathbb{P}^n$. Indeed, if $\varphi = (f_0 : \cdots : f_N)$, and, say, $\psi = (f_0 : \cdots : f_n)$ with $n < N$ then $\psi = \pi \circ \varphi$, where $\pi(x_0 : \cdots : x_N) = (x_0 : \cdots : x_n)$ is the projection, viewed as a rational map.

Let's see how this scheme of things works if we take X to be projective space $\mathbb{P}^m$. We know that $\mathrm{Cl}(\mathbb{P}^m) \cong \mathbb{Z}$, and the class C_k corresponding to an integer k consists of hypersurfaces of degree k. If $k > 0$ and $D \in C_k$ then obviously $\mathcal{L}(-D) = 0$. If $k \leq 0$ then we can take $-D$ to be the divisor kE, where E is the divisor of the hyperplane at infinity $x_0 = 0$. In this case $\mathcal{L}(kE)$ consists of polynomials of degree $\leq k$ in the inhomogeneous coordinates $x_1/x_0, \ldots, x_m/x_0$ (see Ex. 15). If we multiply the formula for the resulting map through by x_0^k we get the Veronese embedding $v_k \colon \mathbb{P}^m \hookrightarrow \mathbb{P}^N$ where $N = \nu_{k,m} = \binom{k+m}{m} - 1$ (see Chap. I, 4.4, Example 2). Thus we see that any rational map from $\mathbb{P}^m$ is obtained by composing the Veronese map with a projection.

Example. Suppose that $X \subset \mathbb{P}^{n+1}$ is an irreducible n-dimensional hypersurface defined by $F = 0$, with $\deg F = k$. We find the space $\mathcal{L}(D)$, where $D = \mathrm{div}\, H$, with H a form of degree m. Since $\mathrm{div}\, H \sim mE$, where $E = \mathrm{div}(x_0)$ is the hyperplane section divisor, we can assume that $D = mE$. Obviously if Φ is any form of degree m then $\Phi/x_0^m \in \mathcal{L}(mE)$. We prove that these functions exhaust $\mathcal{L}(mE)$.

If $\varphi \in \mathcal{L}(mE)$ then $\varphi \in k[U_0]$, where $U_0 \subset X$ is the affine open set given by $x_0 \neq 0$. Let $y_i = x_i/x_0$ for $i = 1, \ldots, n+1$ be inhomogeneous coordinates. We see that $\varphi = P(y_1, \ldots, y_{n+1})$, where P is a polynomial, which can be altered by adding multiples of the defining equation $F_0 = F/x_0^k$ of the hypersurface $U_0 \subset \mathbb{A}^{n+1}$. Our claim is that after adding such a multiple we get a polynomial P of degree $\deg P \leq m$.

By contradiction, suppose that $\deg P = l > m$, and that the degree of P cannot be reduced by adding a multiple of F_0. We choose the coordinate system in such a way that the intersection of X with $x_0 = x_1 = 0$ has dimension $n - 2$. This means that if f_k is the homogeneous component of $F_0(y_1, \ldots, y_{n+1})$ of top degree then f_k is not divisible by y_1.

We pass to the open subset $U_1 \subset X$ with $x_1 \neq 0$, and set $z_1 = x_0/x_1 = 1/y_1$ and $z_i = x_i/x_1 = y_i/y_1$ for $i > 1$. Then $y_1 = 1/z_1$, $y_i = z_i/z_1$ for $i > 1$, and $\varphi = P(y_1, \ldots, y_{n+1}) = z_1^{-l}\widetilde{P}(z_1, \ldots, z_{n+1})$, where $\widetilde{P}$ is a polynomial of degree l. By assumption, $m\,\mathrm{div}\, z_1 + \mathrm{div}\,\varphi > 0$ in U_1, that is, $z_1^m \varphi \in k[U_1]$, or $z_1^{m-l}\widetilde{P} = Q(z_1, \ldots, z_{n+1})$ on U_1, where Q is a polynomial. Let $\deg Q = r$. By assumption, $z_1^{m-l}\widetilde{P} = Q + AF_1$, where $F_1 = F/x_1^k$ is the equation of U_1, and A is a rational function whose denominator does not have F_1 as a factor. Returning to U_0, we get

$$y_1^{-m}P = y_1^{-r}\widetilde{Q} + BF_0, \tag{4}$$

where $\widetilde{Q}(y_1, \ldots, y_{n+1})$ is a polynomial of degree r, and the denominator of B does not have F_0 as a factor. If $m \geq r$ then multiplying (4) by y_1^m gives $P - CF_0 = y_1^{m-r}\widetilde{Q}$, where now C is a polynomial. Since $\deg(y_1^{m-r}\widetilde{Q}) = m < l$, this contradicts the assumption that the degree of P cannot be reduced. If $r \geq m$ then similarly, we get $y_1^{r-m}P - CF_0 = \widetilde{Q}$. Write p_l, q_r, f_k and c for the homogeneous components of top degree in P, $\widetilde{Q}$, F_0 and C. Since $\deg(y_1^{r-m}P) = l + r - m > \deg\widetilde{Q}$, we have $y_1^{r-m}p_l = cf_k$. By the choice of coordinates, f_k is not divisible by y_1, and hence p_l is divisible by f_k, say $p_l = g_{l-k}f_k$. Then $\deg(P - g_{l-m}F_0) < l$, which again contradicts the assumption on P.

This proves the following result.

Proposition. *Let $X \subset \mathbb{P}^{n+1}$ be an irreducible hypersurface defined by $F = 0$, with $\deg F = k$. Then $\mathcal{L}(X, mE)$ is the vector space of forms of degree m, modulo the subspace of multiples of F by forms of degree $m - k$. Therefore $\ell(mE) = \binom{n+1}{m}$ if $m < k$ or $\binom{n+1}{m} - \binom{n+1}{m-k}$ if $m \geq k$.* $\square$

1.6. Pencil of Conics over $\mathbb{P}^1$

We conclude this section with an example that is very pretty, and will be useful later. Let X be a nonsingular projective surface and $\varphi \colon X \to \mathbb{P}^1$ a regular map. Suppose that the point $\infty \in \mathbb{P}^1$ is chosen so that the inverse image $\varphi^{-1}(\infty)$ is nonsingular, $\mathbb{P}^1 \setminus \infty = \mathbb{A}^1$, and the map $\varphi^{-1}(\mathbb{A}^1) \to \mathbb{A}^1$ defines a pencil of conics as in Chap. I, 6.2, Example 1 and Chap. II, 6.4, Example 1. In this situation, X, together with its map $\varphi \colon X \to \mathbb{P}^1$, is called a *pencil of conics* over $\mathbb{P}^1$. The open set $\varphi^{-1}(\mathbb{A}^1)$ is defined in $\mathbb{P}^2 \times \mathbb{A}^1$ by the equation

$$\sum_{i,j=0}^{2} a_{ij}(t)\xi_i\xi_j = 0, \tag{1}$$

where t is a coordinate on $\mathbb{A}^1$. In Chap. II, 6.4, Proposition 1, we saw that the singular fibres of φ correspond to the roots $t = \alpha_1, \ldots, \alpha_m$ of the discriminant $\Delta(t) = \det|a_{ij}(t)|$, that these roots are simple and that the corresponding singular fibres $F_1, \ldots, F_m$ are of the form $F_i = L_i + L_i'$, where L_i and L_i' are distinct lines.

Since $\Delta(t)$ has only simple roots, it is not identically 0, and the conic (1) is nondegenerate. In Chap. I, 6.2, Proposition, Corollary 4, we saw that φ has a section $s \colon \mathbb{A}^1 \to \varphi^{-1}(\mathbb{A}^1)$, a rational map such that $s(\alpha)$ is contained in the fibres $\varphi^{-1}(\alpha)$ for each $\alpha \in \mathbb{A}^1$, that is, $\varphi \circ s = \text{id}$. This rational map extends from $\mathbb{A}^1$ to $\mathbb{P}^1$, and gives a regular map $s \colon \mathbb{P}^1 \to X$. Write S for the curve $s(\mathbb{P}^1)$. We choose some fixed nonsingular fibre F.

Theorem 4. *The divisor class group* $\operatorname{Cl} X$ *is a free Abelian group with* $m + 2$ *generators, the classes defined by* $L_1, \ldots, L_m, F$ *and* S.

Proof. Let C be a prime divisor on X. Then $C \subset X$ is an irreducible curve, and φ either maps it to a point $\gamma \in \mathbb{P}^1$ or onto the whole of $\mathbb{P}^1$. In the first case, C is contained in a fibre $\varphi^{-1}(\gamma)$.

Suppose that $\varphi(C) = \mathbb{P}^1$. Then the map $\varphi \colon C \to \mathbb{P}^1$ defines an inclusion $k(\mathbb{P}^1) \subset k(C)$ of the function fields, and a nonzero function $u \in k(\mathbb{P}^1)$ does not vanish on C; here we identify u and its pullback $\varphi^*(u) \in k(X)$. In other words,

$$v_C(u) = 0 \qquad \text{for any } 0 \ne u \in k(\mathbb{P}^1). \tag{2}$$

Hence v_C defines a function $v \colon \big(k(X) \setminus 0\big) \to \mathbb{Z}$ that satisfies (2) and is a valuation in the sense of 1.1, (2). Chap. I, 6.2, Proposition, Corollary 4 proves that the conic (1) is rational over the field $K = k(\mathbb{P}^1) = k(t)$, that is, $k(X) = K(T)$; the birational map $X \to \mathbb{P}^1_K$ uses the point of the conic (1) corresponding to a section s, and in particular, it can be chosen so that T has a pole of order 1 at this point. Thus v is a function on $K(T) \setminus 0$ that satisfies (2) and 1.1, (2). It is easy to determine all such functions. Suppose that $v(T) \ge 0$. Then it follows from (2) that $v(H) \ge 0$ for every $H \in K[T]$, and if v is not identically 0 then $v(H) > 0$ for some H. Therefore $v(P) > 0$ for some irreducible factor P of H. But then $v(Q) = 0$ for every irreducible polynomial in T not proportional to P: indeed, there exist polynomials $U, V \in K[T]$ such that $PU + QV = 1$; so if $v(P), v(Q) > 0$ it would follow that $v(1) > 0$, whereas $v(u) = 0$ for $u \in K$. It follows that $v(f) = v_P(f) = m$ is the exponent of f when written in the form $f = P^m g$, where P divides neither the numerator nor the denominator of g. In particular, for the divisor C we are considering, there exists an irreducible polynomial $P \in k[T]$ such that $v_P(f) = v_C(f)$, and the divisor C is uniquely determined by P. Hence $v_C(P) = 1$, and since P determines C uniquely, $\operatorname{div}_0 P$ does not contain any irreducible curve except for components of fibres:

$$\operatorname{div}_0 P = C + \sum G_i, \tag{3}$$

where G_i are conics of the pencil or their components.

If $v(T) < 0$ then we set $U = T^{-1}$ and find that v corresponds in the same way to the polynomial $U \in k[U] \subset k(T)$. In terms of $K[T]$, as we see easily, $v(F) = -\deg H$, where $H \in K[T]$, so that there is only one such function v. Since by assumption T has a pole at the point corresponding to the section s, we must have $v = v_S$. As before, $v_S(H) = -\deg H$, and S is the unique curve with this property, so that for any $H \in K[T]$, we have

$$\mathrm{div}_\infty H = (\deg H)S + \sum G'_j, \qquad (4)$$

where G'_i are conics of the pencil or their components. In particular, if P is the irreducible polynomial corresponding to a curve $C \neq S$, we have $\mathrm{div}_\infty P = (\deg P)S + \sum G'_j$ and

$$\mathrm{div}\, P = C - (\deg P)S + \sum G_i - \sum G'_j.$$

Hence $C \sim (\deg P)S + \sum r_l G''_l$, where G''_l are components of conics of the pencil. It remains to consider these. They can either be nondegenerate conics, that is, fibres $\varphi^*(\alpha)$ with $\alpha \in \mathbb{P}^1$. But since all points of $\mathbb{P}^1$ are linearly equivalent, all fibres of $X \to \mathbb{P}^1$ are also linearly equivalent, therefore linearly equivalent to the chosen fibre F, say. Or they can be components L_i or L'_i of reducible fibres. But since $L_i + L'_i \sim F_i \sim F$, we can express L'_i in terms of L_i and F. As a result, we see that every irreducible divisor is linearly equivalent to a linear combination of S, F and $L_1, \ldots, L_m$. Hence the divisor classes of these curves generate $\mathrm{Cl}\, X$.

It remains to check that the classes of S, F and $L_1, \ldots, L_m$ are linearly independent in $\mathrm{Cl}\, X$. Suppose that $nF + lS + \sum_{i=1}^{m} r_i L_i \sim 0$. We consider the restriction of this divisor to various nonsingular curves. It must again be linearly equivalent to 0. Consider the restriction to an irreducible fibre $F' \neq F$. Since $F \cap F' = \emptyset$, $L_i \cap F' = \emptyset$ and the restriction to S gives a point ξ, we must have $l\xi \sim 0$. This is only possible if $l = 0$. Considering the restriction to L'_i we get that $r_i = 0$. The relation that remains is $nF \sim 0$. If $n \neq 0$ then we can assume that $n > 0$. This is impossible: an effective divisor cannot be principal. The theorem is proved.

Exercises to §1

1. Determine the divisor of x/y on the quadric surface $xy - zt = 0$ in $\mathbb{P}^3$.

2. Determine the divisor of the function $x - 1$ on the circle $x_1^2 + x_2^2 = x_0^2$, where $x = x_1/x_0$.

3. Determine the pullback $f^*(D_a)$ where $f(x, y) = x$ is the projection of the circle $x^2 + y^2 = 1$ to the x-axis, and $D_a = a$ is the divisor on $\mathbb{A}^1$ consisting of the point with coordinate a with multiplicity 1.

4. Let X be a nonsingular projective curve and $f \in k[X]$. Viewing f as a regular function $f : X \to \mathbb{P}^1$, prove that $\mathrm{div}\, f = f^*(D)$, where D is the divisor $D = 0 - \infty$ on $\mathbb{P}^1$

5. Let X be a nonsingular affine variety. Prove that $\mathrm{Cl}\, X = 0$ if and only if the coordinate ring $k[X]$ is a UFD.

6. Suppose that $X \subset \mathbb{P}^N$ is a nonsingular projective variety. Let $k[S]$ be the polynomial ring in the homogeneous coordinates of $\mathbb{P}^N$ and $\mathfrak{A}_X \subset k[X]$ the ideal of X. Prove that if $k[S]/\mathfrak{A}_X$ is a UFD then $\operatorname{Cl} X \cong \mathbb{Z}$, and is generated by the class of a hyperplane section.

7. Find $\operatorname{Cl}(\mathbb{P}^n \times \mathbb{A}^m)$.

8. The projection $p \colon X \times \mathbb{A}^1 \to X$ defines a pullback homomorphism $p^* \colon \operatorname{Cl} X \to \operatorname{Cl}(X \times \mathbb{A}^1)$. Prove that p^* is surjective. [Hint: Use the map $q^* \colon \operatorname{Cl}(X \times \mathbb{A}^1) \to \operatorname{Cl} X$, where $q \colon X \to X \times \mathbb{A}^1$ is given by $q(x) = (x, 0)$.]

9. Prove that for any divisor D on $X \times \mathbb{A}^1$ there exists an open set $U \subset X$ such that D is a principal divisor on $U \times \mathbb{A}^1$. [Hint: You can suppose that X is affine, and that D is irreducible. Then it is defined by a prime ideal of $k[X \times \mathbb{A}^1] = k[X][T]$. Use the fact that every ideal in $k(X)[T]$ is principal, and then replace X by some principal affine open set.]

10. Prove that $\operatorname{Cl}(X \times \mathbb{A}^1) \cong \operatorname{Cl} X$. [Hint: Use the results of Ex. 8–9.]

11. Let X be the projective curve defined in affine coordinates by $y^2 = x^2 + x^3$. Prove that every locally principal divisor on X is equivalent to a divisor whose support does not contain the points $(0, 0)$. Using this, together with the normalisation map $\varphi \colon \mathbb{P}^1 \to X$, for which $\varphi^{-1}(0, 0)$ consists of two points $x_1, x_2 \in \mathbb{P}^1$, describe $\operatorname{Pic} X$ as D/P, where D is the group of all divisors on $\mathbb{P}^1$ whose support does not contain x_1, x_2, and P the group of principal divisors $\operatorname{div} f$ such that f is regular at x_1, x_2 and $f(x_1) = f(x_2) \neq 0$. Prove that $\operatorname{Pic} X$ is isomorphic to $\mathbb{Z} \times k^*$, where k^* is the multiplicative group of nonzero elements of k.

12. Determine $\operatorname{Pic} X$ where X is the projective curve $y^2 = x^3$.

13. Let X be a quadratic cone. Using the map $\varphi \colon \mathbb{A}^2 \to X$ described in Chap. II, §5, Ex. 2, determine the image $\varphi^*(\operatorname{Div} X) \subset \operatorname{Div}(\mathbb{A}^2)$. Prove that the principal divisor $D = \operatorname{div} F \in \operatorname{Div}(\mathbb{A}^2)$ is contained in $\varphi^*(\operatorname{Div} X)$ if and only if $F(-u, -v) = \pm F(u, v)$, that is, F is either an odd or an even function. Prove that the principal divisors on X correspond to even functions. Deduce that $\operatorname{Cl} X \cong \mathbb{Z}/2\mathbb{Z}$.

14. Using Theorem 2, determine the points at which the birational map $\varphi \colon X \to \mathbb{P}^2$ is not regular, where X is a surface of degree 2 in $\mathbb{P}^3$ and φ the projection from a point. The same for φ^{-1}.

15. Prove that if E is the hyperplane $x_0 = 0$ in $\mathbb{P}^n$ then the space $\mathcal{L}(kE)$ consists of polynomials of degree $\leq k$ in the inhomogeneous coordinates $x_1/x_0, \ldots, x_n/x_0$. [Hint: $f \in \mathcal{L}(kE)$ implies that $f \in k[\mathbb{A}_0^n]$.]

16. Prove that any automorphism of $\mathbb{P}^n$ takes hyperplane divisors to one another. [Hint: The class of a hyperplane is determined in $\operatorname{Cl}(\mathbb{P}^n)$ by intrinsic properties, and the hyperplane divisors are determined as the effective divisor in this class.]

17. Prove that any automorphism of $\mathbb{P}^n$ is a projective transformation. [Hint: Use the result of Ex. 16.]

18. Suppose that Y is nonsingular, and let $\sigma \colon X \to Y$ be a blowup with centre $y \in Y$. Prove that $\operatorname{Cl} X \cong \operatorname{Cl} Y \oplus \mathbb{Z}$.

2. Divisors on Curves

2.1. The Degree of a Divisor on a Curve

Consider a nonsingular projective curve X. A divisor on X is a linear combination $D = \sum k_i x_i$ of points x_i with coefficients $k_i \in \mathbb{Z}$. The *degree* of D is the number $\deg D = \sum k_i$.

The case $n = 1$ of 1.1, Example 2 shows that when $X = \mathbb{P}^1$, a divisor D is principal if and only if it has degree 0. We prove that the equality $\deg D = 0$ holds for a principal divisor on any nonsingular projective curve. For this we use the notion of the degree $\deg f$ of a map f introduced in Chap. II, 6.3.

Theorem 1. *If $f \colon X \to Y$ is a regular map between nonsingular projective curves and $f(X) = Y$ then $\deg f = \deg(f^*(y))$ for any point $y \in Y$.*

In Theorem 1, $f^*(y)$ is the divisor on X obtained as the pullback of the divisor on Y consisting of y with multiplicity 1. Thus $\deg f$ equals the number of inverse images of any point $y \in Y$, taken with the right multiplicities. This makes the intuitive meaning of the degree of a map f easier to understand: it counts how many times X covers Y under the map f.

Corollary. *The degree of a principal divisor on a nonsingular projective curve equals 0.*

Proof. Indeed, any nonconstant function $f \in k(X)$ defines a regular map $f \colon X \to \mathbb{P}^1$. Moreover, we have $f^*(0) = \mathrm{div}_0\, f$, where the right-hand side is the pullback of the point $0 \in \mathbb{P}^1$, as follows at once from the definition of the two divisors. Similarly, $f^*(\infty) = \mathrm{div}_\infty\, f$. By Theorem 1,

$$\deg(\mathrm{div}\, f) = \deg(\mathrm{div}_0\, f) - \deg(\mathrm{div}_\infty\, f)$$
$$= \deg(f^*(0)) - \deg(f^*(\infty)) = \deg f - \deg f = 0.$$

If X and Y are varieties of the same dimension then a regular map $f \colon X \to Y$ with $f(X)$ dense in Y defines an inclusion $f^* \colon k(Y) \hookrightarrow k(X)$. We use this in what follows to view $k(Y)$ as a subfield of $k(X)$. (That is, for $u \in k(Y)$ we write u instead of $f^*(u)$ when this does not cause confusion.)

Theorem 1 follows from two results. To state these, we introduce the following notation. Let $x_1, \ldots, x_r \in X$ be points of X, and set

$$\tilde{\mathcal{O}} = \bigcap_{i=1}^{r} \mathcal{O}_{x_i}. \tag{1}$$

Thus $\tilde{\mathcal{O}}$ consists of functions that are regular at all the points $x_1, \ldots, x_r$. If $\{x_1, \ldots, x_r\} = f^{-1}(y)$ for $y \in Y$ then the ring $\mathcal{O}_y$, viewed as a subring of $k(X)$ according to the convention just explained, is contained in $\tilde{\mathcal{O}}$.

Theorem 2. $\tilde{\mathcal{O}}$ *is a principal ideal domain with a finite number of prime ideals. There exist elements $t_i \in \tilde{\mathcal{O}}$ such that*

$$v_{x_i}(t_j) = \delta_{ij} \qquad for \ 1 \le i, j \le r \ (Kronecker \ delta). \tag{2}$$

If $u \in \tilde{\mathcal{O}}$ and $u \ne 0$ then

$$u = t_1^{k_1} \cdots t_r^{k_r} v, \tag{3}$$

where $k_i = v_{x_i}(u)$ and v is invertible in $\tilde{\mathcal{O}}$.

Theorem 3. *If $\{x_1, \ldots, x_r\} = f^{-1}(y)$ then $\tilde{\mathcal{O}}$ is a free $\mathcal{O}_y$-module of rank $n = \deg f$, that is, $\tilde{\mathcal{O}} \cong \mathcal{O}_y^{\oplus n}$.*

Proof of Theorem 2 + Theorem 3 $\implies$ Theorem 1. Let t be a local parameter on Y at y, and $\{x_1, \ldots, x_r\} = f^{-1}(y)$. By Theorem 2, $t = t_1^{k_1} \cdots t_r^{k_r} v$, where $k_i = v_{x_i}(t)$ and v is invertible in $\tilde{\mathcal{O}}$. Recalling the definition of the pullback of a divisor, we see that

$$f^*(y) = \sum_{i=1}^{r} k_i x_i \qquad and \qquad \deg f^*(y) = \sum_{i=1}^{r} k_i.$$

Since $t_1, \ldots, t_r$ are pairwise relatively prime in $\tilde{\mathcal{O}}$, it follows that

$$\tilde{\mathcal{O}}/(t) \cong \bigoplus_{i=1}^{r} \tilde{\mathcal{O}}/(t_i^{k_i}).$$

Fix attention on one of the summands $\mathcal{O}/(t_i^{k_i})$ in the direct sum. One sees easily that any element $w \in \tilde{\mathcal{O}}$ can be written in a unique way in the form

$$w \equiv \alpha_0 + \alpha_1 t_i + \cdots + \alpha_{k_i-1} t_i^{k_i - 1} \mod t_i^{k_i}, \tag{4}$$

with $\alpha_i \in k$. Indeed, if we already have an expression

$$w \equiv \alpha_0 + \alpha_1 t_i + \cdots + \alpha_{s-1} t_i^{s-1} \mod t_i^{s},$$

then

$$u = t_i^{-s}(w - \alpha_0 - \cdots - \alpha_{s-1} t_i^{s-1}) \in \tilde{\mathcal{O}} \subset \mathcal{O}_{x_i}.$$

Set $u(x_i) = \alpha_s \in k$. Then $v_{x_i}(u - \alpha_s) > 0$, and it follows from Theorem 2 that $u \equiv \alpha_s$ modulo t_i, that is,

$$w \equiv \alpha_0 + \alpha_1 t_i + \cdots + \alpha_{s-1} t_i^{s-1} + \alpha_s t_i^{s} \mod t_i^{s+1}.$$

This proves (4) by induction. It follows from (4) that $\dim \tilde{\mathcal{O}}/(t_i^{k_i}) = k_i$. Hence

$$\dim \tilde{\mathcal{O}}/(t) = \sum_{i=1}^{r} k_i. \tag{5}$$

Now apply Theorem 3. It follows from this that $\tilde{\mathcal{O}}/(t) \cong (\mathcal{O}_y/(t))^{\oplus n}$. But t is a local parameter at y, and hence

$$\mathcal{O}_y/(t) \cong k \qquad \text{and} \qquad \dim \tilde{\mathcal{O}}/(t) = n = \deg f. \tag{6}$$

The equalities (5) and (6) prove Theorem 1.

Proof of Theorem 2. Write u_i for a local parameter at x_i. Then x_i appears in the divisor $\mathrm{div}(u_i)$ with multiplicity 1, that is, $\mathrm{div}(u_i) = x_i + D$, where x_i does not appear in D. By 1.3, Theorem 1 we can move the support of D away from $x_1, \ldots, x_r$, that is, we can find a function f_i such that none of these points appear in $D + \mathrm{div}(f_i)$. This means that the relations (2) are satisfied by $t_i = u_i f_i$.

Let $u \in \tilde{\mathcal{O}}$. Set $v_{x_i}(u) = k_i$. By assumption, $k_i \geq 0$. The element $v = u t_1^{-k_1} \cdots t_r^{-k_r}$ satisfies $v_{x_i}(v) = 0$ for all $i = 1, \ldots, r$, from which it follows that $v \in \tilde{\mathcal{O}}$ and $v^{-1} \in \tilde{\mathcal{O}}$. This gives an expression (3) for u.

It remains to check that $\tilde{\mathcal{O}}$ is a principal ideal ring. Let $\mathfrak{a}$ be an ideal of $\tilde{\mathcal{O}}$. Set $k_i = \inf_{u \in \mathfrak{a}} v_{x_i}(u)$ and $a = t_1^{k_1} \cdots t_r^{k_r}$. Then $ua^{-1} \in \tilde{\mathcal{O}}$ for any $u \in \mathfrak{a}$, that is, $\mathfrak{a} \subset (a)$. We prove that $\mathfrak{a} = (a)$. For this we denote by $\mathfrak{a}'$ the set of functions ua^{-1} with $u \in \mathfrak{a}$. Obviously $\mathfrak{a}'$ is an ideal of $\tilde{\mathcal{O}}$ and also $\inf_{u \in \mathfrak{a}'} v_{x_i}(u) = 0$. Hence for any $i = 1, \ldots, r$, there exists $u_i \in \mathfrak{a}'$ such that $v_{x_i}(u_i) = 0$, that is $u_i(x_i) \neq 0$. An obvious verification shows that the element $c = \sum_{j=1}^{r} u_j t_1 \cdots \hat{t}_j \cdots t_r \in \mathfrak{a}'$ satisfies $v_{x_i}(c) = 0$ for $i = 1, \ldots, r$. This means that $c^{-1} \in \tilde{\mathcal{O}}$, and hence $\mathfrak{a}' = \tilde{\mathcal{O}}$, and $\mathfrak{a} = (a)$. Theorem 2 is proved.

We proceed to the proof of Theorem 3. We first prove that $\tilde{\mathcal{O}}$ is a finite $\mathcal{O}_y$-module. For this, recall that by Chap. II, 5.3, Theorem 8, the map f is finite. Therefore the result we need follows from the next lemma.

Lemma. *Let $f \colon X \to Y$ be a finite map of curves, with X nonsingular; for $y \in Y$, write $f^{-1}(y) = \{x_1, \ldots, x_r\}$ and $\tilde{\mathcal{O}} = \bigcap \mathcal{O}_{x_i}$. Then $\tilde{\mathcal{O}}$ is a finite $\mathcal{O}_y$-module.*

Proof. Since the assertion is local, we can assume that X and Y are affine. Let $k[X] = A$ and $k[Y] = B$. Then $B \subset A$ and A is a finite B-module. We prove that $\tilde{\mathcal{O}} = A\mathcal{O}_y$.

Indeed, if $\varphi \in \tilde{\mathcal{O}}$ and z_i are the poles of φ on U then $f(z_i) = y_i \neq y$. There exists a function $h \in B$ such that $h(y) \neq 0$ and $h(y_i) = 0$, and moreover $\varphi h \in \mathcal{O}_{z_i}$ and hence $\varphi h \in A$. Since $h^{-1} \in \mathcal{O}_y$, we get $\varphi \in A\mathcal{O}_y$; this proves that $\tilde{\mathcal{O}} \subset A\mathcal{O}_y$. The converse inclusion is obvious.

Obviously, generators of A over $B = k[Y]$ provide at the same time generators of $A\mathcal{O}_y$ over $\mathcal{O}_y$. Hence $\tilde{\mathcal{O}}$ is a finite $\mathcal{O}_y$-module. The lemma is proved.

Proof of Theorem 3. Now it is easy to complete the proof of Theorem 3. By the main theorem on modules over a principal ideal domain, $\tilde{\mathcal{O}}$ is the direct sum of a free module and a torsion module. However, both $\mathcal{O}_y$ and $\tilde{\mathcal{O}}$ are contained in the field $k(X)$, so that it follows that the torsion module is 0, and $\tilde{\mathcal{O}} \cong \mathcal{O}_y^{\oplus m}$ for some m.

It remains to determine m, that is, the rank of $\tilde{\mathcal{O}}$ over $\mathcal{O}_y$. It equals the maximal number of elements of $\tilde{\mathcal{O}}$ that are linearly independent over $\mathcal{O}_y$. Since linear independence over a ring and over its field of fractions is the same thing, and the field of fractions of $\mathcal{O}_y$ is $k(Y)$, our m equals the maximal number of elements of $\tilde{\mathcal{O}}$ that are linearly independent over $k(Y)$.

By assumption $[k(X) : k(Y)] = n$, so that obviously $m \le n$. It remains to prove that $\tilde{\mathcal{O}}$ contains n elements that are linearly independent over $k(Y)$. Suppose that $\alpha_1, \ldots, \alpha_n$ is a basis of the field extension $k(Y) \subset k(X)$. Let t be a local parameter on Y at y, and write k for the maximum order of poles of the α_i at the points x_j. Then obviously the functions $\alpha_i t^k$ are regular at these points, and are hence contained in $\tilde{\mathcal{O}}$. Hence they are linearly independent over $k[Y]$. Theorem 3 is proved.

It follows from Theorem 1, Corollary that on a nonsingular projective curve X, linearly independent divisor have the same degree. Hence it makes sense to talk about the degree of a divisor class. Thus we have a homomorphism

$$\deg \colon \operatorname{Cl} X \to \mathbb{Z},$$

whose image is the whole of $\mathbb{Z}$, and whose kernel consists of divisor classes of degree 0, and is denoted by $\operatorname{Cl}^0 X$. The role of this group is clear already from the following result.

Theorem 4. *A nonsingular projective curve X is rational if and only if $\operatorname{Cl}^0 X = 0$.*

Proof. Indeed, if $X \cong \mathbb{P}^1$ then we are in the case $n = 1$ of 1.1, Example 2. We saw there that $\operatorname{Cl}(\mathbb{P}^1) = \mathbb{Z}$, and hence $\operatorname{Cl}^0 X = 0$.

Conversely, suppose that $\operatorname{Cl}^0 X = 0$. This means that any divisor of degree 0 is principal. In particular, if $x \ne y \in X$ are two points then there exists a function $f \in k(X)$ such that $x - y = \operatorname{div} f$. Viewing f as a map $f \colon X \to \mathbb{P}^1$ we get from Theorem 1 that $k(X) = k(f)$, that is, f is birational. Since X and $\mathbb{P}^1$ are nonsingular projective curves, it follows that f is an isomorphism.

2.2. Bézout's Theorem on a Curve

We now indicate the simplest applications of the theorem on the degree of a principal divisor. They are very special cases of more general theorems, that we will prove in connection with the theory of intersection numbers in Chap. IV. However, it is convenient to treat these simple cases already at this stage, since we will find them useful in 2.3.

Suppose that $X \subset \mathbb{P}^n$ is a nonsingular projective curve and $x \in X$ a point. Let F be a form in the coordinates of $\mathbb{P}^n$, not identically 0 on X; we write $\mathbb{P}^n_F$ for the hypersurface defined by $F = 0$. We introduced in 1.2 the divisor $\operatorname{div} F$ of F on X. Its degree $\deg(\operatorname{div} F)$ is also denoted by XF, and is called the *intersection number* of X and the hypersurface $\mathbb{P}^n_F$.

An important corollary follows at once from Theorem 1: this number $\deg(\operatorname{div} F)$ is the same for all forms of the same degree. Indeed, if $\deg F = \deg F_1$ then $f = F/F_1 \in k(X)$. From the definition of the divisor $\operatorname{div} F$ it follows at once that $\operatorname{div} F = \operatorname{div} F_1 + \operatorname{div} f$, and hence $\operatorname{div} F \sim \operatorname{div} F_1$. By Theorem 1, Corollary, $\deg(\operatorname{div} F) = \deg(\operatorname{div} F_1)$.

To determine how the number XF depends on the degree of F, it is enough to take F to be any form of degree $m = \deg F$. In particular we can take $F = L^m$ where L is a linear form. Then

$$XF = mXL = (\deg F)XL. \tag{1}$$

Finally, we explain the meaning of XL.

Definition. The *degree* of a curve $X \subset \mathbb{P}^N$, denoted by $\deg X$, is the maximum number of points of intersection of X with a hyperplane not containing any component of X.

Since $XL = \sum_{L(x)=0} v_x(\operatorname{div} L)$, we have $\deg XC \le XL$. Here we use the notation $v_x(D)$ for the multiplicity of a divisor D at x, that is, the coefficient k_i in the expression $D = \sum k_i x_i$.

For any form F, we now determine when $v_x(\operatorname{div} F) = 1$. Since the function v_x is additive, it is enough to consider an irreducible form.

Lemma. *Let $X \subset \mathbb{P}^n$ be a curve, F an irreducible form and $Y = \mathbb{P}^n_F \subset \mathbb{P}^n$ the hypersurface given by $F = 0$. Then $v_x(\operatorname{div} F) = 1$ is equivalent to $F(x) = 0$ and $\Theta_{Y,x} \not\supset \Theta_{X,x}$. Here we view both these spaces as vector subspaces of $\Theta_{\mathbb{P}^n,x}$.*

Proof. We obtain a proof by putting together a number of definitions from Chap. II. Let G be a form such that $G(x) \ne 0$ and $\deg G = \deg F$. By definition, $v_x(\operatorname{div} F) = v_x(f)$, where $f = (F/G)_{|X}$. We know that $v_x(\operatorname{div} f) > 1$ is equivalent to $f \in \mathfrak{m}_x^2$, or equivalently, $d_x f = 0$. But $d_x f \in \Theta^*_{X,x}$, and is the restriction to $\Theta_{X,x}$ of the differential $d_x(F/G)$ of the function F/G, which is a rational function on $\mathbb{P}^n$, regular at x. Thus $v_x(\operatorname{div} F) > 1$ is equivalent to $d_x(F/G) = 0$ on $\Theta_{X,x}$. Furthermore, F/G is a local equation of the hypersurface Y in the neighbourhood of x given by $G \ne 0$. Hence $d_x(F/G) = 0$ is the equation of $\Theta_{Y,x}$, and $d_x(F/G) = 0$ on $\Theta_{X,x}$ if and only if $\Theta_{Y,x} \supset \Theta_{X,x}$. The lemma is proved.

We apply this to compute the intersection number XL. Since XL is the same for all linear forms L, the number of points $x \in X$ with $L(x) = 0$ is a maximum when all the $v_x(L) = 1$. By the lemma, this is equivalent to saying that the hyperplane L is not tangent to X at any point. Taking L to be such a linear form, we get

$$\deg X = XL. \tag{2}$$

We need only verify that linear forms with the required property actually exist. This is easy using the dimension counting argument that we have used many times: in the product $X \times \mathbb{P}^{n*}$ (where $\mathbb{P}^{n*}$ is the dual projective space of hyperplanes of $\mathbb{P}^n$), consider the set Γ of points (x, L) such that L is tangent to X at x. A standard application of the theorem on the dimension of fibres of a map then shows that the image of Γ under the projection $X \times \mathbb{P}^{n*} \to \mathbb{P}^{n*}$ has codimension ≥ 1.

Putting together (1) and (2) we get the relation

$$XF = (\deg F)(\deg X), \tag{3}$$

which is called *Bézout's theorem*. Thus we have finally proved this theorem, already stated in Chap. I, 1.6.

2.3. The Dimension of a Divisor

In 1.5, we associated with a divisor D on a nonsingular variety X a vector space $\mathcal{L}(D)$.

Theorem 5. $\mathcal{L}(D)$ *is finite dimensional for any effective divisor D on a nonsingular projective algebraic curve.*

Proof. First of all, the assertion reduces easily to the case $D \geq 0$. Indeed, let $D = D_1 - D_2$ with $D_1, D_2 \geq 0$. Then $\mathcal{L}(D) \subset \mathcal{L}(D_1)$: indeed, $f \in \mathcal{L}(D)$ means that $\operatorname{div} f + D_1 - D_2 = D' \geq 0$, and hence $\operatorname{div} f + D_1 = D' + D_2 \geq 0$, that is, $f \in \mathcal{L}(D_1)$. The required reduction follows from this.

Let $D \geq 0$ and let x be a point appearing in D with multiplicity $r > 0$, that is $D = rx + D_1$. Set $(r-1)x + D_1 = D'$, and let t be a local parameter on X at x. For a function $f \in \mathcal{L}(D)$, set $\lambda(f) = (t^r f)(x)$. Then $\lambda \colon \mathcal{L}(D) \to k$ is obviously a linear function, with kernel equal to $\mathcal{L}(D')$. Carrying out the same construction $\deg D$ times, we see that $\mathcal{L}(0)$ is a vector subspace of $\mathcal{L}(D)$ defined by the vanishing of $\deg D$ linear forms. But we know that $\mathcal{L}(0) = k$ by 1.1 (just before Example 1). It follows from this that $\mathcal{L}(D)$ is finite dimensional, and in fact

$$\ell(D) \leq \deg D + 1. \tag{1}$$

The theorem is proved.

Remark 1. Equality holds in (1) for $X = \mathbb{P}^1$. Indeed, in this case any divisor D is linearly equivalent to rx, where $x \in \mathbb{P}^1$ is the point at infinity. Then $\mathcal{L}(D)$ equals the space of polynomials of degree $\leq r$ and $\ell(D) = r + 1$.

Remark 2. If X is not rational then (1) can be improved. Namely, in this case, for any point $x \in X$ we have $\mathcal{L}(x) = k$. Indeed, if $\mathcal{L}(x)$ contains a nonconstant function, then we would have $\operatorname{div}_\infty f = x$. Then by Theorem 1, Corollary, $\deg(\operatorname{div}_0 f) = 1$, that is, $\operatorname{div} f = y - x$, which contradicts the irrationality of X (see the proof of Theorem 4). Therefore in the process of proving (1) we already get to a divisor x after $\deg D - 1$ steps, for which $\mathcal{L}(x) = 1$, and hence

$$\ell(D) \leq \deg D \qquad \text{if } D > 0. \tag{2}$$

Thus rational curves are characterised by the fact that for them $\ell(D) = \deg D + 1$ for $D > 0$.

Remark 3. The same argument shows that, quite generally, for divisors D_1 and D_2,

$$D_1 < D_2 \implies \ell(D_2) \leq \ell(D_1) + \deg(D_2 - D_1) \tag{3}$$

The inequalities (1) and (2) are particular cases of this, with $D_1 = 0$ and $D_1 = x$ respectively.

Exercises to §2

1. A line l is a *double tangent* or *bitangent* to a plane quartic curve X if l and X are tangent at any point of $l \cap X$. Prove that the set of quartic curves having a given line l as a double tangent has codimension 2 in the space of all quartics. Prove that any irreducible quartic curve has a double tangent.

2. For a singular projective curve X, define the divisor of a form F on the normalisation X^ν using the pullback of functions $\nu^*(F/G)$ as in 1.2, and the intersection number XF as the degree of this divisor on X^ν. Prove that Bézout's theorem continues to hold in this context.

3. Prove that the number of singular points of an irreducible plane curve of degree n is $\leq \binom{n-1}{2}$. [Hint: Pass a curve of degree n through $\binom{n-1}{2} + 1$ singular points, and as many nonsingular ones as possible. Then apply Bézout's theorem.]

4. If X is a nonsingular plane curve and l a line, and the multiplicity of tangency at $x \in X$ is $r \geq 2$, we say that $r - 2$ is the *inflexion multiplicity* of X at x. Prove that the sum of the inflexion multiplicities of a curve of degree n taken over all inflexion points it equal to $3n(n-2)$. [Hint: Prove that the multiplicity of flex points at x is equal to the multiplicity of the zero of the Hessian at x (Chap. I, 6.2).]

5. Let X be a nonsingular curve and $x_1, \ldots, x_m \in X$. Prove that we can take the functions t_i in Theorem 2 to be the equations of hypersurfaces E_i such that $E_i \ni x_i$, $E_i \not\ni x_j$, for $i \neq j$, and $E_i \not\supset \Theta_{X,x_i}$, that is, E_i is not tangent to X at x_i.

6. Prove that a curve of degree n in $\mathbb{P}^n$ not contained in any hyperplane is rational.

3. The Plane Cubic

3.1. The Class Group

We have seen in 2.1, Theorem 4 that $\mathrm{Cl}^0 X = 0$ holds for rational curves X, and for them only. We now work out the simplest example for which $\mathrm{Cl}^0 X \neq 0$. This is the nonsingular plane cubic curve X, one of the most beautiful examples in algebraic geometry, with a wealth of unexpected properties. We proved in Chap. I, 6.2 that X always has an inflexion point, and hence can be put in Weierstrass normal form. It follows from this, as we have seen in Chap. I, 1.6, that X is irrational.

Theorem 1. *Pick any point α_0 of a nonsingular plane cubic curve X, and consider the map $X \to \mathrm{Cl}^0 X$ that sends $\alpha \in X$ to the divisor class C_α containing $\alpha - \alpha_0$. Then $\alpha \mapsto C_\alpha$ defines a one-to-one correspondence between points $\alpha \in X$ and divisor classes $C \in \mathrm{Cl}^0 X$.*

Proof. If $C_\alpha = C_\beta$ then $\alpha - \alpha_0 \sim \beta - \alpha_0$, so that $\alpha \sim \beta$. If $\alpha \neq \beta$, it would follow from the proof of 2.1, Theorem 4 that X is rational, whereas we know that it is not rational.

It remains to prove that any divisor class C of degree 0 contains a divisor of the form $\alpha - \alpha_0$. Suppose first that D is any effective divisor. We show that there exists a point $\alpha \in X$ such that

$$D \sim \alpha + k\alpha_0. \tag{1}$$

If $\deg D = 1$, then (1) holds with $k = 0$. If $\deg D > 1$ then $D = D' + \beta$ with $\deg D' = \deg D - 1$ and $D' > 0$. Using induction, we can assume that (1) is proved for D', that is, $D' \sim \gamma + l\alpha_0$. Then $D \sim \beta + \gamma + l\alpha_0$. If we can find a point α such that

$$\beta + \gamma \sim \alpha + \alpha_0, \tag{2}$$

then (1) will follow. Suppose first that $\beta \neq \gamma$. Pass the line given by $L = 0$ through β and γ. By Bézout's theorem, $LX = 3$, hence

$$\operatorname{div} L = \beta + \gamma + \delta \qquad \text{for some } \delta \in X. \tag{3}$$

Suppose moreover that $\delta \neq \alpha_0$ and pass the line given by $L_1 = 0$ through δ and α_0. In same way as for (3), we get $\operatorname{div} L_1 = \delta + \alpha_0 + \alpha$ for some $\alpha \in X$. Since $\operatorname{div} L \sim \operatorname{div} L_1$ we get $\beta + \gamma + \delta \sim \delta + \alpha_0 + \alpha$, and (2) follows.

We still have to treat the cases with $\beta = \gamma$ or $\delta = \alpha_0$. If $\beta = \gamma$ we pass the tangent to X at β; let $L = 0$ be the equation. According to 1.2, Lemma $v_\beta(L) \geq 2$, and hence $\operatorname{div} L = 2\beta + \delta$. Thus (3) holds in this case. The case $\delta = \alpha_0$ is treated similarly.

Now suppose that D is any nonzero divisor with $\deg D = 0$. Then $D = D_1 - D_2$ with $D_1, D_2 > 0$ and $\deg D_1 = \deg D_2$. Applying (1) to both D_1

and D_2, we get $D_1 \sim \beta + k\alpha_0$ and $D_2 \sim \gamma + k\alpha_0$ with the same k, since $\deg D_1 = \deg D_2$. Hence $D = D_1 - D_2 \sim \beta - \gamma$, and we need only find a point α such that $\beta - \gamma \sim \alpha - \alpha_0$. This relation is equivalent to $\beta + \alpha_0 \sim \alpha + \gamma$, and is the same as (2) up to the notation. The theorem is proved.

The proof of Theorem 1 allows us to determine explicitly the function $\ell(D)$ for divisors D on a nonsingular plane cubic.

Theorem 2. *Let $X \subset \mathbb{P}^2$ be a nonsingular cubic; then*

$$\ell(D) = \deg D \qquad \text{for every effective divisor } D > 0 \text{ on } X. \qquad (4)$$

Conversely, a curve for which (4) holds is isomorphic to a nonsingular cubic. (Compare also 6.6, Corollary 4.)

Proof. By 2.3, Theorem 5, Remark 2, $\ell(D) \le \deg D$ for a divisor $D > 0$ on X, and it is enough to prove that $\ell(D) \ge \deg D$. In the proof of Theorem 1 we proved that $D \sim \alpha + m\alpha_0$. Thus it is enough to prove that $\ell(\alpha + m\alpha_0) > m$ (strict inequality!). If $m = 1$, then $l(\alpha + \alpha_0) > 1$; because $\mathcal{L}(\alpha + \alpha_0)$ contains the nonconstant function L_1/L_0, where L_0 is defining equation of the line through α and α_0 and L_1 any line through the third point of the intersection of L_0 and X (see Figure 12, (a)).

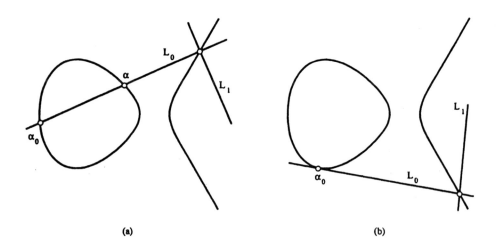

(a)   (b)

Figure 12. Constructing Functions on a Plane Cubic

Hence for $m > 1$ it is sufficient to exhibit a function f_m with $\operatorname{div}_\infty(f) = m\alpha_0$; indeed, then $f_m \in \mathcal{L}(m\alpha_0) \subset \mathcal{L}(\alpha + m\alpha_0)$ and $f_m \notin \mathcal{L}(\alpha + (m-1)\alpha_0)$, whence $\ell(\alpha + m\alpha_0) \geq \ell(\alpha + (m-1)\alpha_0) + 1$, and our assertion is proved by induction. It is easy to find f_m with this property for $m = 2$ or 3. Namely, $f_2 = L_1/L_0$, where L_0 is the tangent line to X at α_0, and L_1 is any line through the third point of the intersection of L_0 and X (see Figure 12, (b)). Similarly, $f_3 = L_1 L_3/L_0 L_2$, where L_0 and L_1 are as before (see Figure 12, (b)), L_2 is the defining equation of the line through α_0 and one of the other points of intersection of L_1 and X, and $L_3 = 0$ a line through the third point of intersection of L_2 and X. Finally, if $m = 2r$ is even, then $f_m = f_2^r$; and if $m = 2r + 3$ is odd and ≥ 3 then $f_m = f_3 f_2^r$. This proves the equality (4).

Conversely, suppose that X is a nonsingular projective curve X such that (4) holds for any divisor $D > 0$. Take any point $p \in X$. Since $\mathcal{L}(2p) > 1$ by (4), there exists a function $x \in k(X)$ with $\mathrm{div}_\infty x = 2p$ (note that $\mathrm{div}_\infty x = p$ is impossible, since then the curve would be rational). By (4) $\mathcal{L}(3p) \neq \mathcal{L}(2p)$, so that there exists a function $y \in k(X)$ with $\dim_\infty y = 3p$. Finally, by (4), $\mathcal{L}(6p) = 6$. But we already know 7 functions belonging to $\mathcal{L}(6p)$, namely $1, x, x^2, x^3, y, xy, y^2$. Hence there must be a linear dependence relation between these

$$a_0 + a_1 x + a_2 x^2 + a_3 x^3 + b_0 y + b_1 xy + b_2 y^2 = 0. \tag{5}$$

Thus the functions x and y define a rational, hence regular map from X to the plane cubic $Y \subset \mathbb{P}^2$ with the equation (5) in inhomogeneous coordinates. This is the rational map defined by the linear system $\mathcal{L}(3p)$.

The map f defines an inclusion of function fields $f^* : k(Y) \hookrightarrow k(X)$. Let us prove that $f^*(k(Y)) = k(X)$. For this, remark that $k(Y) \supset k(x)$ and $k(Y) \supset k(y)$, and the functions x and y each defines a map of $X \to \mathbb{P}^1$. By assumption $\mathrm{div}_\infty x = 2p$, which means that the map g defined by x satisfies $g^*(\infty) = 2p$. From 2.1, Theorem 1 it follows that $\deg g = 2$, that is, $[k(X) : k(f^*(x))] = 2$. Similarly, $[k(X) : k(f^*(y))] = 3$. Since $[k(X) : f^*(k(Y))]$ has to divide both these numbers, $k(X) = f^*(k(Y))$, that is, f is birational. The cubic (5) cannot have singular points, since then it, and X together with it, would be a rational curve, which contradicts (4). Therefore Y is a nonsingular cubic, and hence f is an isomorphism. The theorem is proved.

Thus nonsingular cubic curves in $\mathbb{P}^2$ are characterised by (4) in exactly the same way that rational curves are characterised by $\ell(D) = \deg D + 1$ for $D > 0$.

3.2. The Group Law

Theorem 1 establishes a one-to-one correspondence between the points of a nonsingular cubic curve $X \subset \mathbb{P}^2$ and the elements of the group $\mathrm{Cl}^0 X$, under which a point $\alpha \in X$ corresponds to the class C_α of the divisor $\alpha - \alpha_0$, where α_0 is the fixed point used to define the correspondence.

Using this, we can transfer the group law from $\mathrm{Cl}^0 X$ to X itself. The corresponding operation on points of X is called *addition*, and written $\oplus$, with subtraction denoted by $\ominus$. By definition, $\alpha \oplus \beta = \gamma$ if $C_\alpha + C_\beta = C_\gamma$, that is

$$\alpha + \beta \sim \gamma + \alpha_0. \tag{1}$$

α_0 is obviously the zero element. From now on we denote it by o, so that (1) can be rewritten

$$\alpha + \beta \sim (\alpha \oplus \beta) + o. \tag{2}$$

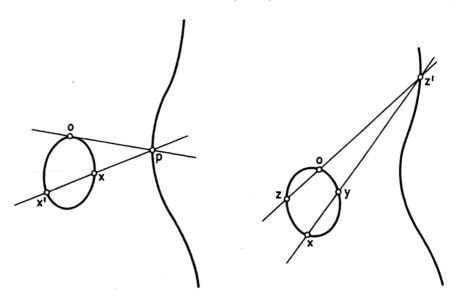

Figure 13. The Group Law on a Plane Cubic

The proof of Theorem 1 allows us to describe the operations $\oplus$ and $\ominus$ in elementary geometric terms. Namely, if the tangent to X at o meets X at π and the line through π and α meets X in a third point α' then

$$2o + \pi \sim \pi + \alpha + \alpha' \qquad \text{so that} \qquad \alpha + \alpha' \sim 2o, \tag{3}$$

which means that $\alpha' = \ominus\alpha$ is the inverse of α in the group law (Figure 13, a). If $\alpha = \pi$, passing a line through α and π should be replaced by drawing the tangent line to X at α.

Similarly, to describe $\oplus$, pass a line through α and β; let γ' be the third point of intersection with X, and γ the third point of intersection of X with the line through o and γ' (Figure 13, b). Then

$$\alpha + \beta + \gamma' \sim \gamma' + \gamma + o$$
$$\alpha + \beta \sim \gamma + o, \qquad \text{that is,} \quad \gamma = \alpha \oplus \beta. \tag{4}$$

If $\alpha = \beta$ or $\gamma' = o$ then passing a secant through α and β, or through γ' and o, should be replaced by drawing the tangent line to X at α or γ'.

This description becomes especially simple if we take o to be an inflexion point of X, which from now on we always assume. Then the section of X by a line is linearly equivalent to $3o$: to see this, take the inflexional tangent line to X at o. If γ_1 is the third point of intersection of X with the line through γ and o then

$$\gamma + \gamma_1 + o \sim 3o, \tag{6}$$

that is, $\gamma_1 \sim \ominus \gamma$ (Figure 14).

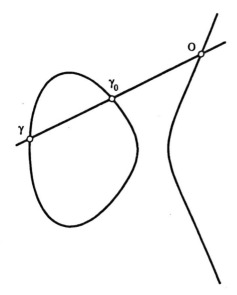

Figure 14. The Inverse Map $\gamma \mapsto \gamma_1$

To describe the operation $\oplus$, pass a line through α and β. Let γ' be the third points of intersection of X, and γ the third point of intersection of X with the line through γ' and o. Then (Figure 13)

$$\begin{aligned} \alpha + \beta + \gamma' &\sim \gamma + \gamma' + o, \\ \alpha + \beta &\sim \gamma + o. \end{aligned} \tag{7}$$

If $\alpha = \beta$ then the secant line through α and β should be replaced by the tangent line to X at α.

Another form of relation (7) is that α, β, γ are collinear if and only if $\alpha \oplus \beta \oplus \gamma = o$. In particular, β lies on the tangent line at α if and only if $2\alpha \oplus \beta = o$ (where $2\alpha = \alpha \oplus \alpha$ in the sense of the group law). Finally, also $\beta = \alpha$ if α is an inflexion point; then $3\alpha = o$. Thus the inflexion points of a cubic are precisely the elements of order 3 in the group law, together with the zero element.

A cubic in Weierstrass normal form has an inflexion point at infinity. We will assume that the characteristic of k is different from 2 or 3. (This is exclusively for the purpose of simplifying the formulas.) Then the equation of X can be written

$$y^2 = x^3 + ax + b \tag{8}$$

and its point at infinity o is on the lines $x = c$ for all $c \in k$. Hence the minus operation of the group law is particularly simple to write down:

$$\ominus(x, y) = (x, -y). \tag{9}$$

To write out the operation $\alpha \oplus \beta$, pass a line through $\alpha = (x_1, y_1)$ and $\beta = (x_2, y_2)$:

$$y - y_1 = \left(\frac{y_2 - y_1}{x_2 - x_1} \right) (x - x_1). \tag{10}$$

The three points of intersection of this line with the cubic (8) are obtained from the equation

$$\left(y_1 + \left(\frac{y_2 - y_1}{x_2 - x_1} \right) (x - x_1) \right)^2 = x^3 + ax + b,$$

that is,

$$x^3 - \left(\frac{y_2 - y_1}{x_2 - x_1} \right)^2 x^2 + \cdots = 0.$$

We know two of the roots x_1 and x_2 of this equation. Therefore the third root is given by

$$x_3 = \left(\frac{y_2 - y_1}{x_2 - x_1} \right)^2 - x_1 - x_2. \tag{11}$$

The coordinate y_3 is given by (10), and finally

$$\alpha \oplus \beta = (x_3, -y_3).$$

When $\alpha = \beta$, we should take the tangent line to X at (x_1, y_1). Similar transformations give its third point of intersection as (x_2, y_2), where

$$x_2 = \frac{(3x_1^2 + a)^2}{4(x_1^3 + ax_1 + b)} - 2x_1, \tag{12}$$

and y_2 is obtained by substituting for x_2 from (12) into the equation of the tangent line. Then $2\alpha = (x_2, -y_2)$.

A remarkable property of the group law we have constructed is that it is given by rational formulas, that is, it defines a rational map $X \times X \to X$. We can even say more.

Theorem 3. *The maps $\varphi: X \to X$ given by $\varphi(\alpha) = \ominus\alpha$ and $\psi: X \times X \to X$ given by $\psi(\alpha, \beta) = \alpha \oplus \beta$ are regular.*

Proof. For φ this is obvious from (9). Similarly, it follows from (10) that φ is regular at a point (α, β) provided that $\alpha = (x_1, y_1)$, $\beta = (x_2, y_2)$ and $x_1 \neq x_2$; or in other words, since $x_2 = x_1$ implies that $y_2 = \pm y_1$, provided that $\alpha \neq \beta$ and $\alpha \neq \ominus \beta$.

Now for any point $\gamma \in X$, consider the reflection map s_γ, that takes a point $\alpha \neq \gamma$ into the third point of intersection of X with the line through α and γ. Obviously $s_\gamma(\alpha) = \ominus(\alpha \oplus \gamma)$. It can be seen explicitly from the formulas that this map is rational, hence regular by Chap. II, 3.1, Theorem 3, Corollary. Moreover, $s_\gamma^2 = \mathrm{id}$, so that s_γ is an automorphism. Let us prove that $s_\gamma(\gamma)$ is the third point of intersection of X with the tangent line at γ. For this we apply s_γ to the relation

$$\alpha + \gamma + s_\gamma(\alpha) \sim 3o.$$

Since s_γ is an automorphism of X it obviously preserves linear equivalence of divisors, and moreover, $s_\gamma(o) = \ominus \gamma$. Hence

$$s_\gamma(\alpha) + s_\gamma(\gamma) + \alpha \sim 3(\ominus \gamma).$$

Substituting in this the expression $s_\gamma = \ominus(\alpha \oplus \gamma)$, we get $s_\gamma(\gamma) = 2(\ominus \gamma)$ (multiplication by 2 in the group law of X), and this is the third point of intersection of X with the tangent line at γ.

Now we can consider the translation automorphism $t_\gamma(\alpha) = \alpha \oplus \gamma$ for $\alpha \neq \gamma$. Obviously t_γ is the composite of the two reflections $t_\gamma = s_0 \circ s_\gamma$, from which it follows that if $\alpha = \gamma$ then $t_\gamma(\alpha) = 2\alpha$. Finally, for any $\alpha, \beta \in X$, we have

$$\psi(\alpha, \beta) = t_{\gamma \oplus \delta}^{-1} \psi(t_\gamma(\alpha), t_\delta(\beta)).$$

Hence if ψ is regular at any point (α_0, β_0), then it is regular at any point $(\alpha, \beta) = (t_\gamma(\alpha_0), t_\delta(\beta_0))$, where $\gamma = \alpha \oplus (\ominus \alpha_0)$ and $\delta = \beta \oplus (\ominus \beta_0)$. But it is regular where $\alpha \neq \beta$ and $\alpha \neq \ominus \beta$, and hence is regular everywhere. The theorem is proved.

The map $\psi: X \times X \to X$ has a differential at $(\alpha, \beta) \in X \times X$,

$$d_{(\alpha, \beta)} \psi: \Theta_{(\alpha, \beta)} \to \Theta_{\alpha \oplus \beta}.$$

Obviously $\Theta_{(\alpha, \beta)} \cong \Theta_\alpha \oplus \Theta_\beta$, and the linear map from the direct sum is determined by the map on the summands. Finally, the composite map $\Theta_\alpha \to \Theta_\alpha \oplus \Theta_\beta \to \Theta_{\alpha \oplus \beta}$ comes from $X \to X \times X \to X$, where the first map is the inclusion $\gamma \mapsto (\gamma, \beta)$, and the second the group law ψ. The composite map $X \to X$ is simply the translation t_β, and hence the restriction of $d\psi$ to Θ_α equals dt_β and finally $d\psi = dt_\alpha + dt_\beta$. We have proved the next result.

Lemma. *The differential* $d\psi: \Theta_{(\alpha, \beta)} \to \Theta_{\alpha \oplus \beta}$ *of the group law* $\psi: X \times X \to X$ *is given by* $d\psi = dt_\alpha + dt_\beta$. *In particular, it is surjective.* $\square$

3.3. Maps

We study regular maps $\lambda\colon X \to X$ of the cubic to itself. An example is the translation t_γ given by $t_\gamma(\alpha) = \alpha \oplus \gamma$. If $\lambda(o) = \gamma$ then $t_{\ominus\gamma} \circ \lambda = \lambda'$ fixes o. From now on we always assume this, that is, $\lambda(o) = o$. We prove in 4.3, Theorem 3 below that then λ is a homomorphism of the group law on X, but we do not use this at present.

Just as with any maps to a group, we can add maps, defining $\lambda + \mu$ by $(\lambda + \mu)(\alpha) = \lambda(\alpha) \oplus \mu(\alpha)$. Obviously all regular maps $\lambda\colon X \to X$ with $\lambda(o) = o$ form a group. If $\lambda(X)$ is not just a point then $\lambda(X) = X$. Then the degree $\deg \lambda$ is defined, and is positive; we write $n(\lambda)$ for $\deg \lambda$. If $\lambda(X) = o$ then we set $n(\lambda) = 0$.

The basic result is the following theorem, which has many applications.

Theorem 4. *There exists a scalar product* (λ, μ) *on the group of regular maps* $\lambda\colon X \to X$ *with* $\lambda(o) = o$ *such that* $(\lambda, \lambda) = n(\lambda)$.

Here by scalar product, we mean that a number $(\lambda, \mu) \in \mathbb{Q}$ is defined for any elements λ, μ, with the properties

$$(\lambda, \mu) = (\mu, \lambda) \qquad \text{and} \qquad (\lambda_1 + \lambda_2, \mu) = (\lambda_1, \mu) + (\lambda_2, \mu).$$

For any $\mathbb{Q}$-valued function $n(\lambda)$ with $n(\lambda) > 0$ for $\lambda \neq 0$ and $n(\lambda) = 0$ for $\lambda = 0$, there exists a scalar product (λ, μ) with $(\lambda, \lambda) = n(\lambda)$ if and only if

$$n(\lambda + \mu) + n(\lambda - \mu) = 2\big(n(\lambda) + n(\mu)\big). \tag{1}$$

This is an elementary and purely algebraic fact (see Appendix, §1, Proposition 1). Thus to prove the theorem, it is enough to check the relation (1) for $n(\lambda) = \deg \lambda$.

We write $\Delta \subset X \times X$ for the diagonal subvariety of pairs (α, α) with $\alpha \in X$, and $\Sigma \subset X \times X$ for the set of pairs $(\alpha, \ominus\alpha)$. Obviously these are both nonsingular irreducible subvarieties isomorphic to X. For Δ, compare Chap. I, 2.3, Example 10; and $\Sigma = \iota(\Delta)$ where $\iota = (\mathrm{id}, \ominus)$ is the involution $(\alpha, \beta) \mapsto (\alpha, \ominus\beta)$. Consider the regular map $\psi\colon X \times X \to X$ given by $\psi(x, y) = x \oplus y$, and the divisor $\psi^*(o)$. By 3.2, Lemma, $d\psi$ is surjective. It follows from this that $\psi^*(o) = \Sigma$ is the prime divisor Σ with multiplicity 1. Indeed, if t_o is a local parameter on X at o then $\psi^*(t_o)$ is a local equation of Σ. Since $d\psi$ is surjective, the dual map $\mathfrak{m}_o/\mathfrak{m}_o^2 \to \mathfrak{m}_{\alpha,\ominus\alpha}/\mathfrak{m}_{\alpha,\ominus\alpha}^2$ is injective. Hence $\psi^*(t_o) \notin \mathfrak{m}_{\alpha,\ominus\alpha}^2$, and it follows that $\psi^*(o)$ is nonsingular. In the same way, the map $\psi_1\colon X \times X \to X$ defined by $\psi_1(\alpha, \beta) = \alpha \ominus \beta$ differs from ψ by the involution ι, and a similar argument gives $\psi_1^*(0) = \Delta$. Finally, set $p_1(\alpha, \beta) = \alpha$ and $p_2(\alpha, \beta) = \beta$. Obviously $p_1^*(o) = o \times X$ and $p_2^*(o) = X \times o$. The identity (1) follows easily from the next assertion.

Lemma. *The linear equivalence*

$$\Delta + \Sigma \sim 2(o \times X + X \times o) = 2\big(p_1^*(o) + p_2^*(o)\big) \tag{2}$$

holds on the surface $X \times X$.

Proof. To prove this, we exhibit a function f on $X \times X$ for which $\operatorname{div}_0 f$ equals the left-hand side of (2) and $\operatorname{div}_\infty f$ the right-hand side. We assume that an affine piece of X is given in the Weierstrass normal form 2.2, (8); then x defines a map $x \colon X \to \mathbb{P}^1$ such that $x(\alpha) = x(\beta)$ only if $\alpha = \beta$ or $\alpha = \ominus\beta$. Moreover, $\deg x = 2$, and since $x^{-1}(\infty) = \{o\}$, we have $\operatorname{div}_\infty x = x^*(\infty) = 2o$.

The affine product variety $X \times X$ is the subset of $\mathbf{A}^4$ (with coordinates x_1, y_1, x_2, y_2) defined by $y_1^2 = f(x_1)$, $y_2^2 = f(x_2)$. Inside $X \times X$, Δ is defined by $x_1 = x_2$, $y_1 = y_2$, and Σ by $x_1 = x_2$, $y_1 = -y_2$. The function f we require to establish the linear equivalence (2) is $f = x_1 - x_2$. The verification is an almost tautological calculation. The divisor defined on the affine surface by $x_1 - x_2$ is $\Delta + \Sigma$. In fact

$$(y_1 - y_2)(y_1 + y_2) = (y_1^2 - y_2^2) = (x_1 - x_2)g(x_1, x_2),$$

where $g(x_1, x_2) = (f(x_1) - f(x_2))/(x_1 - x_2)$. Now $(y_1 - y_2)$ is invertible outside Δ, so $x_1 - x_2$ is a local equation for Σ there, and similarly $(y_1 + y_2)$ is invertible outside Σ, so $x_1 - x_2$ is a local equation for Δ there. Therefore $\operatorname{div}_0(x_1 - x_2) = \Delta + \Sigma$.

Now since $\operatorname{div}_\infty x = 2o$, it's clear that on the projective variety $X \times X$ we have $\operatorname{div}_\infty(x_1 - x_2) = 2(o \times X + X \times o)$, so that finally,

$$\operatorname{div}(x_1 - x_2) = \Delta + \Sigma - 2(o \times X + X \times o).$$

This proves the lemma.

Proof of Theorem 4. Consider the map $f \colon X \to X \times X$ given by $f(\alpha) = \big(\lambda(\alpha), \mu(\alpha)\big)$. Obviously $p_1 \circ f = \lambda$ and $p_2 \circ f = \mu$, so that $f^*(p_1^*(o)) = \lambda^*(o)$ and $f^*(p_2^*(o)) = \mu^*(o)$ for $\lambda, \mu \neq 0$. Similarly, $\psi \circ f = \lambda + \mu$, so that, since $\Sigma = \psi^*(o)$, we get $f^*(\Sigma) = (\lambda + \mu)^*(o)$ for $\lambda + \mu \neq 0$, and similarly $f^*(\Delta) = (\lambda - \mu)^*(o)$ for $\lambda - \mu \neq 0$. Therefore applying f^* to (2) gives

$$(\lambda + \mu)^*(o) + (\lambda - \mu)^*(o) \sim 2\big(\lambda^*(o) + \mu^*(o)\big)$$

(linear equivalence of divisor on X), provided that $\lambda, \mu, \lambda + \mu, \lambda - \mu \neq 0$. Since linearly equivalent divisors have the same degree, and $\deg \lambda^*(o) = \deg \lambda = n(\lambda)$, and similarly for $\mu, \lambda + \mu, \lambda - \mu$, (1) follows, provided that $\lambda, \mu, \lambda + \mu, \lambda - \mu \neq 0$. If $\lambda = 0$ or $\mu = 0$ then (1) is obvious. If, say, $\lambda + \mu = 0$ then we need only use the assertion in 1.3, Example, together with $n(\lambda + \mu) = 0$; similarly if $\lambda - \mu = 0$. This proves (1) and the theorem.

3.4. Applications

Example 1. m-torsion points of X. Consider the homomorphism $\delta_m : X \to X$ of multiplication by m in X, that is,

$$\delta_m(\alpha) = \underbrace{\alpha \oplus \cdots \oplus \alpha.}_{m \text{ times}}$$

By 3.3, (1) with $\lambda = \mu = \delta_1$ it follows that $n(\delta_2) = 4$, and, by induction on m, one sees that $n(\delta_m) = m^2$. Suppose that k has characteristic 0. By Chap. II, 6.3, Theorem 4, there exists a nonempty open set $U \subset X$ such that points $\alpha \in U$ have exactly m^2 inverse images under δ_m. But δ_m is a group homomorphism, so that the number of inverse images of any point is equal to the order of the kernel. We deduce that the number of solutions in X of the equation $m\alpha = 0$ is equal to m^2.

Suppose now that k has characteristic $p > 0$, but m is not divisible by p. In order to be able to apply Chap. II, 6.3, Theorem 4, we need to prove that δ_m is separable. If this were not the case, then by Chap. II, 6.4, Theorem 6, we could write δ_m in the form $g \circ \varphi$, where φ is the Frobenius map of X, and then $\deg \delta_m$ would be divisible by p, whereas we know that $\deg \delta_m = m^2$ and $p \nmid m$. Thus the number of solutions of $m\alpha = 0$ equals m^2 provided that m is not divisible by char k.

In particular, the equation $3\alpha = 0$ has 9 solutions if char $k \neq 3$. We saw in 3.2 that the points satisfying $3\alpha = 0$ are the inflexion points of X. Hence a nonsingular plane cubic has 9 inflexion points. These enjoy a number of remarkable properties. For example, the line through any two of them intersects X again in an inflexion point. This follows at once from the fact that the sum of two solutions of $3\alpha = 0$ in a group is again a solution.

Example 2. Hasse–Weil estimates. Suppose now that $X \subset \mathbb{P}^2$ is a nonsingular plane cubic curve defined by an equation with coefficients in the field $\mathbb{F}_p$ with p elements. In Chap. I, 2.3, Example 6, we defined the Frobenius map $\varphi : (\alpha_1, \ldots, \alpha_n) \mapsto (\alpha_1^p, \ldots, \alpha_n^p)$ for affine varieties defined over $\mathbb{F}_p$. This definition extends automatically to arbitrary quasiprojective varieties. By Chap. II, 6.4, Theorem 6, $\deg \varphi = p$.

We apply Theorem 4 to maps $\lambda : X \to X$ of the form $a + b\varphi$ with $a, b \in \mathbb{Z}$, given by

$$a + b\varphi : \alpha \mapsto \underbrace{\alpha \oplus \cdots \oplus \alpha}_{a \text{ times}} \oplus \underbrace{\varphi(\alpha) \oplus \cdots \oplus \varphi(\alpha)}_{b \text{ times}}.$$

By 3.3, Theorem 4, we know that $n(a + b\varphi) = (a + b\varphi, a + b\varphi)$, and hence

$$
\begin{aligned}
n(a + b\varphi) &= a^2 n(1) + 2ab(1, \varphi) + b^2 n(\varphi) \\
&= a^2 + 2ab(1, \varphi) + b^2 p.
\end{aligned}
\tag{1}
$$

By definition $n(a + b\varphi) \geq 0$ for all a and b, so that, viewed as a quadratic form in a, b, $n(a + b\varphi)$ has positive discriminant, and therefore

$$|(1, \varphi)| \le \sqrt{p}. \tag{2}$$

On the other hand, it follows from (1) that $2(1, \varphi) = n(1 - \varphi) - p - 1$, and hence (2) gives

$$|n(1 - \varphi) - p - 1| \le 2\sqrt{p}. \tag{3}$$

Moreover, $n(1-\varphi) = \deg(1-\varphi)^*(o)$, and $\text{Supp}\big((1-\varphi)^*(o)\big)$ consists of points $\alpha \in X$ with $(1 - \varphi)(\alpha) = o$, that is, $\alpha = \varphi(\alpha)$. These are the points of X with coordinates in $\mathbb{F}_p$. We prove that all these points appear in the divisor $(1 - \varphi)^*(o)$ with multiplicity 1. As in the preceding example, referring to Chap. II, 6.3, Theorem 4, it is enough to prove that the map $1-\varphi$ is separable. For this, by Chap. II, 6.4, Theorem 6, we need to prove that $1 - \varphi \ne \mu\varphi$ for any map $\mu \colon X \to X$. But it would follow from this that $1 = (1 + \mu)\varphi$, and this contradicts $\deg \varphi = p > 1$.

Thus (3) can be rewritten

$$|N - p - 1| \le 2\sqrt{p}, \tag{3'}$$

where N is the number of points of X with coordinates in $\mathbb{F}_p$ (including the point at infinity). In other words, the number N_0 of solutions of the congruence

$$y^2 \equiv x^3 + ax + b \pmod{p} \tag{4}$$

satisfies the inequality

$$|N_0 - p| \le 2\sqrt{p}. \tag{5}$$

This result has the following interpretation: for a given residue x modulo p, the congruence (4) has

$$\begin{cases} \text{no solutions} & \text{if } \left(\frac{x^3+ax+b}{p}\right) = -1, \\ \text{2 solutions} & \text{if } \left(\frac{x^3+ax+b}{p}\right) = 1, \end{cases}$$

where $\left(\frac{u}{p}\right)$ is the Legendre symbol. Hence $N_0 - p = \sum_{x=0}^{p-1} \left(\frac{x^3+ax+b}{p}\right)$, and (5) gives the estimate

$$\left| \sum_{x=0}^{p-1} \left(\frac{x^3 + ax + b}{p}\right) \right| \le 2\sqrt{p}. \tag{6}$$

The estimates (5) and (6) have many applications in number theory.

3.5. Algebraically Nonclosed Field

Suppose that the coefficients a, b in the equation 3.2, (8) belong to some field k_0, not necessarily algebraically closed. Write k for the algebraic closure of k_0. The definition, or the explicit formulas for the group law on X, show that the points of X with coordinates in k_0 form a subgroup, denoted by $X(k_0)$.

Example 1. The Mordell–Weil theorem. Suppose that $k_0 = \mathbb{Q}$ is the rational number field. In this case, the theorem known as the *Mordell theorem* asserts that the group $X(\mathbb{Q})$ is finitely generated. In principle, this is a description of the set of rational solutions of equation 3.2, (8) in finite terms, just as that provided in the case of conics by the parametrisation of Chap. I, 1.2.

Example 2. Divisors and rationality. In the more general situation, when X is a nonsingular projective algebraic curve defined by equations with coefficients in a field k_0, we can apply the preceding theory to the field extension $k_0 \subset k$. In what follows we assume that k_0 is a perfect field (just to simplify the arguments somewhat). With X we associate a function field $k_0(X) \subset k(X)$ over k_0 consisting of rational functions in the coordinates with coefficients in k_0. Suppose that $D = \sum n_i x_i$ is a divisor with $x_i \in X$ such that the coordinates of the points x_i are contained in an intermediate field k_1 with $k_0 \subset k_1 \subset k$; we can assume here that $k_0 \subset k_1$ is a Galois extension. An automorphism σ of the field extension $k_0 \subset k_1$ applied to the coordinates of a point $x_i \in X(k_1)$ obviously takes it into a point $\sigma(x_i) \in X(k_1)$. If for every point x_i, and every automorphism $\sigma \in \mathrm{Gal}(k_1/k_0)$, the conjugate points $\sigma(x_i)$ appears in D with the same multiplicity, then we say that the divisor D is *rational* over k_0. This applies, in particular, to the divisor of a function $f \in k_0(x)$.

We write $\mathcal{L}_{k_0}(D)$ for the subspace of functions $f \in k_0(X)$ such that $\mathrm{div}\, f + D > 0$. This is a vector space over k_0. Set $\ell_{k_0}(D) = \dim_{k_0} \mathcal{L}_{k_0}(D)$. The automorphisms of k_1 over k_0 take functions of $\mathcal{L}(D)$ to one another and preserve the subspace $\mathcal{L}_{k_0}(D)$. They are not, however, linear maps: $\sigma(\alpha f) = \sigma(\alpha)\sigma(f)$ for $\alpha \in k$ and $f \in \mathcal{L}(D)$; transformations of this type are said to be *quasilinear*. The so-called main theorem on quasilinear maps (Appendix, §3, Proposition 1) asserts that $\mathcal{L}(D)$ is generated over k by the k_0-vector subspace $\mathcal{L}_{k_0}(D)$ of invariant elements, that is $\mathcal{L}(D) = k \cdot \mathcal{L}_{k_0}(D)$. In particular,

$$\ell_{k_0}(D) = \ell(D). \tag{1}$$

If divisors D and D' are rational over k_0 and linearly equivalent then there exists a function $f \in k_0(X)$ such that $D - D' = \mathrm{div}\, f$. To prove this one must apply the main theorem on quasilinear maps to the 1-dimensional space of functions $g \in k_1(X)$ for which $\mathrm{div}\, g = D - D'$.

Example 3. Zeta functions and the Weil Riemann hypothesis. We return to the cubic X. In Chap I, 2.3, we defined the zeta functions $Z_X(t)$ and $\zeta_X(s)$ for an affine variety defined by equations with coefficients in $\mathbb{F}_p$. The definition obviously extends to arbitrary quasiprojective varieties. If X is a curve then the cycles defined in Chap. I, 2.3 are rational divisors over $\mathbb{F}_p$, and as such, are obviously irreducible, that is, cannot be expressed as sums of other rational divisors. The Euler product Chap. I, 2.3, (2) can then be rewritten, as in the case of the Riemann zeta function, in the form

$$Z_X(t) = \sum_{D \geq 0} t^{\deg D},$$

where D runs through all rational divisors over $\mathbb{F}_p$. In other words,

$$Z_X(t) = \sum a_n t^n,$$

where a_n is the number of effective rational divisors D of degree n. We are now in a position to determine this number explicitly when X is a cubic curve.

We first find the number of linear equivalence classes of rational divisors of degree n. We proved in 3.1 that if $\deg D = n$ then $D \sim x + (n-1)o$. From the fact that D is rational over $\mathbb{F}_p$, it follows that $x \in X(\mathbb{F}_p)$. Indeed, if the coordinates of x are contained in a Galois extension k_1 of k_0 then $\sigma(x) \sim x$ for any automorphism of this extension, and hence $\sigma(x) = x$. Thus the number of divisor classes of degree n equals N, where N is the number of points $x \in X(\mathbb{F}_p)$. Now we find the number of divisors in a given class, that is, the number of divisors $D \sim D_0$ where D_0 is given. They correspond to nonzero functions $f \in \mathcal{L}_{\mathbb{F}_p}(D)$, considered up to constant factors in $\mathbb{F}_p$. Thus the number of rational divisors with $D \sim D_0$ and $D > 0$ is equal to the number of points of $\mathbb{P}^{\ell_{k_0}(D)-1}$, that is, $(p^{\ell_{k_0}(D)} - 1)/(p - 1)$. By (1) $\ell_{k_0}(D) = \ell(D)$, and by 3.1, Theorem 2, $\ell(D) = n$. Therefore $a_n = \left(\frac{p^n-1}{p-1}\right)N$, and we get that

$$Z_X(t) = 1 + N \sum_{n=1}^{\infty} \left(\frac{p^n - 1}{p - 1}\right) t^n$$

$$= 1 + \frac{N}{p-1}\left(\frac{pt}{1-pt} - \frac{t}{1-t}\right)$$

$$= \frac{1 + (N - p - 1)t + pt^2}{(1-t)(1-pt)}.$$

We see that the zeta function $Z_X(t)$ is a rational function of t. Moreover, the inequality 3.4, (3′) shows that the roots α_1 and α_2 of the quadratic polynomial $1 + (N - p - 1)t + pt^2$ are complex conjugate algebraic integers. Since their product equals $1/p$, we have $|\alpha_i| = p^{-1/2}$. For the zeta function $\zeta_X(s) = Z_X(p^{-s})$ this gives that the zeros β_1 and β_2 lie on the line $\operatorname{Re} s = 1/2$. We get in this way an analogue of the Riemann hypothesis.

Analogous results hold for arbitrary nonsingular projective varieties, but their proofs are very much harder (see for example Hartshorne [35], Appendix C for a survey).

Exercises to §3

1. Find all points of order 2 on a cubic curve in Weierstrass normal form.

2. Prove that if two cubic curves intersect in exactly 9 points then any cubic through 8 of these points also passes through the 9th.

3. Prove that the x-coordinates of the inflexion points of the cubic curve 3.2, (8) satisfy $f(x) = x^4 + 2ax^2 + 4bx + a^2$. Prove that if $a, b \in \mathbb{R}$ then not all four of these points can be real. [Hint: Use the fact that $f'(x) = 4(x^3 + ax + b)$.] Prove that a real cubic has one or 3 real inflexion point. In the latter case, they are collinear.

4. Prove that there are 4 tangent lines to a cubic X through every point of X. (We only count $x \in T_x X$ if x is an inflexion point.)

5. Prove that the points of tangency of the 4 tangent lines to a cubic X through a point $a \in X$ lie on a conic tangent to X at a.

6. Prove that if two cubics X_1 and X_2 with equation $y^2 = x^3 + a_i x + b_i$ for $i = 1$ and 2 are isomorphic, then there exists an isomorphism that takes their points at infinity to one another.

7. Under the assumptions of Ex. 6, prove that an isomorphism between X_1 and X_2 that takes their points at infinity to one another is given by a linear map.

8. Under the assumptions of Ex. 6–7, prove that if $b_1, b_2 \neq 0$, then the two cubics X_1 and X_2 are isomorphic if and only if $a_1^3/b_1^2 = a_2^3/b_2^2$.

9. Prove that the zeta function $\zeta_X(s)$ associated with a cubic satisfies the functional equation $\zeta_X(1 - s) = \zeta_X(s)$.

10. Prove that over a field of characteristic p, any map of the cubic $\alpha \colon X \to X$ with $\alpha(X) = X$ can be written in the form $\alpha = \varphi^r \beta$ where φ is the Frobenius map, $r \geq 0$ and β is separable.

4. Algebraic Groups

The results of the preceding sections lead to an interesting topic in algebraic geometry, the theory of algebraic groups. We will not go very deeply into this theory, but in order to give the reader at least an impression, we discuss some of its basic results, leaving out most of the proofs.

4.1. Algebraic Groups

The plane cubic curves of the preceding section are one of the most important examples of a general notion that we now introduce.

Definition. An *algebraic group* is an algebraic variety G which is at the same time a group, in such a way that the following conditions are satisfied: the maps $\varphi\colon G \to G$ given by $\varphi(g) = g^{-1}$ and $\psi\colon G \times G \to G$ given by $\psi(g_1, g_2) = g_1 g_2$ are regular maps (here g^{-1} and $g_1 g_2$ are the inverse and product in the group G).

Examples of algebraic groups

Example 1. A nonsingular plane cubic curve with the group law $\oplus$. 3.2, Theorem 3 asserts that the conditions in the definition of algebraic group are satisfied.

Example 2. The affine line $\mathbf{A}^1$, with the group law defined by addition of coordinates of points. This is called the *additive group*, and denoted by $\mathbf{G}_a$.

Example 3. The variety $\mathbf{A}^1 \setminus 0$, where 0 is the origin, with the group law defined by multiplication of coordinates of points. This is called the *multiplicative group*, and denoted by $\mathbf{G}_m$.

Example 4. The open subset of the space $\mathbf{A}^{n^2}$ of $n \times n$ matrixes consisting of nondegenerate matrixes, with the usual matrix multiplication. This is called the *general linear group*.

Example 5. The closed subset of the space $\mathbf{A}^{n^2}$ of $n \times n$ matrixes consisting of orthogonal matrixes, with the usual matrix multiplication. This is called the *orthogonal group*.

We give a very simple example illustrating how being an algebraic group affects the geometry of the variety G.

Theorem 1. *The variety of an algebraic group is nonsingular.*

Proof. It follows from the definition of an algebraic group that for any $h \in G$ the map
$$t_h\colon G \to G \quad \text{given by} \quad t_h = hg$$
is an automorphism of the variety G. For any g_1, g_2, we have $t_h(g_1) = g_2$, where $h = g_2 g_1^{-1}$, and the property that a point is singular is invariant under isomorphism, so that if any point of G is singular, then so are all its points. But this contradicts the fact that the singular points of any algebraic variety form a proper closed subvariety. Therefore G does not have singular points. The theorem is proved.

A generalisation of this situation is the case when a variety X has a group G of automorphisms, with the property that for any two points $x_1, x_2 \in X$ there exists $g \in G$ such that $g(x_1) = x_2$. In this case we say that X is *homogeneous.* The argument just given shows that a homogeneous variety is nonsingular. An example is the Grassmannian (Chap. I, 4.1, Example 1).

4.2. Quotient Groups and Chevalley's Theorem

This section contains statements of some of the basic theorems on algebraic groups. Theorems are labelled with letters (Theorem A, etc.) to indicate that proofs are omitted.[9]

Definition. An *algebraic subgroup* of an algebraic group G is a subgroup $H \subset G$ that is a closed subset in G. As in the theory of abstract groups, a subgroup $H \subset G$ is a *normal subgroup* if $g^{-1}Hg = H$ for every $g \in G$. Finally, a *homomorphism* $\varphi\colon G_1 \to G_2$ of algebraic groups is a regular map that is a homomorphism of abstract groups.

The problem of constructing the quotient group by a given normal subgroup is quite delicate. The difficulty, of course, is how to turn G/N into an algebraic variety.

Theorem A. *The abstract group G/N can be made into an algebraic variety in such a way that the following conditions are satisfied:*

1. The natural map $\varphi\colon G \to G/N$ is a homomorphism of algebraic groups.

2. For every homomorphism of algebraic groups $\psi\colon G \to G_1$ whose kernel contains N, there exists a homomorphism of algebraic groups $f\colon G/N \to G_1$ such that $\psi = f \circ \varphi$. $\square$

The algebraic group G/N is obviously uniquely determined by conditions 1 and 2. It is called the *quotient group* of G by N.

An algebraic group G is *affine* if the algebraic variety G is affine, and is an *Abelian variety* if G is projective and irreducible. The general linear group (4.1, Example 4) is obviously an affine group. Indeed, it is the principal open set (Chap. I, 4.2) of $\mathbf{A}^{n^2}$ defined by $\det M \neq 0$. Hence any algebraic subgroup of the general linear group is affine.

Theorem B. *An affine algebraic group is isomorphic to an algebraic subgroup of the general linear group.* $\square$

Theorem C (Chevalley's theorem). *Every algebraic group G has a normal subgroup N such that N is an affine group, and G/N an Abelian variety. The subgroup N is uniquely determined by these properties.* $\square$

---

[9] For a modern introduction to algebraic groups, including proofs of Theorems A–B, see Humphreys [38].

4.3. Abelian Varieties

In the definition of Abelian variety, the projectivity condition on the algebraic group G contains a surprising amount of information; many unexpected properties of Abelian varieties flow from it. We deduce here the simplest of these, which only require application of simple theorems already proved in Chap. I.

We need a property of arbitrary projective varieties. Define a *family of maps* $X \to Z$ between varieties to be a map $f: X \times Y \to Z$, where Y is some algebraic variety, the *base* of the family. Obviously for any $y \in Y$ we have a map $f_y: X \to Z$ defined by $f_y(x) = f(x, y)$, which justifies the terminology.

Lemma. *Suppose that X and Y are irreducible varieties with X projective, and let $f: X \times Y \to Z$ be a family of maps from X to a variety Z with base Y. Suppose that for some point $y_0 \in Y$, the image $f(X \times y_0) = z_0 \in Z$ is a point. Then $f(X \times y)$ is a point for every $y \in Y$.*

Proof. Consider the graph Γ of f. Obviously $\Gamma \subset X \times Y \times Z$ and Γ is isomorphic to $X \times Y$. Write $p: X \times Y \times Z \to Y \times Z$ for the projection to the second and third factors, and $\overline{\Gamma} = p(\Gamma)$. Since X is projective, $\overline{\Gamma}$ is closed by Chap. I, 5.2, Theorem 3. Now let $q: \overline{\Gamma} \to Y$ be the restriction of the first projection $Y \times Z \to Y$. The fibre of q over y is of the form $y \times f_y(X \times y)$, and so is nonempty, so that $q(\overline{\Gamma}) = Y$. On the other hand, by assumption, the fibre over y_0 consists of a single point $y_0 \times z_0$. By the theorem on dimension of fibres, Chap. I, 6.3, Theorem 7, we see that $\dim \overline{\Gamma} = \dim Y$.

Choose any point $x_0 \in X$. We obviously have $\{(y, f(x_0, y)) \mid y \in Y\} \subset \overline{\Gamma}$, and this is a variety isomorphic to Y. Now since both of these varieties are irreducible and of the same dimension, they must be equal, and therefore $f(X \times y) = f(x_0, y)$. The lemma is proved.

Remark. Without the assumption that X is projective, the lemma is false, as shown by the family of maps $f_y: \mathbf{A}^1 \to \mathbf{A}^1$ given by $f(x, y) = xy$. The reason is that the set $\overline{\Gamma}$ is not closed, and Chap. I, 6.3, Theorem 7 is not applicable to it. In the example $\overline{\Gamma} \subset \mathbf{A}^1 \times \mathbf{A}^1 = \mathbf{A}^2$ consists of all points (u, v) except those with $u = 0, v \neq 0$. That is, it is the plane with the line $u = 0$ deleted, but the origin $u = v = 0$ kept (Figure 15). Chap. I, 6.3, Theorem 7 is really false for the projection $q: (u, v) \mapsto u$: the domain has dimension 2, the image dimension 1, but the fibre over 0 has dimension 0.

Theorem 2. *An Abelian variety is an Abelian group.*

Proof. Consider the family of maps from G to G with base G given by $f(g, h) = g^{-1}hg$. Obviously when $h = e$ is the identity element we have $f(g, e) = e$, and hence by the lemma, $f(G, h)$ is a point for every h. Hence $f(g, h) = f(e, h) = h$, and this means that G is Abelian.

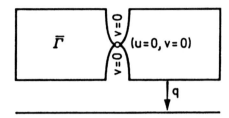

Figure 15. The Map $f(x, y) = xy$

Theorem 3. *If $\psi\colon G \to H$ is a regular map of an Abelian variety G to an algebraic group H, then $\psi(g) = \psi(e)\varphi(g)$, where $e \in G$ is the identity element, and $\varphi\colon G \to H$ a group homomorphism.*

Proof. We set $\varphi(g) = \psi^{-1}(e)\psi(g)$ and prove that φ is a homomorphism. For this, consider the following family of maps $G \to H$ with base G:

$$f\colon G \times G \to H \quad \text{given by} \quad f(g', g) = \varphi(g')\varphi(g)\varphi(g'g)^{-1}.$$

Then since $\varphi(e) = e'$ is the identity element of H, we have $f(G, e) = e'$. By the lemma, $f(G, g)$ is a single point for every $g \in G$, that is, $f(g', g)$ does not depend on g'. Setting $g' = e$ we get $f(g', g) = f(e, g) = e'$, and this means that φ is a homomorphism.

Corollary. *If two Abelian varieties are isomorphic as algebraic varieties, they are isomorphic as groups; that is, "the geometry determines the algebra".*

In particular, the maps of the cubic curve $\lambda\colon X \to X$ with $\lambda(o) = o$ considered in 3.3 are homomorphisms.

4.4. The Picard Variety

The only examples of Abelian varieties appearing so far are the plane cubic curves considered in §3. The group law on these was defined starting out from the study of their divisor class groups. This example is typical of a much more general situation. Starting from an arbitrary nonsingular projective variety X, we can construct an Abelian variety whose group of points is isomorphic to a certain subgroup of the divisor class group $\operatorname{Cl} X$, corresponding to $\operatorname{Cl}^0 X$ in the case of the cubic curve. We give this definition, omitting proofs of all but the simplest assertions.

We now define a new equivalence relation for divisors, algebraic equivalence. It is a coarser relation than the linear equivalence of divisors considered up to now (that is, linear equivalence implies algebraic equivalence). Our aim is to study divisors on nonsingular varieties, but divisors on arbitrary varieties will appear at intermediate stages of the argument. In this case, we always take divisors to mean locally principal divisors.

Let X and T be two arbitrary irreducible varieties. For any point $t \in T$ the map $j_t \colon x \mapsto (x,t)$ defines an embedding $X \hookrightarrow X \times T$. Every divisor C on $X \times T$ with $\operatorname{Supp} C \not\supset X \times t$ defines a pullback divisor $j_t^*(C)$ on X. In this case we say that $j_t^*(C)$ is defined.

Definition. A *family of divisors* on X with base T is any map $f \colon T \to \operatorname{Div} X$. We say that the family f is an *algebraic family of divisors* if there exists a divisor $C \in \operatorname{Div}(X \times T)$ such that $j_t^*(C)$ is defined for each $t \in T$ and $j_t^*(C) = f(t)$. Divisors D_1, D_2 on X are *algebraically equivalent* if there exists an algebraic family of divisors f on X with base T, and two points $t_1, t_2 \in T$ such that $f(t_1) = D_1$ and $f(t_2) = D_2$. This equivalence relation is denoted by $D_1 \equiv D_2$.

Thus algebraic equivalence of divisors means that they can be "algebraically deformed" into one another. Algebraic equivalence is obviously reflexive and symmetric. It is easy to prove that it is transitive: if an algebraic equivalence between D_1 and D_2 is realised by a divisor C on $X \times T$ and an algebraic equivalence between D_2 and D_3 by a divisor C' on $X \times T'$, then to prove that D_1 and D_3 are equivalent we need to consider the divisor $(C \times T') + (C' \times T) - D_2 \times T \times T'$ on $X \times T \times T'$. We leave the detailed verification to the reader.

Finally, one sees easily that algebraic equivalence is compatible with addition in $\operatorname{Div} X$: divisors D with $D \equiv 0$ form a subgroup. We denote this by $\operatorname{Div}^a X$.

Linear equivalence of divisors implies algebraic equivalence. It is enough to prove this for equivalence to 0. Suppose that $D \in \operatorname{Div} X$ and $D \sim 0$, that is, $D = \operatorname{div} g$ with $g \in k(X)$. Consider the variety $T = \mathbb{A}^2 \setminus (0,0)$, and write u, v for coordinates on $\mathbb{A}^2$. We can view g, u, v as functions on $X \times T$, meaning the pullbacks $p^*(g)$, $q^*(u)$ and $q^*(v)$, where as usual $p \colon X \times T \to X$ and $q \colon X \times T \to T$ are the projections. Set $C = \operatorname{div}(u + vg)$ and consider the algebraic family of divisors f defined by the divisor C on $X \times T$. One checks that $f(1,0) = 0$ (the zero divisor) and $f(0,1) = D$, and hence $D \equiv 0$.

Finally, consider algebraic equivalence of divisors in the example of a nonsingular projective curve X. Then $x \equiv y$ for any two points $x, y \in X$. For this, it is enough to consider the family of divisors f parametrised by X itself, and defined by the diagonal $\Delta \subset X \times X$; it is easy to check that $f(x) = x$ for every $x \in X$. Hence for every divisor $D = \sum n_i x_i$ and any point $x_0 \in X$ we have $D \equiv (\sum n_i) x_0$, that is, any two divisors of the same degree are algebraically equivalent.

It is slightly more complicated to prove the converse implication, that algebraically equivalent divisors have the same degree. We do not give the proof here.[10] Thus for divisors on a nonsingular projective curve X, algebraic

---

[10]The so-called "Principle of conservation of number" (roughly, algebraically equivalent cycles have the same numerical properties) is discussed in Fulton [27],

equivalence of divisors is equivalent to them having equal degree. Therefore

$$\operatorname{Div} X/\operatorname{Div}^{\mathrm{a}} X = \operatorname{Cl} X/\operatorname{Cl}^0 X = \mathbb{Z}.$$

A generalisation of this is the following theorem proved by Severi (for fields of characteristic 0) and Néron (in the general case).

Theorem D. (The Néron–Severi Theorem). *For X a nonsingular projective variety, the group $\operatorname{NS} X = \operatorname{Div} X/\operatorname{Div}^{\mathrm{a}} X$ is finitely generated.* $\square$

One can show that algebraic and linear equivalence of divisors coincide on $X = \mathbb{P}^{n_1} \times \cdots \times \mathbb{P}^{n_k}$. This example shows that $\operatorname{Div} X/\operatorname{Div}^{\mathrm{a}} X$ can be more complicated that $\mathbb{Z}$.

When X is a plane cubic curve, the quotient $\operatorname{Cl}^0 X = \operatorname{Div}^{\mathrm{a}} X/P(X)$, where $P(X)$ is the group of principal divisors, is a 1-dimensional Abelian variety. In a similar way, for any nonsingular projective variety X there exists an Abelian variety G whose group of points is isomorphic to $\operatorname{Div}^{\mathrm{a}} X/P(X)$, that is, divisors algebraically equivalent to 0 modulo divisors linearly equivalent to 0, and having the following property: for any algebraic family of divisors f on X over a base T there exists a regular map $\varphi \colon T \to G$ such that $f(t) - f(t_0) \in \varphi(t)$, where t_0 is some fixed point of T. (Here G is identified with $\operatorname{Div}^{\mathrm{a}} X/P(X)$, so that $\varphi(t)$ is considered as a divisor class.)

The Abelian variety G is uniquely determined by this property. It is called the *Picard variety* of X. The Picard variety of a nonsingular projective curve X is also called the *Jacobian* of X.

Exercises to §4

1. Let G be an algebraic group, $\psi \colon G \times G \to G$ the regular map defined by the group law, and Θ_e and $\Theta_{e'}$ the tangent spaces to G and $G \times G$ at their respective identity elements. Prove that $\Theta_{e'} = \Theta_e \oplus \Theta_e$ and that $d_e\psi \colon \Theta_e \oplus \Theta_e \to \Theta_e$ is given by addition of vectors.

2. In the notation of Ex. 1, suppose that G is an Abelian group, and define $\varphi_n \colon G \to G$ by $\varphi_n(g) = g^n$. Supposing that the ground field has characteristic 0, prove that $d_e\varphi_n$ is a nondegenerate linear map. Deduce from this that in a Abelian algebraic group the number of elements of order n is finite, and that every element has an nth root. ˙

Chap. 10, although the proof given there is very high-powered and abstract. §19.3 of the same book also contains a condensed discussion of the Néron–Severi theorem over $\mathbb{C}$.

5. Differential Forms

5.1. Regular Differential 1-forms

We introduced the notion of the differential $d_x f$ of a regular function f at a point $x \in X$ of a variety in Chap. II, 1.3. By definition, $d_x f$ is a linear form on the tangent space Θ_x to X at x, that is, $d_x f \in \Theta_x^*$. We now study how this notion depends on x.

Fix a function f regular everywhere on X. Then, as a function of x, the differential $d_x f$ is an object of a new type that we have not met before: it sends each point $x \in X$ to a vector $d_x f \in \Theta_x^*$ in the dual space of the tangent space at x. Objects of this nature will appear all the time in what follows. Perhaps the following explanation will be helpful. In linear algebra, we deal with constants, but also with other quantities, such as vectors, linear forms, and arbitrary tensors. In geometry, the analogue of a constant is a function, which takes constant values at points. The analogue of a vector, linear form or whatever, is a "field", or "function" that sends each point x of an algebraic variety (or differentiable manifold) into a vector, linear form or whatever, of the tangent space Θ_x at the point x.

Consider the set $\Phi[X]$ of all possible functions φ sending each point $x \in X$ to a vector $\varphi(x) \in \Theta_x^*$. This set is of course much too big to be of any interest, just as the set of all k-valued function on X is too big. Now, just as we distinguished the regular functions among all k-valued functions on X, we now distinguish in $\Phi[X]$ a subset that is more closely related to the structure of X. For this we note that $\Phi[X]$ is an Abelian group, if we set $(\varphi + \psi)(x) = \varphi(x) + \psi(x)$. Moreover, $\Phi[X]$ is a module over the ring of all k-valued functions on X, if we set $(f\varphi)(x) = f(x)\varphi(x)$ for $f \colon X \to k$ and $\varphi \in \Phi[X]$. In particular we may view $\Phi[X]$ as a module over the ring $k[X]$ of regular functions on X.

As we have seen, a regular function f on X defines a differential $d_x f \in \Theta_x^*$ at x. Thus any function $f \in k[X]$ defines an element $\varphi \in \Phi[X]$ by $\varphi(x) = d_x f$. We denote this function by df.

Definition. An element $\varphi \in \Phi[X]$ is a *regular differential form* on X if every point $x \in X$ has a neighbourhood U such that the restriction of φ to U belongs to the $k[U]$-submodule of $\Phi[U]$ generated by the elements df with $f \in k[U]$.

All the regular differential forms on X obviously form a module over $k[X]$; we denote it by $\Omega[X]$. Thus $\varphi \in \Omega[X]$ if it can be written in the form

$$\varphi = \sum_{i=1}^{m} f_i dg_i, \qquad (1)$$

in a neighbourhood of every point $x \in X$, where $f_1, \ldots, f_m, g_1, \ldots, g_m$ are regular functions in a neighbourhood of x.

Taking the differential of functions defines a map $d: k[X] \to \Omega[X]$. The properties Chap. II, 1.3, (1) then take the form

$$d(f + g) = df + dg \quad \text{and} \quad d(fg) = f\,dg + g\,df. \tag{2}$$

From these formulas one deduces easily an identity that holds for any polynomial $F \in k[T_1, \ldots, T_m]$ and any functions $f_1, \ldots, f_m \in k[X]$:

$$d\big(F(f_1, \ldots, f_m)\big) = \sum_{i=1}^{m} \frac{\partial F}{\partial T_i}(f_1, \ldots, f_m)\,df_i. \tag{3}$$

To obtain this, using (2), one reduces the proof to the case of a monomial, and then, using (2) again, proves it by induction of the degree of the monomial. We leave the details of the verification to the reader.

Once (3) is proved for polynomials, it generalises immediately to rational functions F. Here we should note that if a rational function F is regular at x, then so are all the $\partial F/\partial T_i$; indeed, then $F = P/Q$, where P, Q are polynomials and $Q(x) \neq 0$, and so

$$\frac{\partial F}{\partial T_i} = \frac{1}{Q^2}\left(Q\frac{\partial P}{\partial T_i} - P\frac{\partial Q}{\partial T_i}\right),$$

whence regularity.

Example 1. $X = \mathbf{A}^n$. Since the differentials $d_x t_1, \ldots, d_x t_n$ of the coordinates form a basis of the vector space Θ_x^* at any point $x \in \mathbf{A}^n$, any element $\varphi \in \Phi[\mathbf{A}^n]$ can be written uniquely in the form $\varphi = \sum_{i=1}^{n} \psi_i dt_i$, where the ψ_i are k-valued functions on $\mathbf{A}^n$.

If $\varphi \in \Omega[\mathbf{A}^n]$ then φ must have an expression (1) in a neighbourhood of any point $x \in \mathbf{A}^n$. Applying the relations (3) to g_i gives an expression $\varphi = \sum h_i dt_i$, in which the h_i are rational functions regular at x. Since such an expression is unique, the ψ_i must be regular at every $x \in \mathbf{A}^n$, that is, $\psi_i \in k[\mathbf{A}^n]$. Therefore

$$\Omega[\mathbf{A}^n] = \bigoplus k[\mathbf{A}^n]dt_i.$$

Example 2. Let $X = \mathbb{P}^1$ with coordinate t. Then $X = \mathbf{A}_0^1 \cup \mathbf{A}_1^1$, where $\mathbf{A}_0^1 \cong \mathbf{A}_1^1 \cong \mathbf{A}^1$. By the result of Example 1, any element $\varphi \in \Omega[\mathbb{P}^1]$ can be written $\varphi = P(t)dt$ on $\mathbf{A}_0^1$ and $\varphi = Q(u)du$ on $\mathbf{A}_1^1$, where $u = t^{-1}$. The final relation gives $du = -dt/t^2$, so that if $n = \deg Q$, we have in $\mathbf{A}_0^1 \cap \mathbf{A}_1^1$

$$P(t)dt = -\frac{1}{t^2}Q(1/t)dt, \qquad \text{that is,} \qquad P(t) = \frac{-Q^*(t)}{t^{n+2}},$$

where $Q^*(t) = t^n Q(1/t)$, so that $Q^*(0) \neq 0$. A relation of this kind between polynomials is only possible if $P = Q = 0$, and hence $\Omega[\mathbb{P}^1] = 0$.

Example 3. Suppose that $X \subset \mathbb{P}^2$ is given by the equation $x_0^3 + x_1^3 + x_2^3 = 0$, and that char $k \neq 3$. Write U_{ij} for the open set in which $x_i, x_j \neq 0$. Then $X = U_{01} \cup U_{12} \cup U_{20}$. We set

$$x = \frac{x_1}{x_0}, \quad y = \frac{x_2}{x_0} \quad \text{and} \quad \varphi = \frac{dy}{x^2} \quad \text{in } U_{01};$$

$$u = \frac{x_2}{x_1}, \quad v = \frac{x_0}{x_1} \quad \text{and} \quad \psi = \frac{dv}{u^2} \quad \text{in } U_{12};$$

$$s = \frac{x_0}{x_2}, \quad t = \frac{x_1}{x_2} \quad \text{and} \quad \chi = \frac{dt}{s^2} \quad \text{in } U_{20}.$$

Obviously $\varphi \in \Omega[U_{01}]$, $\psi \in \Omega[U_{12}]$ and $\chi \in \Omega[U_{20}]$. It is easy to check that $\varphi = \psi = \chi$ in $U_{01} \cap U_{12} = U_{01} \cap U_{20} = U_{12} \cap U_{20}$. Hence φ, ψ and χ define a global form $\omega \in \Omega[X]$. This example is interesting, in that $\Omega[X] \neq 0$, whereas X is a projective variety, and so has no everywhere regular functions other than the constants.

In the general case one can prove a weaker version of the result of Example 1.

Theorem 1. *Any nonsingular point x of an algebraic variety X has an affine neighbourhood U such that $\Omega[U]$ is a free $k[U]$-module of rank $\dim_x X$.*

Proof. Let $X \subset \mathbb{A}^N$, and suppose that $F_1, \ldots, F_m$ are polynomials forming a basis of the ideal $\mathfrak{A}_X$. Then $F_i = 0$ on X, and hence by (3) we have

$$\sum_{j=1}^{N} \frac{\partial F_i}{\partial T_j} dt_j = 0. \tag{4}$$

If $x \in X$ is a nonsingular point and $\dim_x X = n$ then the matrix $\left((\partial F_i / \partial T_j)(x)\right)$ has rank $N - n$. Suppose, for example, that $t_1, \ldots, t_n$ are local parameters at x. Then it follows from (4) that all the dt_j can be expressed in terms of $dt_1, \ldots, dt_n$ with coefficients that are rational functions, regular at x.

Consider a neighbourhood U of x in which all these functions are regular. Then $d_y t_1, \ldots, d_y t_n$ form a basis of Θ_y^* for every $y \in U$. Let $\varphi \in \Omega[U]$. By what we have said above, there exists a unique expression

$$\varphi = \sum_{i=1}^{n} \psi_i dt_i, \tag{5}$$

where ψ_i are k-valued functions on U. By the expression (1) and formula (3) it follows that φ can be written in a neighbourhood of any point $y \in U$ as a linear combination of $dt_1, \ldots, dt_N$ with coefficients functions that are regular at y. As we have seen, $dt_1, \ldots, dt_N$ can in turn be expressed in the same way in terms of $dt_1, \ldots, dt_n$. Hence $\varphi = \sum_{i=1}^{n} g_i dt_i$, where the g_i are regular in a neighbourhood of y. Since the expression (5) is unique, it follows that $\psi_i = g_i$ in a neighbourhood of each $y \in U$, and hence $\psi_i \in k[U]$. We see that $\Omega[U] = \sum_{i=1}^{n} k[U] dt_i$.

Suppose that $dt_1, \ldots, dt_n$ are related by $\sum_{i=1}^{n} g_i dt_i = 0$, with, say, $g_n \neq 0$. Then $dt_1, \ldots, dt_n$ are linearly dependent in the open set $g_n \neq 0$, which contradicts the linear independence of $d_y t_i \in \Theta_y^*$ for all $y \in U$. Therefore $\Omega[U] = \bigoplus_{i=1}^{n} k[U] dt_i$. The theorem is proved.

Corollary. *If $u_1, \ldots, u_n$ is any system of local parameters at a point x, then $du_1, \ldots, du_n$ generate $\Omega[U]$ as a $k[U]$-module for some affine neighbourhood U of x.*

Proof. Let $dt_1, \ldots, dt_n$ be the basis of the free module $\Omega[U]$ in a neighbourhood U of x that exists by Theorem 1. Then $du_i = \sum_{j=1}^{n} g_{ij} dt_j$, and since the u_i are local parameters, $\det |g_{ij}(x)| \neq 0$. Therefore in the neighbourhood U' where $\det |g_{ij}| \neq 0$, the elements $du_1, \ldots, du_n$ generate $\Omega[U']$.

5.2. Algebraic Definition of the Module of Differentials

We saw in Chap. I that the category of affine varieties is equivalent to the category of rings of a special type. Thus the whole theory of affine varieties can be seen from a purely algebraic point of view; in particular, we can try to understand the algebraic meaning of the module of differential forms.

Consider an affine variety X, and write $A = k[X]$ for its coordinate ring and $\Omega = \Omega[X]$ for the module of differentials. Taking the differential of a function defines a k-linear homomorphism $d \colon A \to \Omega$.

Proposition 1. *Ω is generated as an A-module by the elements df with $f \in A$.*

Proof. This is analogous to Chap. I, 3.2, Theorem 4, and the proof is entirely similar. If $\omega \in \Omega$ then, by definition, for any $x \in X$ we can write $\omega = \sum f_{i,x} dg_{i,x}$ with $f_{i,x}, g_{i,x} \in \mathcal{O}_x$. Every function $u \in \mathcal{O}_x$ can be expressed as $u = v/w$ with $v, w \in A$ and $w(x) \neq 0$. Using such expressions for $f_{i,x}, g_{i,x}$ and taking a common denominator for all the fractions, we obtain a function p_x such that $p_x(x) \neq 0$ and

$$p_x \omega = \sum r_{i,x} dh_{i,x} \qquad \text{with } r_{i,x}, h_{i,x} \in A.$$

Now because $p_x(x) \neq 0$ for each $x \in X$, there exist a finite set of x and functions $q_x \in A$ such that $\sum p_x q_x = 1$. Therefore $\omega = \sum_x \sum_i q_x r_{i,x} dh_{i,x}$. This proves Proposition 1.

Proposition 1 suggests the idea of trying to describe Ω in terms of its generators df with $f \in A$. The following relations obviously hold:

$$d(f + g) = df + dg, \qquad \text{and} \qquad d\alpha = 0 \quad \text{for } \alpha \in k. \qquad (1)$$
$$d(fg) = fdg + gdg,$$

Proposition 2. *If X is a nonsingular affine variety and $A = k[X]$ then the A-module Ω is defined by the relations (1).*

Proof. Write R for the A-module having generators df in one-to-one correspondence with elements $f \in A$, and relations (1). There is an obvious homomorphism $\xi \colon R \to \Omega$, and Proposition 1 shows that ξ is surjective.

It remains to prove that $\ker \xi = 0$. Suppose that $\varphi \in R$ and $\xi(\varphi) = 0$. We observe that the arguments in the proof of Theorem 1 only used the relations (1). Hence they are applicable also to R and show that for any $x \in X$ there exists a function $p \in A$ such that $p(x) \neq 0$ and $p\varphi = \sum g_i dt_i$ with $g_i \in A$, where now the local parameters t_i are chosen as elements of A. If $\xi(\varphi) = 0$, then $\sum g_i dt_i = 0$ in the module Ω, and it follows from Theorem 1 that all the $g_i = 0$. Thus $p\varphi = 0$. We see that for every $x \in X$ there exists a function $p \in A$ such that $p(x) \neq 0$ and $p\varphi = 0$. Arguing as in the proof of Proposition 1, we see that $\varphi = 0$. The proposition is proved.

Remark. It follows easily from what we have just said that for any A-module M we have

$$\mathrm{Hom}_A(\Omega_{A/A_0}, M) = \mathrm{Der}_{A_0}(A, M),$$

where $\mathrm{Der}_{A_0}(A, M)$ is the module of derivations of A into M, that is, of A_0-linear maps $D \colon A \to M$ satisfying $D(ab) = aD(b) + bD(a)$ for all $a, b \in A$.

Thus in this case $\Omega[X]$ can be described purely algebraically starting from the ring $k[X]$. This suggests the idea of considering a similar module for any ring A, with A an algebra over a subring A_0. The module defined by generators df for $f \in A$ and relations (1) (where of course $\alpha \in A_0$ in the final relation) is called the *module of differentials* or *module of Kähler differentials* of A over A_0 and denoted by Ω_A or Ω_{A/A_0}.

If X is singular, then the module Ω_A defined in this purely algebraically way no longer coincides with $\Omega[X]$ in general (see Ex. 9). Proposition 1, which still holds for a singular variety X, shows that Ω_A contains more information on X than $\Omega[X]$. However, in what follows, we mostly deal with nonsingular varieties, and the distinction will not be important for us.

5.3. Differential p-forms

The differential forms considered in 5.1 were functions sending each point $x \in X$ to an element of Θ_x^*. We now consider more general differential forms, that send $x \in X$ to a skewsymmetric multilinear form on Θ_x, that is, to an element of the rth exterior power $\bigwedge^r \Theta_x^*$ of Θ_x^*.

The definition is entirely analogous to that considered in 5.1. We write $\Phi^r[X]$ for the set of all possible functions sending each point $x \in X$ to an element of $\bigwedge^r \Theta_x^*$. Thus if $\omega \in \Phi^r[X]$ and $x \in X$ then $\omega(x) \in \bigwedge^r \Theta_x^*$. In particular, $\Phi^0[X]$ is the ring of all k-valued functions on X and $\Phi^1[X]$ is the set $\Phi[X]$ considered in the preceding section. Hence $df \in \Phi^1[X]$ for $f \in k[X]$.

We recall the operation of *exterior product* $\wedge$, which is defined for any vector space L. If $\varphi \in \bigwedge^r L$ and $\psi \in \bigwedge^s L$, then $\varphi \wedge \psi \in \bigwedge^{r+s} L$. Moreover, the exterior product is distributive, associative and satisfies $\varphi \wedge \psi = (-1)^{rs} \psi \wedge \varphi$. If $e_1, \ldots, e_n$ is a basis of L then a basis of $\bigwedge^r L$ is given by all the products $e_{i_1} \wedge \cdots \wedge e_{i_r}$ with $1 \le i_1 < \cdots < i_r \le n$. Hence $\dim \bigwedge^r L = \binom{n}{r}$ (the binomial coefficient), and in particular $\dim \bigwedge^n L = 1$ and $\bigwedge^r L = 0$ for $r > n$.

We define the exterior product on the sets $\Phi^r[X]$. If $\omega_r \in \Phi^r[X]$ and $\omega_s \in \Phi^s[X]$ then we define $\omega = \omega_r \wedge \omega_s$ by $\omega(x) = \omega_r(x) \wedge \omega_s(x)$ for all $x \in X$. Obviously $\omega \in \Phi^{r+s}[X]$. For $r = 1$ and $s = 0$ we return to the multiplication of elements of $\Phi^1[X] = \Phi[X]$ by k-valued functions on X. Taking any r and $s = 0$, we see that elements of $\Phi^r[X]$ can be multiplied by functions on X. In particular all the $\Phi^r[X]$ are modules over $k[X]$.

Definition. An element $\varphi \in \Phi^r[X]$ is a *regular differential r-form* on X if any point $x \in X$ has a neighbourhood U such that φ on U belongs to the submodule of $\Phi^r[U]$ generated over $k[U]$ by the elements $df_1 \wedge \cdots \wedge df_r$ with $f_1, \ldots, f_r \in k[U]$. In terms of this definition, the differential forms considered in 5.1–2 are regular differential 1-forms.

All regular differential r-forms on X form a $k[X]$-module, denoted by $\Omega^r[X]$. Thus an element $\omega \in \Omega^r[X]$ can be written in a neighbourhood of any point $x \in X$ in the form

$$\omega = \sum g_{i_1 \ldots i_r} df_{i_1} \wedge \cdots \wedge df_{i_r}, \tag{1}$$

where $g_{i_1 \ldots i_r}$ and $f_{i_1}, \ldots, f_{i_r}$ are regular functions in a neighbourhood of x. The exterior product is defined on regular differential forms, and for $\omega_r \in \Omega^r[X]$ and $\omega_s \in \Omega^s[X]$, we obviously have $\omega_r \wedge \omega_s \in \Omega^{r+s}[X]$.

5.2, Theorem 1 has an analogue for the forms $\Omega^r[X]$ for any r.

Theorem 2. *Any nonsingular point $x \in X$ of an n-dimensional variety has a neighbourhood U such that $\Omega^r[U]$ is a free $k[U]$-module of rank $\binom{n}{r}$.*

Proof. In the proof of Theorem 1, we saw that a nonsingular point x has a neighbourhood U on which there are n regular functions $u_1, \ldots, u_n$ such that $d_y u_1, \ldots, d_y u_n$ form a basis of Θ_y^* for any $y \in U$. It follows from this that any element $\varphi \in \Phi^r[U]$ is of the form

$$\varphi = \sum \psi_{i_1 \ldots i_r} du_{i_1} \wedge \cdots \wedge du_{i_r},$$

where the $\psi_{i_1 \ldots i_r}$ are k-valued functions on U.

If $\varphi \in \Omega^r[U]$ then φ can be expressed in the form (1) in a neighbourhood of any point $y \in U$. Applying Theorem 1 to the forms df_i we see that $\psi_{i_1 \ldots i_r}$ are regular at y; but since y is any point of U, they are regular functions on U. Thus the forms $du_{i_1} \wedge \cdots \wedge du_{i_r}$ for $1 \leq i_1 < \cdots < i_r \leq n$ generate $\Omega^r[U]$. It remains to see that they are linearly independent over $k[U]$. But any dependence relation

$$\sum g_{i_1 \ldots i_r} du_{i_1} \wedge \cdots \wedge du_{i_r} = 0$$

gives a relation

$$\sum g_{i_1 \ldots i_r}(x) d_x u_{i_1} \wedge \cdots \wedge d_x u_{i_r} = 0 \qquad (2)$$

at any point $x \in U$. Since $d_x u_1, \ldots, d_x u_n$ form a basis of Θ_x^*, the elements $d_x u_{i_1} \wedge \cdots \wedge d_x u_{i_r}$ form a basis of $\bigwedge^r \Theta_x^*$. Hence from (2) it follows that $g_{i_1 \ldots i_r}(x) = 0$ for all $x \in U$, that is, $g_{i_1 \ldots i_r} = 0$. The theorem is proved.

Of special importance is the module $\Omega^n[U]$, which under the assumptions of Theorem 2 is of rank 1 over $k[U]$. [11] Thus if $\omega \in \Omega^n[U]$, we have

$$\omega = g du_1 \wedge \cdots \wedge du_n \qquad \text{with } g \in k[U]. \qquad (3)$$

This expression for ω depends in an essential way on the choice of the local parameters $u_1, \ldots, u_n$. We determine what this dependence is. Let $v_1, \ldots, v_n$ be another n regular functions on X such that $v_1 - v_1(x), \ldots, v_n - v_n(x)$ are local parameters at any point $x \in U$. Then also

$$\Omega^1[U] = k[U] dv_1 \oplus \cdots \oplus k[U] dv_n.$$

and in particular the du_i can be expressed

$$du_i = \sum_{j=0}^{n} h_{ij} dv_j \qquad \text{for } i = 1, \ldots, n. \qquad (4)$$

Since $d_x u_1, \ldots, d_x u_n$ form a basis of the vector space Θ_x^* for each $x \in U$, it follows from (4) that $\det |h_{ij}(x)| \neq 0$. By analogy with what happens in analysis, $\det |h_{ij}(x)|$ is called the *Jacobian determinant* of the functions $u_1, \ldots, u_n$ with respect to $v_1, \ldots, v_n$. We denote it by $J\left(\frac{u_1, \ldots, u_n}{v_1, \ldots, v_n}\right)$. As we have seen,

$$J\left(\frac{u_1, \ldots, u_n}{v_1, \ldots, v_n}\right) \in k[U], \qquad \text{and} \qquad J\left(\frac{u_1, \ldots, u_n}{v_1, \ldots, v_n}\right)(x) \neq 0 \qquad (5)$$

for all $x \in U$.

Substituting (4) in the expression for ω and simple calculations in the exterior algebra shows that

---

[11] Elements of $\Omega^n[U]$ are called *canonical differentials*, following a suggestion of Mumford.

$$\omega = gJ\left(\frac{u_1, \ldots, u_n}{v_1, \ldots, v_n}\right) dv_1 \wedge \cdots \wedge dv_n. \tag{6}$$

Thus although $\omega \in \Omega^n[U]$ is specified by a function $g \in k[X]$, this specification is only possible once local coordinates have been chosen, and depends in an essential way on this choice.

We recall that the expression (3) is in general only possible locally (see the statements of Theorems 1–2). If $X = \bigcup U_i$ is an open cover, and in each U_i an expression (3) is possible, we still cannot associate with ω a global function g on the whole of X: the functions g_i obtained on different U_i are not compatible. We have already seen an example of this in 5.1, Example 2.

5.4. Rational Differential Forms

5.1, Example 2 shows that there may be very few regular differential forms on an algebraic variety X (for example, $\Omega^1[\mathbb{P}^1] = 0$) whereas there are lots on its open subsets (for example, $\Omega^1[U] = k[u]du$). A similar thing happened in connection with regular functions, and it was precisely these considerations that led us to introduce the notion of rational functions, as functions regular on some open subset. We now introduce the analogous notion for differential forms.

Consider a nonsingular irreducible quasiprojective variety X. Let ω be a differential r-form on X. Recall that it makes sense to speak of ω being 0 at a point $x \in X$; for $\omega(x) \in \bigwedge^r \Theta_x^*$, and in particular, it can be 0 there.

Lemma. *The set of points at which a regular differential form ω is 0 is closed.*

Proof. Since closed is a local property, we can restrict ourselves to a sufficiently small neighbourhood U of any point $x \in X$. In particular we can choose U so that 5.1, Theorem 1 and 5.3, Theorem 2 hold for it. Then there exist functions $u_1, \ldots, u_n \in k[U]$ such that $\Omega^r[U]$ is the free $k[U]$-module based by

$$du_{i_1} \wedge \cdots \wedge du_{i_r} \qquad \text{for } 1 \le i_1 < \cdots < i_r \le n.$$

Hence ω has a unique expression in the form $\omega = \sum g_{i_1 \ldots i_r} du_{i_1} \wedge \cdots \wedge du_{i_r}$, and the conditions $\omega(x) = 0$ is equivalent to $g_{i_1 \ldots i_r} = 0$, which define a closed set. The lemma is proved.

It follows in particular from the lemma that if $\omega(x) = 0$ for all points x of an open set U then $\omega = 0$ on the whole of X.

We now introduce a new object, consisting of an open set $U \subset X$ and a differential r-form $\omega \in \Omega^r[U]$. On pairs (ω, U) we introduce the equivalence relation $(\omega, U) \sim (\omega', U')$ if $\omega = \omega'$ on $U \cap U'$. By the above remark, it is enough to require that $\omega = \omega'$ on some open subset of $U \cap U'$. The transitivity of the equivalence relation follows from this. An equivalence class under this relation is called a *rational differential r-form* on X. The set of all rational differential r-forms on X is denoted by $\Omega^r(X)$. Obviously $\Omega^0(X) = k(X)$.

Algebraic operations on representatives of equivalence classes carry over to the classes, and define the exterior product: if $\omega_r \in \Omega^r(X)$ and $\omega_s \in \Omega^s(X)$ then $\omega_r \wedge \omega_s \in \Omega^{r+s}(X)$. When $s = 0$ we see that $\Omega^r(X)$ is a $k(X)$-module.

If a rational differential form ω (an equivalence class of pairs) contains a pair (ω, U) then we say that ω is *regular* in U. The union of all open sets on which ω is regular is an open set U_ω, called the *domain of regularity* of ω. Obviously ω defines a regular form belonging to $\Omega^r[U_\omega]$. If $x \in U_\omega$ then we say that ω is regular at x. Obviously $\Omega^r(X)$ does not change if we replace X by an open subset, that is, it is a birational invariant.

We determine the structure of $\Omega^r(X)$ as a module over the field $k(X)$.

Theorem 3. *$\Omega^r(X)$ is a vector space over $k(X)$ of dimension $\binom{n}{r}$.*

Proof. Consider any open set $U \subset X$ for which $\Omega^r[U]$ is free over $k[U]$, as in Theorems 1–2. Then there exist n functions $u_1, \ldots, u_n \in k[U]$ such that the products
$$du_{i_1} \wedge \cdots \wedge du_{i_r} \qquad \text{for } 1 \leq i_1 < \cdots < i_r \leq n \tag{1}$$
form a basis of $\Omega^r[U]$ over $k[U]$. Any form $\omega \in \Omega^r(X)$ is regular in some open subset $U' \subset U$, over which (1) still gives a basis of $\Omega^r[U']$ over $k[U']$. Hence ω' can be uniquely written in the form
$$\sum_{1 \leq i_1 < \cdots < i_r \leq n} g_{i_1 \ldots i_r} du_{i_1} \wedge \cdots \wedge du_{i_r},$$
where $g_{i_1 \ldots i_r}$ are regular in some open set $U' \subset U$, that is, are rational functions on X. This just means that the forms (1) are a basis of $\Omega^r(X)$ over $k(X)$. The theorem is proved.

For which n-tuples of functions $u_1, \ldots, u_n \in k(X)$ is $du_{i_1} \wedge \cdots \wedge du_{i_r}$ for $1 \leq i_1 < \cdots < i_r \leq n$ a basis of $\Omega^r(X)$ over $k(X)$? We give a sufficient condition for this – in fact it is also necessary, but we do not need this.

Theorem 4. *If $u_1, \ldots, u_n$ is a separable transcendence basis of $k(X)$ then the forms $du_{i_1} \wedge \cdots \wedge du_{i_r}$ for $1 \leq i_1 < \cdots < i_r \leq n$ form a basis of $\Omega^r(X)$ over $k(X)$.*

Proof. Since $\Omega^r(X)$ and $k(X)$ are birational invariants, we can assume that X is affine, $X \subset \mathbf{A}^N$. Let $u_1, \ldots, u_n$ be a separable transcendence basis of $k(X)$. Then any element $v \in k(X)$ satisfies a relation $F(v, u_1, \ldots, u_n) = 0$ that is separable in v. In particular for each of the coordinates t_i of $\mathbf{A}^N$, there is a relation $F_i(t_i, u_1, \ldots, u_n) = 0$, for $i = 1, \ldots, N$. It follows from these that the relations

$$\frac{\partial F_i}{\partial t_i} dt_i + \sum_{j=1}^{n} \frac{\partial F_i}{\partial u_j} du_j = 0 \qquad \text{for } i = 1, \ldots, N$$

hold on X. Since F_i is separable in t_i it follows that $\partial F_i/\partial t_i \neq 0$ on X. Hence

$$dt_i = \sum_{j=1}^{n} -\frac{(\partial F_i/\partial u_j)}{(\partial F_i/\partial t_i)} du_j. \tag{2}$$

All the function $(\partial F_i/\partial u_j) \big/ (\partial F_i/\partial t_i)$ and u_i are regular on some open set $U \subset X$, and then (2) shows that at any point $y \in U$, the differentials $d_y u_j$ generate Θ_y^*. Since the number of these differentials is equal to $\dim X = \dim \Theta_y^*$, they form a basis. Hence the du_i form a basis of $\Omega^1[U]$ as a $k[U]$-module, and the products (1) a basis of $\Omega^r[U]$ over $k[U]$, hence a fortiori of $\Omega^r(X)$ over $k(X)$. The theorem is proved.

Exercises to §5

1. Prove that the rational differential form dx/y is regular on the affine circle X defined by $x^2 + y^2 = 1$. We suppose that the ground field has characteristic $\neq 2$.

2. In the notation of Ex. 1, prove that $\Omega^1[X] = k[X](dx/y)$. [Hint: Write any form $\omega \in \Omega^1[X]$ in the form $\omega = f dx/y$, and use the fact that $dx/y = -dy/x$.]

3. Prove that $\dim \Omega^1[X] = 1$ in 5.1, Example 3.

4. Prove that $\Omega^n[\mathbb{P}^n] = 0$.

5. Prove that $\Omega^1[\mathbb{P}^n] = 0$.

6. Prove that $\Omega^r[\mathbb{P}^n] = 0$ for any $r > 0$.

7. Let $\omega = \left(P(t)/Q(t)\right) dt$ be a rational form on $\mathbb{P}^1$, with coordinate t, where P and Q are polynomials with $\deg P = m$, $\deg Q = n$. At what points $x \in \mathbb{P}^1$ is ω not regular?

8. Prove that for a nonsingular variety X, the tangent fibre space introduced in Chap. II, 1.4 is birational to the product $X \times \mathbb{A}^n$. [Hint: For the open set U in 5.1, Theorem 1, construct an isomorphism of the tangent fibre space of U to $U \times \mathbb{A}^n$ by $(x, \xi) \mapsto x, (d_x u_1)(\xi), \ldots, (d_x u_n)(\xi)$, for $\xi \in \Theta_x$.]

9. Compute the module $R = \Omega_A$ constructed in the proof of 5.2, Proposition 2 when X is the curve $y^2 = x^3$, and prove that $3y\,dx - 2x\,dy$ is a nonzero element of Ω_A, but $\xi(3y\,dx - 2x\,dy) = 0 \in \Omega^1[X]$ (where $\xi: R = \Omega_A \to \Omega[X]$ is as in the proof of 5.2, Proposition 2). Show also that

$$y(3y\,dx - 2x\,dy) = x^2(3y\,dx - 2x\,dy) = 0.$$

[Hint: Use the fact that $k[X] = k[x] + k[x]y$. The point is that on a singular variety Kähler differentials (5.2) and regular differentials are different notions.]

10. Let K be an extension field of k. A *derivation* of K over k is a k-linear map $D: K \to K$ satisfying the conditions $D(xy) = yD(x) + xD(y)$ for $x, y \in K$. Prove that if $u \in K$ and D is a derivation, then the map $D_1(x) = uD(x)$ is also a derivations, so that all the derivations of K over k form a vector space over k. This is denoted by $\operatorname{Der}_k(K)$.

11. Let D be a derivation of $K = k(X)$ over k, and $\omega = \sum f_i dg_i \in \Omega^1(X)$. Prove that the function $(D, \omega) = \sum f_i D(g_i)$ is independent of the representation of ω in the form $\sum f_i dg_i$. Prove that it is a scalar product, and establishes an isomorphism $\operatorname{Der}_k(K) \cong (\Omega^1(X))^* = \operatorname{Hom}_{k(X)}(\Omega^1(X), k(X))$.

6. Examples and Applications of Differential Forms

6.1. Behaviour Under Maps

We first study the behaviour of differential forms under regular maps. If $\varphi: X \to Y$ is a regular map, $x \in X$ and $y = \varphi(x) \in Y$ then $d_x\varphi$ is a map $d_x\varphi: \Theta_{X,x} \to \Theta_{Y,y}$, and its dual a map $(d_x\varphi)^*: \Theta_{Y,y}^* \to \Theta_{X,x}^*$. Hence for $\omega \in \Phi[Y]$ we have a pullback $\varphi^*(\omega) \in \Phi[X]$ defined by $\varphi^*(\omega)(x) = (d_x\varphi)^*(\omega(y))$.

It follows easily from the definition that $(d_x\varphi)^*$ is compatible with taking the differential, that is, $(d_x\varphi)^*(d_yf) = d_x(\varphi^*(f))$ for $f \in k[Y]$. It follows that if $\omega \in \Omega^1[Y]$ then $\varphi^*(\omega) \in \Omega^1[X]$, and φ^* defines a homomorphism $\varphi^*: \Omega^1[Y] \to \Omega^1[X]$ that is compatible with taking the differential of $f \in k[Y]$.

Finally, it is known from linear algebra that a linear map $\varphi: L \to M$ between vector spaces determines a linear map $\bigwedge^r \varphi: \bigwedge^r L \to \bigwedge^r M$. Applying this to $(d_x\varphi)^*$, we get a map $\bigwedge^r (d_x\varphi)^*: \bigwedge^r \Theta_{Y,y}^* \to \bigwedge^r \Theta_{X,x}^*$, hence maps $\Phi^r[Y] \to \Phi^r[X]$ and $\Omega^r[Y] \to \Omega^r[X]$. These maps will also be denoted φ^*.

From what we have said above, it follows that the effective computation of the action of φ^* on differential forms is very simple: if

$$\omega = \sum g_{i_1\ldots i_r} du_{i_1} \wedge \cdots \wedge du_{i_r},$$

then

$$\varphi^*(\omega) = \sum \varphi^*(g_{i_1\ldots i_r}) d(\varphi^*(u_{i_1})) \wedge \cdots \wedge d(\varphi^*(u_{i_r})). \tag{1}$$

Now suppose that X is irreducible, and $\varphi: X \to Y$ a rational map such that $\varphi(X)$ is dense in Y. Since φ is a regular map of an open set $U \subset X$ to

Y, and any open set $V \subset Y$ intersects $\varphi(U)$, the preceding arguments define a map $\varphi^* \colon \Omega^r(Y) \to \Omega^r(X)$. This is again given by (1).

We know that for $r = 0$, in other words, for functions, the map $\varphi^* \colon k(Y) \to k(X)$ is an inclusion. For differential forms this is not always so. Suppose, for example, that $X = Y = \mathbb{P}^1$, with respective coordinates t and u, so that $k(X) = k(t)$ and $k(Y) = k(u)$. Suppose that k has finite characteristic p and that φ is given by the formula $u = t^p$. Then $\varphi^*(f(u)) = f(t^p)$, and $\varphi^*(df) = d(f(t^p)) = 0$ for all $f \in k(u)$ (because $dt^p = pt^{p-1}dt = 0$), so that $\varphi^*(\Omega^1(Y)) = 0$. The situation is clarified by the following result.

Theorem 1. *If $k(X)$ has a separable transcendence basis over $k(Y)$ then $\varphi^* \colon \Omega^r(Y) \to \Omega^r(X)$ is an inclusion. Here we identify $k(Y)$ with the subfield $\varphi^*(k(Y)) \subset k(X)$.*

Proof. Suppose that the extension $k(Y) \subset k(X)$ has a separable transcendence basis $v_1, \dots, v_s$. This means that $v_1, \dots, v_s$ are algebraically independent over $k(Y)$, and $k(X)$ is a finite separable extension of the subfield $k(Y)(v_1, \dots, v_s)$. The field $k(Y)$ has a separable transcendence basis over k (see Chap. I, 3.3, Theorem 5, Remark 1). Denote this by $u_1, \dots, u_t$. Then $u_1, \dots, u_t, v_1, \dots, v_s$ is a separable transcendence basis of $k(X)$ over k.

We write any differential form $\omega \in \Omega^r(Y)$ in the form

$$\omega = \sum g_{i_1 \dots i_r} du_{i_1} \wedge \cdots \wedge du_{i_r}, \tag{2}$$

and apply (1) to it, giving an expression for $\varphi^*(\omega)$ as a linear combination of elements $d\varphi^*(u_{i_1}) \wedge \cdots \wedge d\varphi^*(u_{i_r})$, which, by 4.4, Theorem 4, are a subset of a basis of $\Omega^r(X)$ over $k(X)$, since the $\varphi^*(u_i)$ are part of the separable transcendence basis $u_1, \dots, u_t, v_1, \dots, v_s$. Hence $\varphi^*(\omega) = 0$ only if all $\varphi^*(g_{i_1 \dots i_r}) = 0$, and this is only possible if all $g_{i_1 \dots i_r} = 0$, that is $\omega = 0$. The theorem is proved.

So far everything has been more or less obvious. We now arrive at an unexpected fact.

Theorem 2. *If X and Y are nonsingular varieties, with Y projective, and $\varphi \colon X \to Y$ a rational map such that $\varphi(X)$ is dense in Y, then*

$$\varphi^* \Omega^r[Y] \subset \Omega^r[X].$$

In other words, φ^* takes regular differential forms to regular differential forms. Since φ is only a rational map, this seems quite implausible, even for functions, that is, the case $r = 0$. In this case we are saved by the fact that, since Y is projective, the only regular functions on Y are constant, and the theorem is vacuous. In the general case, the theorem is less obvious.

Proof. We use the fact that by Chap. II, 3.1, Theorem 3, φ is regular on $X \setminus Z$, where $Z \subset X$ is a closed subset and $\operatorname{codim}_X Z \geq 2$. If $\omega \in \Omega^r[Y]$ then $\varphi^*(\omega)$ is regular on $X \setminus Z$. Let us prove that regularity on the whole of X follows from this. For this, we write $\varphi^*(\omega)$ in some open set U in the form

$$\varphi^*(\omega) = \sum g_{i_1 \ldots i_r} du_{i_1} \wedge \cdots \wedge du_{i_r},$$

where $u_1, \ldots, u_n$ are regular functions on U such that $du_{i_1} \wedge \cdots \wedge du_{i_r}$ is a basis for $\Omega^r[U]$ over $k[U]$. Then from the fact that $\varphi^*(\omega)$ is regular on $X \setminus Z$ it follows that all the functions $g_{i_1 \ldots i_r}$ are regular on $U \setminus (U \cap Z)$. But $\operatorname{codim}_U(U \cap Z) \geq 2$, and this means that the set of points where $g_{i_1 \ldots i_r}$ is not regular has codimension ≥ 2. On the other hand, this set is a divisor $\operatorname{div}_\infty(g_{i_1 \ldots i_r})$. This is only possible if $\operatorname{div}_\infty(g_{i_1 \ldots i_r}) = 0$, and hence $g_{i_1 \ldots i_r}$ are regular functions. The theorem is proved

Corollary. *If two nonsingular projective varieties X and Y are birational then the vector space $\Omega^r[X]$ and $\Omega^r[Y]$ are isomorphic.* $\square$

The significance of Theorem 2 and its corollary is enhanced by the fact that for projective varieties X, the vector spaces $\Omega^r[X]$ are finite dimensional over k. This result is a consequence of a general theorem on coherent sheaves proved in Chap. VI, 3.4, Theorem. For the case of curves we prove it in 6.3 below. Set $h^r = \dim \Omega^r[X]$. The corollary means that the numbers h^r for $r = 0, \ldots, n$ are birational invariants of a nonsingular projective variety X.

6.2. Invariant Differential Forms on a Group

Let X be an algebraic variety, ω a differential form on X, and g an automorphism of X. We say that ω is invariant under g if $g^*(\omega) = \omega$. Suppose in particular that G is an algebraic group (see 4.1 for the definition). It follows at once from the definition that for any element $g \in G$, the translation map $t_g : G \to G$ given by

$$t_g(x) = gx$$

is regular and is an automorphism of G as an algebraic variety. A differential form on G is *invariant* if it is invariant under all the translations t_g.

An invariant rational differential form is regular. Indeed, if ω is regular at a point $x_0 \in G$ then $t_g^*(\omega)$ is regular at $g^{-1}x_0$. But $t_g^*(\omega) = \omega$, so that ω is regular at all points gx_0 for $g \in G$, and these are all the points of G.

We show how to find all invariant differential forms on an algebraic group. For this, consider the vector spaces $\Phi^r[G]$ as in 5.1 and 5.3, and their automorphisms t_g^* corresponding to the translations t_g. We determine first the set of elements $\varphi \in \Phi^r[G]$ that are invariant under all t_g^* for $g \in G$. This set contains in particular all the invariant regular differential r-forms.

The condition $t_g^*(\varphi) = \varphi$ means that for any point $x \in G$,

$$\varphi(x) = \Big(\bigwedge^r dt_g^*\Big)(\varphi(gx)). \tag{1}$$

In particular, for $g = x^{-1}$,

$$\Big(\bigwedge^r dt_{x^{-1}}^*\Big)(\varphi(e)) = \varphi(x). \tag{2}$$

This formula shows that φ is uniquely determined by the element $\varphi(e)$ of the finite dimensional vector space $\bigwedge^r \Theta_e^*$. Conversely, if we specify an arbitrary element $\eta \in \bigwedge^r \Theta_e^*$, we can use (2) to construct the element $\varphi \in \varPhi^r[G]$ given by

$$\varphi(x) = \Big(\bigwedge^r dt_{x^{-1}}^*\Big)(\varphi(\eta)).$$

A simple substitution shows that it also satisfies (1), that is, is invariant under t_g^*. Thus the subspace of elements of $\varphi \in \varPhi^r[G]$ that are invariant under t_g^* is isomorphic to $\bigwedge^r \Theta_e^*$, with the isomorphism defined by

$$\varphi \mapsto \varphi(e).$$

Now we show that all the φ just constructed are regular differential forms, that is, are elements of $\Omega^r[G]$. In view of their invariance, it is enough to show that the forms are regular at any point, for example at the identity element e. Moreover, it is enough to restrict ourselves to the case $r = 1$. Indeed, if $\eta = \sum \alpha_{i_1} \wedge \cdots \wedge \alpha_{i_r}$, with $\alpha_j \in \bigwedge^1 \Theta_e^*$, and we prove that the forms φ_j corresponding by (2) to the α_j are regular, then the form $\varphi = \sum \varphi_{i_1} \wedge \cdots \wedge \varphi_{i_r}$ is regular, and it corresponds by (2) to η.

We choose an affine neighbourhood V of e such that $\Omega^1[V]$ is free over $k[V]$, and write $du_1, \ldots, du_n$ for a basis. Then there exists an affine neighbourhood U of e such that $\mu(U \times U) \subset V$, where $\mu : G \times G \to G$ is the multiplication map of G. Just as any function of $k[U \times U]$, the elements $\mu^*(u_l)$ can be written in the form

$$\mu^*(u_l)(g_1, g_2) = \sum v_{lj}(g_1) w_{lj}(g_2) \qquad \text{for } (g_1, g_2) \in U \times U \subset G \times G,$$

where $v_{lj}, w_{lj} \in k[U]$. By definition, $t_h = \mu \circ s_h$, where s_h is the embedding $G \hookrightarrow G \times G$ given by $s_h(g) = (h, g)$. Hence $t_h^*(u_l)(g) = \sum v_{lj}(h) w_{lj}(g)$, and since $(t_h^*(du_l))(g) = d_{hg}(t_h^*(u_l))$, we have $(t_h^*(du_l))(g) = \sum v_{lj}(h) d_{hg}(w_{lj})$. In particular, setting $h = g^{-1}$, we get

$$\big(t_{g^{-1}}^*(du_l)\big)(g) = \sum v_{lj}(g^{-1}) d_e w_{lj}.$$

Expressing dw_{lj} in terms of du_k, we obtain the relations

$$t_{g^{-1}}^*(du_l) = \sum_k c_{kl}(g) du_k, \qquad \text{with } c_{kl} \in k[U], \tag{3}$$

where

$$c_{kl}(g) = \sum_j v_{lj}(g^{-1})\frac{\partial w_{lj}}{\partial u_k}(e). \tag{4}$$

Now write the invariant form φ in the form $u = \sum \psi_k du_k$ and consider the relation $t_g^*(\varphi) = \varphi$ at e. Substituting (3) and equating coefficients of du_k, we get

$$\sum c_{kl}\psi_l = \psi_k(e). \tag{5}$$

Since $\big(c_{kl}(e)\big)$ is the identity matrix, we get $\det|c_{kl}|(e) \neq 0$, and it follows from the system of equations (5) that $\psi_k \in \mathcal{O}_e$.

We state the result we have proved:

Proposition. *The map $\omega \mapsto \omega(e)$ establishes an isomorphism from the vector space of invariant regular differential r-forms on G to $\bigwedge^r \Theta_e^*$.* $\square$

6.3. The Canonical Class

We now consider more particularly rational differential n-forms on an n-dimensional nonsingular variety X (compare 5.3). In some neighbourhood of a point $x \in X$, such a form can be written $\omega = gdu_1 \wedge \cdots \wedge du_n$. We cover X by affine sets U_i such that on each U_i we have such an expression $\omega = g^{(i)}du_1^{(i)} \wedge \cdots \wedge du_n^{(i)}$. On the intersection $U_i \cap U_j$, by 5.3, (6), we get

$$g^{(j)} = g^{(i)} J\left(\frac{u_1^{(i)},\ldots,u_n^{(i)}}{u_1^{(j)},\ldots,u_n^{(j)}}\right).$$

Since the Jacobian determinant J is regular and nowhere zero in $U_i \cap U_j$ (see 5.3, (5)), the system of functions $g^{(i)}$ on U_i is a compatible system of functions in the sense of 2.1, and hence defines a divisor on X. This divisor is called the divisor of ω, and is denoted by $\operatorname{div}\omega$.

The following properties of the divisor of a rational differential n-form on a nonsingular n-dimensional variety follow at once from the definition:

(a) $\operatorname{div}(f\omega) = \operatorname{div} f + \operatorname{div}\omega$ for $f \in k(X)$.

(b) $\operatorname{div}\omega \geq 0$ if and only if $\omega \in \Omega^n[X]$.

By the case $r = n$ of 5.4, Theorem 3, $\Omega^n(X)$ is a 1-dimensional vector space over $k(X)$. Hence if $\omega_1 \in \Omega^n(X)$ and $\omega_1 \neq 0$, then any form $\omega \in \Omega^n(X)$ can be written $\omega = f\omega_1$. Hence property (a) shows that the divisors of all forms $\omega \in \Omega^n(X)$ are linearly equivalent, and form one divisor class on X. This class is called the *canonical class* of X, and is denoted by K or K_X.

Let $\omega_1 \in \Omega^n(X)$ be a fixed n-form, so that any other can be written $f\omega_1$. (b) shows that ω is regular on X if and only if $\operatorname{div} f + \operatorname{div}(\omega_1) \geq 0$. In other words, in terms of the notion of vector space associated with a divisor introduced in 1.5, we have $\Omega^n[X] = \mathcal{L}(\operatorname{div}(\omega_1))$. Thus $h^n = \dim_k \Omega^n[X] = \ell(K_X)$. We see that the invariant h^n introduced in 6.1 is equal to the dimension of the canonical class.

Example. Suppose that X is the variety of an algebraic group. We showed in 6.2 that the vector space of invariant differential r-forms on X is isomorphic to $\bigwedge^r \Theta_e^*$, where Θ_e is the tangent space to X at the identity element e. In particular, the space of invariant differential n-forms is 1-dimensional, since $\bigwedge^n \Theta_e^* \cong k$. If ω is a nonzero invariant form then $\omega \in \Omega^n[X]$, that is div $\omega \geq 0$. But if $\omega(x) = 0$ for some point $x \in X$, then by invariance also $\omega(y) = 0$ for every $y \in X$. Hence $\omega(x) \neq 0$ for all $x \in X$, that is, ω is regular and nowhere vanishing on X. This means that div $\omega = 0$, that is $K_X = 0$.

In 2.3, Theorem 5, we proved that the number $\ell(D)$ is finite for any divisor D on a nonsingular projective algebraic curve. It follows in particular that the number $h^1 = \dim_k \Omega^1[X] = \ell(K_X)$ is finite for any nonsingular projective algebraic curve X. This number is called the *genus* of the curve X, and denoted by $g = g(X)$; that is, for curves, $h^1 = g$. There are several other characterisations of the genus of a curve, see for example 6.6, Corollary 1 and Chap. VII, 3.3.

In the case $\dim X = 1$ we know that all divisors in one linear equivalence class have the same degree, so that it makes sense to speak of the degree $\deg C$ of a divisor class C. In particular, the degree $\deg K_X$ of the canonical class is a birational invariant of a curve X.

The invariants $g(X)$ and $\deg K_X$ we have introduced are not independent. It can be proved that the relation $\deg K_X = 2g(X) - 2$ holds between them; see 6.6, Corollary 1. In particular, if a nonsingular projective curve X is an algebraic group then $K_X = 0$, as we have just seen. Hence $g(X) = 1$, that is, of all projective curves, only the curves of genus 1 can have an algebraic group law defined on them. We will see in 6.6 below that the curves of genus 1 are exactly the nonsingular cubic curves.

6.4. Hypersurfaces

We now compute the canonical class and the invariant $h^n(X) = \ell(K_X)$ in the case that $X \subset \mathbb{P}^N$ is a nonsingular n-dimensional hypersurface, with $N = n + 1$. Suppose that X is defined by the equation $F(x_0 : \cdots : x_N) = 0$ with $\deg F = \deg X = m$. Consider the affine open set U with $x_0 \neq 0$. Our X is defined in U by $G(y_1, \ldots, y_N) = 0$, where $y_i = x_i/x_0$ and $G(y_1, \ldots, y_N) = F(1, y_1, \ldots, y_N)$.

Define the open subset $U_i \subset U$ by $\partial G/\partial y_i \neq 0$; then $y_1, \ldots, \widehat{y_i}, \ldots, y_N$ (with y_i omitted) are local parameters in U_i, and the form

$$dy_1 \wedge \cdots \wedge \widehat{dy_i} \wedge \cdots \wedge dy_N$$

is a basis of $\Omega^n[U_i]$ over $k[U_i]$. However, it is more convenient to take as basis the form

$$\omega_i = \frac{(-1)^i}{\partial G/\partial y_i} dy_1 \wedge \cdots \wedge \widehat{dy_i} \wedge \cdots \wedge dy_N,$$

which is permissible, since $\partial G/\partial y_i \neq 0$ in U_i. The advantage is that then the forms $\omega_1, \ldots, \omega_N$ are equal: multiplying the relation

$$\sum_{i=1}^{N} \frac{\partial G}{\partial y_i} dy_i = 0$$

by the product $dy_1 \wedge \cdots \wedge \widehat{dy_i} \wedge \cdots \wedge \widehat{dy_j} \wedge \cdots \wedge dy_N$, and using the fact that $dy_i \wedge dy_i = 0$ we see that

$$\omega_j = \omega_i. \tag{1}$$

Since X is nonsingular, $U = \bigcup U_i$, and it follows from (1) that the ω_i fit together to give a form ω that is regular and everywhere nonzero on U, so that $\operatorname{div} \omega = 0$ in U.

It remains to study points not in U. Consider, say, the open subset V in which $x_1 \neq 0$. This affine space has coordinates $z_1, \ldots, z_N$ with

$$z_1 = \frac{1}{y_1} \quad \text{and} \quad z_i = \frac{y_i}{y_1} \quad \text{for } i = 2, \ldots, N;$$

$$y_1 = \frac{1}{z_1} \quad \text{and} \quad y_i = \frac{z_i}{z_1} \quad \text{for } i = 2, \ldots, N.$$

Hence

$$dy_1 = -\frac{dz_1}{z_1^2} \quad \text{and} \quad dy_i = \frac{z_1 dz_i - z_i dz_1}{z_1^2} \quad \text{for } i = 2, \ldots, N.$$

We substitute these expressions in ω_N, and use the fact that $dz_1 \wedge dz_1 = 0$, obtaining

$$\omega = -\frac{(-1)^N}{z_1^N (\partial G/\partial y_N)} dz_1 \wedge \cdots \wedge dz_{N-1}.$$

The equation of X in V is

$$H(z_1, \ldots, z_N) = 0, \quad \text{where } H = z_1^m G\left(\frac{1}{z_1}, \frac{z_2}{z_1}, \ldots, \frac{z_N}{z_1}\right).$$

From the relation

$$\frac{\partial H}{\partial z_N} = z_1^{m-1} \frac{\partial G}{\partial y_N}\left(\frac{1}{z_1}, \frac{z_2}{z_1}, \ldots, \frac{z_N}{z_1}\right) = z_1^{m-1} \frac{\partial G}{\partial y_N}(y_1, \ldots, y_N)$$

it follows that

$$\omega = -\frac{(-1)^N}{z_1^{N-m+1}(\partial H/\partial z_N)} dz_1 \wedge \cdots \wedge dz_{N-1}. \tag{3}$$

All the arguments that we carried out for U are also valid for V, and show that

$$\Omega^n[V] = k[V] \frac{1}{\partial H/\partial z_N} dz_1 \wedge \cdots \wedge dz_{N-1}. \tag{4}$$

Hence in V we have $\operatorname{div}\omega = -(N - m + 1)\operatorname{div}(z_1)$. Obviously $\operatorname{div}(z_1)$ in V is the divisor of the form x_0 on X, as defined in 1.2. Finally, we get that the relation $\operatorname{div}\omega = (m - N - 1)\operatorname{div}(x_0) = (m - n - 2)\operatorname{div}(x_0)$. Thus K_X is the divisor class containing $(m - n - 2)L$, where L is the hyperplane section of X.

We now determine $\Omega^n[X]$. Writing any form $\eta \in \Omega^n(X)$ as $\eta = \varphi\omega$, we see that $\eta \in \Omega^n[X]$ if and only if $\varphi \in \mathcal{L}\big((m - n - 2)\operatorname{div}(x_0)\big)$. By 1.5, Example, this is equivalent to $\varphi = P(z_1, \ldots, z_n)$, where P is a polynomial, and

$$\deg P \leq m - n - 2. \tag{5}$$

From this it is easy to compute the dimension of $\Omega^n[X]$. Namely two different polynomials $P, Q \in k[y_1, \ldots, y_N]$ satisfying (5) define different elements of $k[X]$, since otherwise $P - Q$ is divisible by Q, which contradicts (5). Thus the dimension of $\Omega^n[X]$ equals the dimension of the space of polynomials P satisfying (5). This dimension is equal to the binomial coefficient $\binom{m-1}{N} = \frac{(m-1)\cdots(m-N)}{N!}$. Thus

$$h^n(X) = \ell(K_X) = \binom{m - 1}{n + 1}. \tag{6}$$

The simplest case of this formula is when $N = 2$, that is, $n = 1$. We get the formula

$$g(X) = \binom{m - 1}{2} = \frac{(m - 1)(m - 2)}{2}$$

for the genus of a nonsingular plane curve of degree m. Compare Chap. IV, 2.3, Example 1.

We can make an important deduction at once from (6). Namely, interpreting the binomial coefficients as the number of combinations, we get at once that

$$\binom{m - 1}{n + 1} > \binom{m' - 1}{n + 1} \qquad \text{for } m > m' \geq n + 1.$$

Since by 6.1, Theorem 2, $h^n(X)$ is a birational invariant, (6) thus implies that hypersurfaces of different degrees $m, m' \geq n + 1$ are not birational. We see that there exist infinitely many algebraic varieties of any given dimension not birational to one another.

In particular, when $N = 2$, $m = 3$ we get $g(X) = 1$, and since $g(\mathbb{P}^1) = 0$, we see once again that a nonsingular cubic curve in $\mathbb{P}^2$ is not rational.

It follows from (6) that $h^n(X) = 0$ if $m \leq N$. In particular, $h^n(\mathbb{P}^n) = 0$. We verified this directly for $n = 1$ in 5.1, Example 2.

Consider the case $m \leq N$ in more detail. If $N = 2$ then this means $m = 1$ or 2. For $m = 1$ we have $X = \mathbb{P}^1$, and we already know that $h^1(\mathbb{P}^1) = 0$. For $m = 2$ we have a nonsingular plane curve of degree 2, which is isomorphic to $\mathbb{P}^1$, so that in this case also $h^1(X) = 0$ tells us nothing new.

Suppose that $N = 3$. For $m = 1$ we have $\mathbb{P}^2$, and we already know that $h^2 = 0$. If $m = 2$ then X is a nonsingular surface of degree 2, which is birational to $\mathbb{P}^2$, so that $h^2(X) = 0$ is a consequence of $h^2(\mathbb{P}^2) = 0$ and 6.1, Theorem 2. If $m = 3$ then X is a nonsingular cubic surface. If such a surface contains two skew lines then it is birational to $\mathbb{P}^2$ (Chap. I, 3.3, Example 2). One can show that every nonsingular cubic surface contains two skew lines, so that $h^2(X) = 0$ is again a consequence of $h^2(\mathbb{P}^2) = 0$ and 6.1, Theorem 2.

The examples we have considered lead to interesting questions on nonsingular hypersurfaces of small degree, $X \subset \mathbb{P}^N$ with $\deg X = m \leq N$. We see that for $N = 2, 3$, X is birational to projective space $\mathbb{P}^n$, with $n = N - 1$, which "explains" the equality $h^n(X) = 0$.

For $N = 4$ we run into a new phenomenon. For $m = 3$, for example, already for the cubic hypersurface X given by

$$x_0^3 + x_1^3 + x_2^3 + x_3^3 + x_4^3 = 0, \qquad (7)$$

the question of whether X is birational to $\mathbb{P}^3$ is very subtle. However, one can show that there exists a rational map $\varphi \colon \mathbb{P}^3 \to X$ such that $\varphi(\mathbb{P}^3)$ is dense in X and $k(X) \subset k(\mathbb{P}^3)$ is a separable extension (see Ex. 13). Already this, together with $h^3(\mathbb{P}^3)$ and 6.1, Theorems 1–2, implies $h^3(X) = 0$. The following terminology arises in connection with this: we say that a variety is *rational* if it birational to $\mathbb{P}^n$ where $n = \dim X$, and *unirational* if there exists a rational map $\varphi \colon \mathbb{P}^n \to X$ such that $\varphi(\mathbb{P}^n)$ is dense in X and $k(X) \subset k(\mathbb{P}^n)$ is separable. It follows from §5, Ex. 6 and 6.1, Theorems 1–2 that all $h^i = 0$ for a unirational variety.

The question of whether the notions of rational and unirational varieties are the same is typical of a series of problems arising in algebraic geometry. This question is called the *Lüroth problem*. It can obviously be stated as a problem in the theory of fields: if K is a subfield of the rational function field $k(T_1, \ldots, T_n)$ such that $K \subset k(T_1, \ldots, T_n)$ is a finite separable extension, then is it true that K is isomorphic to the rational function field?

For $n = 1$ the answer is positive, in fact even without the assumptions that k is algebraically closed and $K \subset k(T)$ is separable. For $n = 2$ the answer is negative without these assumptions, but positive with them, but the proof is very delicate. It is given, for example, for fields of characteristic 0 in Shafarevich [67], Chap. III or Barth, Peters and van de Ven [8], and in the general case in Bombieri and Husemoller [12].

For $n \geq 3$ the answer is negative even if $k = \mathbb{C}$. This is a delicate result of the theory of 3-folds. One of the examples of a unirational but not rational variety is the nonsingular cubic 3-fold, in particular the hypersurface (7) (see Clemens and Griffiths [21]). Another example of an irrational variety is the nonsingular quartic hypersurface of $\mathbb{P}^4$ (see Iskovskikh and Manin [41]); some of these hypersurfaces are unirational. For another type of example, see Artin and Mumford [6].

Whereas for 3-folds the Lüroth problem is a subtle geometric problem, in higher dimension it turns out to be more algebraic in spirit, and its solution more elementary. For example, there are examples of finite group G of linear transformations of variable $x_1, \ldots, x_n$ such that the subfield of invariants $k(x_1, \ldots, x_n)^G$ of this group is not isomorphic to the field of rational functions (see Bogomolov [11] or Saltman [65]).

6.5. Hyperelliptic Curves

As a second example we consider one type of curves. Write Y for the affine plane curve with equation $y^2 = F(x)$, where $F(x)$ is a polynomial with no multiple roots and of odd degree $n = 2m + 1$ (we proved in Chap. I, 1.4 that the case of even degree reduces to the odd degree case). We suppose that char $k \neq 2$. The nonsingular projective model X of Y is called a *hyperelliptic curve*. We compute the canonical class and the genus of X.

The rational map $Y \to \mathbf{A}^1$ given by $(x, y) \mapsto x$ defines a regular map $f: X \to \mathbf{P}^1$. Obviously $\deg f = 2$, so that, by 2.1, Theorem 1, if $\alpha \in \mathbf{P}^1$ and u is a local parameter at α, the inverse image $f^{-1}(\alpha)$ either consists of two points z', z'' with $v_{z'}(u) = v_{z''}(u) = 1$, or $f^{-1}(\alpha) = z$ with $v_z(u) = 2$.

It is easy to check that the affine curve Y is nonsingular. If $\overline{Y}$ is its projective closure in $\mathbf{P}^2$ then X is the normalisation of $\overline{Y}$, and we have a map $\varphi: X \to \overline{Y}$ which is an isomorphism of $\varphi^{-1}(Y)$ and Y. It follows that if $\xi \in \mathbf{A}^1$ has coordinate α then

$$f^{-1}(\xi) = \begin{cases} \{z', z''\} & \text{if } F(\alpha) \neq 0; \\ z & \text{if } F(\alpha) = 0. \end{cases}$$

Now consider the point at infinity $\alpha_\infty \in \mathbf{P}^1$. If x denotes the coordinate on $\mathbf{P}^1$ then a local parameter at α_∞ is $u = x^{-1}$. If $f^{-1}(\alpha_\infty) = \{z', z''\}$ consisted of 2 points then u would be a local parameter at z', say. It would follow that $v_{z'}(u) = 1$ and hence $v_{z'}(F(x)) = -n$; but since $y^2 = F(x)$, we have $v_{z'}(F(x)) = 2v_{z'}(y)$, and this contradicts n odd. Thus $f^{-1}(\alpha_\infty)$ consists of just one point z_∞, and $v_{z_\infty}(x) = -2$, $v_{z_\infty}(y) = -n$. It follows that $X = \varphi^{-1}(Y) \cup z_\infty$.

We proceed to differential forms on X. Consider, for example, the form $\omega = dx/y$. At a point $\xi \in Y$, if $y(\xi) \neq 0$ then x is a local parameter, and $v_\xi(\omega) = 0$. If $y(\xi) = 0$ then y is a local parameter, and $v_\xi(x) = 2$, so that it again follows that $v_\xi(\omega) = v_\xi(dx) - v_\xi(y) = 1 - 1 = 0$. Thus $\operatorname{div} \omega = kz_\infty$, and it remains to determine the value of k. For this, it is enough to recall that if t is a local parameter at z_∞ then $x = t^{-2}u$ and $y = t^{-n}v$, where $u, v, u^{-1}, v^{-1} \in \mathcal{O}_{z_\infty}$. Hence $\omega = t^{n-3}wdt$ with $w, w^{-1} \in \mathcal{O}_{z_\infty}$, and therefore $\operatorname{div} \omega = (n - 3)z_\infty = (2m - 2)z_\infty$.

Now we determine $\Omega^1[X]$. As we have seen, ω is a basis of the module $\Omega^1[Y]$, that is, $\Omega^1[Y] = k[Y]\omega$, so that any form in $\Omega^1[X]$ is of the form $u\omega$, where $u \in k[Y]$, and hence u is of the form $P(x) + Q(x)y$ with $P, Q \in k[X]$. It remains to see when these forms are regular at z_∞. This happens if and only if

$$v_{z_\infty}(u) \geq -(n - 3). \tag{1}$$

We find all such $u \in k[Y]$. Since $v_{z_\infty}(x) = -2$, it follows that $v_{z_\infty}(P(x))$ is always even and since $v_{z_\infty}(y) = -n$, that $v_{z_\infty}(Q(x)y)$ is always odd. Hence

$$v_{z_\infty}(u) = v_{z_\infty}(P(x) + Q(x)y) \leq \min\{v_{z_\infty}(P(x)), v_{z_\infty}(Q(x)y)\}$$

and so if $Q \neq 0$ we have $v_{z_\infty}(u) \leq -n$. Thus $u = P(x)$ and (1) gives $2 \deg P \leq n - 3$, that is, $\deg P \leq m - 1$, where $n = 2m + 1$.

We have found that $\Omega^1[X]$ consists of forms $P(x)dx/y$ where the degree of $P(x)$ is $\leq m-1$. It follows from this that $g(X) = h^1(X) = \dim \Omega^1[X] = m$.

It is interesting to compare the results of 6.4 when $N = 2$ and of 6.5. In the second case we saw that there exist algebraic curves of any given genus. In the first case, the genus of a nonsingular plane curve is $\binom{n-1}{2}$, that is, is a long way from giving an arbitrary integer. Thus not every nonsingular projective curve is isomorphic to a plane curve. For example, a hyperelliptic curve of genus 4, with $n = 9$ is not.

6.6. The Riemann–Roch Theorem for Curves

One of the central results of the theory of algebraic curves is the *Riemann–Roch theorem*. This is the equality

$$\ell(D) - \ell(K - D) = 1 - g + \deg D, \tag{1}$$

where D is an arbitrary divisor on a nonsingular projective curve, K the canonical class and g the genus. We omit the proof of this theorem, which would requires us to go into the details of the theory of algebraic curves in some depth. We prove instead a weaker statement, the *Riemann inequality*, which, together with the inequality 2.2, (1), determines the order of growth of the function $\ell(D)$.

Lemma. *Let X be a nonsingular projective curve X. Then there exists a constant $\gamma = \gamma(X)$ such that*

$$\ell(D) \geq \deg D - \gamma \tag{2}$$

for every divisor D on X.

Proof. The point is of course to construct lots of functions $\varphi \in \mathcal{L}(D)$. We do this first for certain divisors of a special form. As in 2.1, Theorem 1, we consider any map $f \colon X \to \mathbb{P}^1$ defined by a nonconstant rational function $f \in k(X)$, and set $\deg f = n$. Write $\mathbb{P}^1 = \mathbb{A}^1 \cup \infty$ and let t be the coordinate on $\mathbb{A}^1$. We set $D_\infty = f^*(\infty)$, and prove (2) for the divisors rD_∞ with $r \gg 0$.

Write $D_\infty = \sum n_i x_i$ with $n_i > 0$. Then $n_i = -v_{x_i}(t)$, where we use the inclusion $f^* \colon k(t) \hookrightarrow k(X)$ to identify t with $f^*(t) = f \in k(X)$. By 2.1, Theorem 1, $\sum n_i = n$, where $n = \deg f = [k(X) : k(t)]$. Let $w_1, \ldots, w_n$ be a basis of $k(X)$ over $k(t)$. If some w_j has a pole at a point $x \notin \operatorname{Supp} D_\infty$ with

$t(x) = \alpha$ then $v_x((t-\alpha)^k w_j) \geq 0$ for sufficiently large k. Therefore, multiplying each w_j by a suitable polynomial in t, we get a new basis $u_1, \ldots, u_n$ of $k(X)$ over $k(t)$ such that the poles of all the u_j are concentrated in $\operatorname{Supp} D_\infty$.
Set $v_{x_i}(u_j) = -m_{ij}$. Then for a polynomial $p \in k[t]$,

$$v_{x_i}(pu_j) = -n_i k - m_{ij}, \qquad \text{where } k = \deg p.$$

Hence $pu_j \in \mathcal{L}(rD_\infty)$ provided that $n_i k + m_{ij} \leq rn_i$, that is, $k + m_{ij}/n_i \leq r$. Set $m = \max\{m_{ij}/n_i\}$. Then $\sum p_j u_j \in \mathcal{L}(rD_\infty)$ for all $p_j \in k[t]$ for $j = 1, \ldots, n$ with $\deg p_j \leq r - m$. This is a subspace of $\mathcal{L}(rD_\infty)$ of dimension $(r - m + 1)n = rn - (m - 1)n$, which implies (2) for the divisors rD_∞, with the constant $\gamma = (m - 1)n$.

Now let $D = \sum k_i y_i$ be an arbitrary effective divisor. For $y_i \notin \operatorname{Supp} D_\infty$ with $k_i > 0$, suppose that $t(y_i) = \alpha_i$, and set $u = \prod (t - \alpha_i)^{k_i}$, where the product runs over all i with $y_i \notin \operatorname{Supp} D_\infty$ with $k_i > 0$. Then $D' = D - \operatorname{div} u$ is a divisor with $D' \sim D$ and by construction $v_y(D') \leq 0$ for every $y \notin \operatorname{Supp} D_\infty$. Therefore $D' < rD_\infty$ for some large r. Now by 2.3, Remark 3, inequality (3), $\ell(rD_\infty) \leq \ell(D') + \deg(rD_\infty - D')$, so that

$$\ell(D') \geq \deg D' \ell(rD_\infty) - \deg(rD_\infty) \geq \deg D' - \gamma.$$

Since $\ell(D') = \ell(D)$ and $\deg D' = \deg D$, the lemma follows.

We now point out some of the consequences of the Riemann–Roch formula (1), which clarify its significance for the theory of curves.

Corollary 1. *Setting $D = K$, since $\ell(K) = g$ and $\ell(K - K) = \ell(0) = 1$, we get that $\deg K = 2g - 2$. This equality was mentioned in 6.3, and verified for hyperelliptic curves in 6.5.* $\square$

Corollary 2. *If $\deg D > 2g - 2$ then $\ell(D) = 1 - g + \deg D$.*

This follows because $\deg D > 2g - 2$ implies that $\deg(K - D) < 0$, and hence $\ell(K - D) = 0$; indeed, $K - D \sim D' \geq 0$, which would contradict $\deg D' < 0$. $\square$

Corollary 3. *$g(X) = 0$ is a necessary and sufficient condition for $X \cong \mathbb{P}^1$.*

Proof. We have seen that $g(\mathbb{P}^1) = 0$ in 5.1, Example 2. If $g = 0$ and $D = x$ is a point of X then $\ell(D) \geq 2$ by (1). This means that $\mathcal{L}(D)$ contains a function f with $\operatorname{div}_\infty f = x$ in addition to the constants. That is, if we interpret f as a map $f \colon X \to \mathbb{P}^1$ then $\deg f = 1$ by 2.1, Theorem 1. It follows that $X \cong \mathbb{P}^1$. The corollary is proved.

Corollary 4. *If $g = 1$ then X is isomorphic to a cubic in $\mathbb{P}^2$.*

Proof. Indeed, for $g = 1$, Corollary 2 gives $\ell(D) = \deg D$ for $D > 0$, and the assertion follows from 3.1, Theorem 2. $\square$

Corollary 5. *Consider a basis $f_0, \ldots, f_n$ of the space $\mathcal{L}(D)$ with $D \geq 0$, and the corresponding rational map $\varphi = (f_0 : \cdots : f_n) \colon X \to \mathbb{P}^n$. Then φ is an embedding provided that*

$$\ell(D - x) = \ell(D) - 1 \quad and \quad \ell(D - x - y) = \ell(D) - 2 \quad for\ all\ x, y \in X. \quad (2)$$

In particular, (2) holds if $\deg D \geq 2g + 1$ by Corollary 2, so that in this case φ is an embedding.

Proof. Note first that, in the terminology of 1.4, the first condition of (2) implies that $-D = \text{hcd}\{\text{div}(f_i)\}$. Indeed, $\text{hcd}\{\text{div}(f_i)\} \geq -D$ by definition of $\mathcal{L}(D)$. If equality does not hold, then there exists a point x such that $\text{div}(f_i) \geq -D + x$, that is, $\mathcal{L}(D) = \mathcal{L}(D - x)$, so that $\ell(D) = \ell(D - x)$, which contradicts (2). Thus by the remark at the end of 1.4, the divisors $D_\lambda = \text{div}(\sum \lambda_i f_i) + D$ are the pullbacks of hyperplanes under the map φ.

To prove that φ is an isomorphic embedding, we use Chap II, 5.3, Theorem 8 and Chap. II, 5.4, Lemma, the assumptions of which we verify using the above remark. If $\varphi(x) = \varphi(y)$ then every hyperplane E through $\varphi(x)$ also passes through $\varphi(y)$. This means that if $D_\lambda - x \geq 0$ then also $D_\lambda - x - y \geq 0$, that is, $\ell(D_\lambda - x) = \ell(D_\lambda - x - y)$, which contradicts the second condition of (2).

We prove that the tangent space at a point is mapped isomorphically. This is equivalent to saying that

$$\varphi^* \colon \mathfrak{m}_{\varphi(x)} / \mathfrak{m}_{\varphi(x)}^2 \to \mathfrak{m}_x / \mathfrak{m}_x^2$$

is surjective. If this does not hold then $\varphi^*(\mathfrak{m}_{\varphi(x)}) \subset \mathfrak{m}_x^2$, since $\dim \mathfrak{m}_x / \mathfrak{m}_x^2 = 1$. In other words, for any function $u \in \mathfrak{m}_{\varphi(x)}$ we have $v_x(\varphi^*(u)) \geq 2$. Applied to linear functions, this shows that if $D_\lambda - x \geq 0$ then $D_\lambda - 2x \geq 0$. We again get $\ell(D_\lambda - x) = \ell(D_\lambda - 2x)$, which contradicts the second condition of (2). This completes the proof of Corollary 5.

Obviously, changing the choice of basis of $\mathcal{L}(D)$ changes φ by composing it with a projective transformation of $\mathbb{P}^n$. On the other hand, changing D to another divisor $D + \text{div} f$ leads to an isomorphism $\mathcal{L}(D) \xrightarrow{\sim} \mathcal{L}(D + \text{div} f)$ given by $u \mapsto uf$, and so does not change φ. Thus it makes sense to talk of the map φ associated with a divisor class.

Suppose, for example, that X is a curve of genus 1, and $x_0 \in X$. The conditions of Corollary 5 are satisfied for $3x_0$. Hence the map φ corresponding to $3x_0$ is an isomorphism of X to a curve $Y \subset \mathbb{P}^2$ (since $\ell(3x_0) = 3$ by Corollary 2). As we have seen, $3x_0$ is the pullback of a section of Y by a line, and since $\deg 3x_0 = 3$, also $\deg Y = 3$. Thus every curve of genus 1 is isomorphic to a plane cubic. (For more details, compare the proof of the converse part of 3.1, Theorem 2.)

The most interesting maps φ are those corresponding to classes intrinsically related to X, for example, the multiples nK of the canonical class. Corollary 1 shows that $\deg(nK) \geq 2g+1$ if $n \geq 2$ for $g > 2$, and if $n \geq 3$ for $g = 2$. Thus for $g > 1$ the class $3K$ always satisfies the conditions of Corollary 4. The corresponding map φ_{3K} maps X to $\mathbb{P}^m$, where $m = \ell(3K) - 1 = 5g - 6$ (by Corollary 2). Moreover, two curves X and X' are isomorphic if and only if their images $\varphi_{3K}(X)$ and $\varphi_{3K}(X')$ can be obtained from one another by a projective transformation. The question of birational classification thus reduces to the projective classification.

The map φ corresponding to the canonical class itself is not always an embedding. However, one can enumerate all cases where this fails (see Ex. 11–12).

As a simple application of these ideas, consider plane curves of degree 4. By the result of 6.4, the canonical class of $X \subset \mathbb{P}^2$ equals the class of the intersection of X with a line of $\mathbb{P}^2$. Hence the map φ_K corresponding to this class is just the natural embedding $X \hookrightarrow \mathbb{P}^2$. It follows from what we have said above that two such curves are isomorphic if and only if they are projectively equivalent. This leads to an extremely important conclusion. The set of plane quartics can be identified with the projective space $\mathbb{P}^{14}$, where $14 = \binom{6}{2} - 1$, as in Chap. I, 4.4, Example 2. On the other hand, the group of projective linear transformations of $\mathbb{P}^2$ has dimension 8 (it is the group of nondegenerate 3×3 matrixes up to constant multiple). Using the theorem on the dimension of fibres one can deduce from this that $\mathbb{P}^{14}$ contains an open set U and a map $f\colon U \to M$ such that two points $u_1, u_2 \in U$ correspond to projectively equivalent curves only if they lie in the same fibre of f. The dimension of the fibre is therefore equal to 8, so that $\dim M = 14 - 8 = 6$.

Thus two plane curves of degree 4 are by no means always isomorphic: for this they must correspond to the same point of a 6-dimensional variety M. This shows that the genus is not a complete set of birational invariants of curves. In addition to their integer invariant, the genus, curves also have continuous invariants, called *moduli*. It can be proved that the set of all curves of given genus $g > 1$ form (in a sense that we do not make precise) a single irreducible variety of dimension $3g - 3$. In the case of plane quartic curves, $g = 3$ and $3g - 3 = 6 = \dim M$. A similar thing happens for curves of genus 1 (see §3, Ex. 8). Only for $g = 0$ are all curves of the same genus isomorphic.

6.7. Projective Embedding of a Surface

We discuss here how to generalise to surfaces the facts proved in the previous section for algebraic curves. We do not give any proofs. The reader can find them in Shafarevich [67], Bombieri and Husemoller [12] or Barth, Peters and van de Ven [8]. We restrict ourselves, moreover, to the case of a field of characteristic 0.

The analogue of curves of genus > 1 are surfaces for which some multiple of the canonical class defines a birational embedding. These are called *surfaces of general type*, and their classification reduces in a certain sense to a projective classification. The main result on surfaces of general type is that already $5K$ defines a regular map that is a birational embedding.

It remains to enumerate the surfaces not of general type. They play the role of curves of genus 0 and 1 and are given by analogous constructions.

The analogue of rational curves are first the rational surfaces, that is, surfaces birational to $\mathbb{P}^2$, and then ruled surfaces. These are the surfaces that can be mapped to a curve C such that all the fibres of the map are isomorphic to $\mathbb{P}^1$. These are thus algebraic families of lines.

There are three types of surfaces that play the role of curves of genus 1. The first type are the *Abelian surfaces*, that is, 2-dimensional Abelian varieties. The surfaces of the second type, called *K3 surfaces* have the property in common with Abelian varieties that their canonical class is equal to 0. However, in distinction to Abelian varieties, they have no regular differential 1-forms. According to 6.2, Proposition, Abelian varieties have invariant (and therefore regular) differential 1-forms. The third type are *elliptic surfaces*, that is, families of elliptic curves. These surfaces have maps $f\colon X \to C$ to a curve C such that for all $y \in C$ for which $f^{-1}(y)$ is a nonsingular curves (that is, for all but finitely many y), this nonsingular fibre is a curve of genus 1.

The main theorem asserts that all surfaces not of general type are exhausted by the above 5 types, the rational, ruled, Abelian, K3 and elliptic surfaces.

To get a better idea of these classes of surfaces, it is convenient to classify them by the invariant κ, the maximal dimension of the image of X under the rational map given by a divisor class nK for $n = 1, 2, \ldots$. If $\ell(nK) = 0$ for all n then there are no such maps, and we[12] set $\kappa = -\infty$. Here is the result of the classification. The surfaces of general type are the surfaces with $\kappa = 2$. Surfaces with $\kappa = 1$ are all elliptic; more precisely, these are all the elliptic surfaces for which $nK \neq 0$ for all $n > 0$. For an elliptic surface X, the order of the canonical class in $\operatorname{Cl} X$ is either infinite or a divisor of 12. The surfaces with $\kappa = 0$ are characterised by $12K = 0$. Thus these are the elliptic surfaces for which $12K = 0$, the K3 surfaces and 2-dimensional Abelian varieties. Surfaces with $\kappa = -\infty$ are rational or ruled.

Each of these 5 types of surfaces can be characterised by invariants, in the same way that $g = 0$ characterises rational curves. We only give such a characterisation for the two first types. For this we use the result of Ex. 7, according to which the numbers $\ell(mK)$ for $m \geq 0$ are birational invariants

---

[12] $\kappa = -1$ also occurs in the literature. The invariant κ is usually called the *Kodaira dimension*, although it was introduced in different contexts by the Shafarevich seminar [67] and by Iitaka.

of nonsingular projective varieties. They are called *plurigenera*, and denoted by P_m. In particular, $P_1 = h^n = \dim \Omega^n[X]$ where $n = \dim X$.

Rationality criterion. *A surface X is rational if and only if $\Omega^1[X] = 0$ and $P_1 = P_2 = 0$.* $\square$

The positive solution of the Lüroth problem (discussed in 6.4) for surfaces follows at once from this criterion.

Ruledness criterion. *A surface X is ruled if and only if $P_3 = P_4 = 0$.* $\square$

Generalisations of the results of this section to varieties of dimension ≥ 3 are not known, although there has been a lot of progress on this question in recent years. For this see the surveys Esnault [24], Kawamata [43] and Wilson [77], and for the relation with minimal models, Kawamata, Matsuda and Matsuki [44].

Exercises to §6

1. Prove that an element $f \in k(X)$ satisfies $df = 0$ if and only if $f \in k$ (in the case of a field of characteristic 0), or $f = g^p$ (in the case char $k = p > 0$). [Hint: Use 6.1, Theorem 1 and the following lemma: if $L \subset K$ is a finite separable field extension in characteristic $p > 0$, and $x \in K$ has the property that its minimal polynomial is of the form $\sum a_i^p x^i$ with $a_i \in L$ then $x = y^p$ with $y \in K$.]

2. Let X and Y be nonsingular projective curves and $\varphi \colon X \to Y$ a regular map such that $\varphi(X) = Y$ and $x \in X$, $y = f(x) \in Y$, and let t be a local parameter on Y at y. Prove that the number $e_x = v_x(f^*(dt))$ does not depend on the choice of the local parameter t and that $e_x > 0$ if and only if x is a branch point of φ. The number e_x is the *ramification multiplicity* of φ at x. (Compare Chap. VII, 3.1.)

3. In the notation of Ex. 2, suppose that $\varphi^*(y) = \sum l_i x_i$ where y is a divisor consisting of the single point y. Suppose that the characteristic of k is equal either to 0, or to a prime $p > l_i$. Prove that $e_{x_i} = l_i - 1$.

4. In the notation of Ex. 2–3, suppose that $Y = \mathbb{P}^1$. Prove that $g(X)$ is given by $2g(X) - 2 = -2 \deg \varphi + \sum_{x \in X} e_x$ (the *Hurwitz ramification formula*). Generalise this relation to the case of Y an arbitrary curve.

5. Suppose that $\varphi \colon X \to Y$ satisfies the conditions of Ex. 2. Prove that a rational differential $\omega \in \Omega^1(Y)$ is regular if and only if $\varphi^*(\omega) \in \Omega^1[X]$.

6. Write Ψ_m for the set of all functions ψ of mn vectors $x_{ij} \in L$ for $i = 1, \ldots, m$ and $j = 1, \ldots, n$, where L is an n-dimensional vector space, satisfying the conditions: (a) ψ is linear in each argument; (b) ψ is skewsymmetric as a function of $x_{i_0 j}$, for any fixed i_0 and $j = 1, \ldots, n$; (c) ψ is symmetric as a function of $x_{i j_0}$, for any fixed j_0 and $i = 1, \ldots, m$. Suppose that char $k > m$. Prove that every function $\psi \in \Psi$ is determined by its values $\psi_{y_1 \ldots y_n}$ at vectors $x_{ij} = y_j$, and that $\psi_{y_1 \ldots y_n} = d^m \psi_{e_1 \ldots e_n}$, where d is the determinant of the vectors $y_1, \ldots, y_n$ in the basis $e_1, \ldots, e_n$. Suppose that $\xi_1, \ldots, \xi_n \in L^*$. The function ψ for which $\psi_{y_1 \ldots y_n} = \left(\det |\xi_i(y_j)| \right)^m$ is written $\left(\xi_1 \wedge \cdots \wedge \xi_n \right)^m$. Prove that $\dim \Psi_m = 1$ and that $\left(\xi_1 \wedge \cdots \wedge \xi_n \right)^m$ is a basis.

7. Generalise the construction of regular and rational differential n-forms, replacing throughout the space $\bigwedge^r \Theta_x^*$ by Ψ_m. The resulting object is called a *differential form of weight m*. Prove that in the analogue of 5.3, (6) we should replace J by J^m. Prove that a differential form of weight m has a divisor, that all these divisors belong to one divisor class, and that this class is mK. Generalise 6.1, Theorem 2.

8. Compute the space of regular differential forms of weight 2 on a hyperelliptic curve. [Hint: Write in the form $f(dx)^2/y^2$.]

9. Verify the relation $\deg K = 2g - 2$ for hyperelliptic curves and nonsingular curves in the plane.

10. Prove that for a hyperelliptic curve, the ratio between regular differential forms generate a subfield of $k(X)$ isomorphic to the field of rational functions. From this deduce that a nonsingular plane curve $X_m \subset \mathbb{P}^2$ of degree $m > 3$ is not hyperelliptic.

11. Prove that for a hyperelliptic curve, the rational map corresponding to the canonical class is not an embedding.

12. Prove that if the map corresponding to .the canonical class of a curve X is not an embedding then X is rational or hyperelliptic. [Hint: If one or other of the conditions 6.6, (2) fails then the Riemann–Roch theorem gives $\ell(x) \geq 2$ or $\ell(x + y) \geq 2$.]

13. Prove that a nonsingular cubic 3-fold $X_3 \subset \mathbb{P}^4$ is unirational. [Hint: Use Chap. I, 6.4, Theorem 10 to show that X contains a line l. Using §5, Ex. 8, prove that there exists an open set $U \subset X$ with $U \cap l \neq \emptyset$ such that the tangent fibre space to U is isomorphic to $U \times \mathbb{A}^3$. Write $\mathbb{P}^2$ for the projective plane consisting of lines through the origin of $\mathbb{A}^3$. For a point $\xi = (u, \alpha)$ with $u \in l \cap U$ and $\alpha \in \mathbb{P}^2$, denote by $\varphi(\xi)$ the point of intersection of the line α lying in $\Theta_{X,u}$ with X. Prove that φ defines a rational map $\mathbb{P}^1 \times \mathbb{P}^2 \to X$.]

14. Let o be a point of an algebraic curve X of genus g. Using the Riemann–Roch theorem, prove that any divisor D with $\deg D = 0$ is equivalent to a divisor of the form $D_0 - go$, where $D_0 > 0$, $\deg D_0 = g$. This is a generalisation of 3.1, Theorem 1.

15. Let $X \subset \mathbb{P}^2$ be an irreducible nonsingular plane curve with equation $F = 0$, and suppose that $\alpha = (\alpha_0 : \alpha_1 : \alpha_3) \notin X$ and $x \in X$. The multiplicity c_x of x in the divisor of the form $\sum_{i=0}^{2} \partial F / \partial x_i$ is called the *multiplicity of tangency* at x. Prove that $c_x = e_x$ is the ramification multiplicity of x with respect to the map $\varphi : X \to \mathbb{P}^1$ given by projecting from α. Deduce that $c = \sum_{x \in X} c_x$ is the number of tangent lines to X through α, counted with multiplicities. It does not depend on α. It is called the *class* of X. Prove that $c = n(n-1)$ where $n = \deg X$.

16. Prove that if X is a nonsingular affine hypersurface then $K_X = 0$.

Chapter IV. Intersection Numbers

1. Definition and Basic Properties

1.1. Definition of Intersection Number

The theorems proved in Chap. I, 6.2 on the dimension of intersection of varieties often allow us to assert that some system of equations has solutions. However, they do not say anything about the number of solutions if this number is finite. The distinction is the same as that between the theorem that roots of a polynomial exist, and the theorem that the number of roots of a polynomial equals its degree. The latter result is only true if we count each root with its multiplicity. In the same way, to state general theorems on the number of points of intersection of varieties, we must assign certain intersection multiplicities to these points. This will be done in the present section.

We will consider intersection of codimension 1 subvarieties on a nonsingular variety X. We are interested in the case that the number of points of intersection is finite. If $\dim X = n$ and $C_1, \ldots, C_k$ are codimension 1 subvarieties with nonempty intersection, then by Chap. I, 6.2, Theorem 4, Corollary 5, we have $\dim(C_1 \cap \cdots \cap C_k) > 0$ if $k < n$. Hence it is natural to consider the case $k = n$. The theory that we apply in the following is simpler if we consider arbitrary divisors in place of codimension 1 subvarieties. Thus we consider n divisors $D_1, \ldots, D_n$ on an n-dimensional variety X. If $x \in X$ with $x \in \bigcap \operatorname{Supp} D_i$ and $\dim_x \bigcap \operatorname{Supp} D_i = 0$ then we say that $D_1, \ldots, D_n$ are in *general position* at x. The condition means that in some neighbourhood of x, the intersection $\bigcap \operatorname{Supp} D_i$ consists of x only. If $D_1, \ldots, D_n$ are in general position at all points of the subvariety $\bigcap \operatorname{Supp} D_i$ then this subvariety either consists of a finite number of points, or is empty. We then say that $D_1, \ldots, D_n$ are in general position.

We define intersection numbers first of all for effective divisors in general position. Suppose that $D_1, \ldots, D_n$ are effective and in general position at x, and have local equations $f_1, \ldots, f_n$ in some neighbourhood of x. Then there exists a neighbourhood U of x in which $f_1, \ldots, f_n$ are regular and have no common zeros on U other than x. It follows from the Nullstellensatz that the ideal generated by $f_1, \ldots, f_n$ in the local ring $\mathcal{O}_x$ of x contains some power of the maximal ideal $\mathfrak{m}_x$. Suppose that

$$(f_1, \ldots, f_n) \supset \mathfrak{m}_x^k. \tag{1}$$

We consider the quotient $\mathcal{O}_x/(f_1, \ldots, f_n)$ as a vector space over k; it is finite dimensional. Indeed, in view of (1), for this it is enough to prove that $\dim_k \mathcal{O}_x/\mathfrak{m}_x^k < \infty$. This last condition follows at once from the theorem on power series expansion (Chap. II, 2.2): $\dim_k \mathcal{O}_x/\mathfrak{m}_x^k$ equals the dimension of the space of polynomials of degree $< k$ in n variables.

From now on we write $\ell(E)$ for the dimension of a k-vector space E.

Definition 1. If $D_1, \ldots, D_n$ are effective divisors on an n-dimensional non-singular variety X, in general position at a point $x \in X$, and having local equations $f_1, \ldots, f_n$ in some neighbourhood of x, then the number

$$\ell(\mathcal{O}_x/(f_1, \ldots, f_n)) \tag{2}$$

is the *intersection multiplicity* or *local intersection number* of $D_1, \ldots, D_n$ at x. We denote it by $(D_1 \cdots D_n)_x$.

The number (2) actually only depends on the divisors $D_1, \ldots, D_n$ and not on the choice of local equations $f_1, \ldots, f_n$: if $f_1', \ldots, f_n'$ are other local equations then $f_i' = f_i g_i$ with g_i a unit of $\mathcal{O}_x$, and hence $(f_1, \ldots, f_n) = (f_1', \ldots, f_n')$.

Now suppose that $D_1, \ldots, D_n$ are not necessarily effective divisors. Write D_i in the form $D_i = D_i' - D_i''$, with $D_i', D_i'' \geq 0$ having no common components; this expression is unique. Suppose that $D_1, \ldots, D_n$ are in general position at x. Then, since $\operatorname{Supp} D_i = \operatorname{Supp} D_i' \cup \operatorname{Supp} D_i''$, it follows that the divisors $D_{i_1}', \ldots, D_{i_k}', D_{i_{k+1}}'', \ldots, D_{i_n}''$ are in general position at x for any permutation $i_1, \ldots, i_n$ and any k.

We now define the intersection number of $D_1, \ldots, D_n$ at x by multi-linearity, that is, we set

$$
\begin{aligned}
(D_1 \cdots D_n)_x &= \left(\prod_{i=1}^n (D_i' - D_i'') \right)_x \\
&= \sum_{i_1 \ldots i_n} \sum_{k=0}^n (-1)^{n-k} (D_{i_1}' \cdots D_{i_k}' D_{i_{k+1}}'' \cdots D_{i_n}'')_x.
\end{aligned}
\tag{3}
$$

Definition 2. If divisors $D_1, \ldots, D_n$ on an n-dimensional variety X are in general position, then the number

$$D_1 \cdots D_n = \sum_{x \in \bigcap \operatorname{Supp} D_i} (D_1 \cdots D_n)_x$$

is called their *intersection number*.

We can formally extend the sum over all points $x \in X$, although of course only the terms with $x \in \bigcap \operatorname{Supp} D_i$ are nonzero.

Remark. We can also define intersection numbers without requiring X to be a nonsingular variety; however, we then have to restrict attention to locally principal divisors (Chap. III, 1.2). All the definitions given above preserve their meaning.[13]

We now give some examples, with the aim of showing that the definition of intersection multiplicity just introduced agrees with geometric intuition.

Example 1. Suppose that $\dim X = 1$, and that t is a local parameter at a point x. Let f be a local equation of a divisor D with $v_x(f) = v_x(D) = k$. Then $(D)_x = \ell(\mathcal{O}_x/(f)) = \ell(\mathcal{O}_x/(t^k)) = k$. Thus in this case the local multiplicity $(D)_x$ is just the multiplicity of x in the divisor D.

In the following examples we will assume that D_i are prime divisors, that is, irreducible codimension 1 subvarieties of X.

Example 2. If $x \in D_1 \cap \cdots \cap D_n$ then $(D_1 \cdots D_n)_x \geq 1$ by definition. Let us determine when $(D_1 \cdots D_n)_x = 1$.

Now $f_i \in \mathfrak{m}_x$, so that $(f_1, \ldots, f_n) \subset \mathfrak{m}_x$, and since $\ell(\mathcal{O}_x/\mathfrak{m}_x) = 1$, the condition $(D_1 \cdots D_m)_x = 1$ is equivalent to $(f_1, \ldots, f_n) = \mathfrak{m}_x$. In other words, $f_1, \ldots, f_n$ should form a local system of parameters at x. We saw in Chap. II, 2.1 that this holds if and only if the subvarieties $D_1, \ldots, D_n$ intersect transversally at x, that is, x is a nonsingular point on each D_i, and $\bigcap \Theta_{D_i, x} = 0$.

Example 3. Suppose that $\dim X = 2$, and that the point x is nonsingular on both curves D_1 and D_2. By Example 2, $(D_1 D_2)_x > 1$ if and only if the two tangent lines $\Theta_{D_1, x}$ and $\Theta_{D_2, x}$ coincide. Let u, v be local parameters at x and f_1, f_2 local equations of D_1, D_2, and write $f_i \equiv \alpha_i u + \beta_i v \bmod \mathfrak{m}_x^2$. Then for $i = 1, 2$, the tangent line $\Theta_{D_i, x}$ is given by the equation $\alpha_i \xi + \beta_i \eta = 0$, where $\xi = \mathrm{d}_x u$ and $\eta = \mathrm{d}_x v$ are coordinates in $\Theta_{X, x}$. Hence $\Theta_{D_1, x} = \Theta_{D_2, x}$ if and only if $\alpha_2 u + \beta_2 v = \gamma(\alpha_1 u + \beta_1 v)$ for some nonzero $\gamma \in k$, or in other words, $f_2 \equiv \gamma f_1 \bmod \mathfrak{m}_x^2$. It is thus natural to define the *order of tangency* of D_1 and D_2 at x to be the number k such that there exists an invertible element $g \in \mathcal{O}_x$ such that $f_2 \equiv g f_1 \bmod \mathfrak{m}_x^{k+1}$, and no such g exists for greater values of the exponent $k + 1$. We now show that the intersection multiplicity is one plus the order of tangency of the curves D_1 and D_2 at x, that is, $(D_1 D_2)_x = k + 1$.

For this note that, because x is a nonsingular point of D_1, we can assume that f_1 is one element of a system of local parameters at x. On the other hand, $g^{-1} f_2$ is a local equation of D_2. Hence we can assume that u, v are

---

[13]Although the prime divisors Γ_i that are components of a Cartier divisor $D = \sum a_i \Gamma_i$ are not necessarily Cartier, it is still true that any locally principal divisor D can be written $D = D' - D''$ with D' and D'' effective; this follows in a neighbourhood of any point $x \in X$ simply because the rational function field $k(X)$ is the field of fractions of $\mathcal{O}_x$.

local parameters, the local equation of D_1 is u, that of D_2 is f, and $f \equiv u \bmod \mathfrak{m}_x^{k+1}$. Then $f \equiv u + \varphi(u,v) \bmod \mathfrak{m}_x^{k+2}$, where φ is a form of degree $k+1$. Moreover, φ is not divisible by u, since otherwise D_1 and D_2 would have order of tangency $> k$. Hence

$$\varphi(0,v) = cv^{k+1}, \qquad \text{with } c \neq 0. \tag{4}$$

By definition of intersection multiplicity,

$$(D_1 D_2)_x = \ell(\mathcal{O}_x/(u,f)) = \ell\Big((\mathcal{O}_x/(u)) \Big/ ((u,f)/(u))\Big).$$

Now obviously, $\mathcal{O}_x/(u) = \overline{\mathcal{O}}$ is the local ring of the point x on D_1, and the quotient map $\mathcal{O}_x \to \overline{\mathcal{O}}$ is restriction of functions from X to D_1. Moreover, $(u,f)/(u) = (\overline{f})$, where $\overline{f}$ is the image of f in $\overline{\mathcal{O}}$. Since, as an element of $\overline{\mathcal{O}}$, we have $\overline{f} \in (\overline{\mathfrak{m}}_x)^{k+1}$ and $\overline{f} \equiv \overline{\varphi} \bmod (\overline{\mathfrak{m}}_x)^{k+2}$, and by (4) $\overline{\varphi} \notin (\overline{\mathfrak{m}}_x)^{k+2}$, therefore $v_x(\overline{f}) = k+1$ and $\ell(\overline{\mathcal{O}}/(\overline{f})) = k+1$. Thus $(D_1 D_2)_x = k+1$.

Example 4. Suppose again that $\dim X = 2$, and that the point x is singular on D. This means that $f \in \mathfrak{m}_x^2$, where f is the local equation of D. Hence it is natural to define the *multiplicity* of the singularity $x \in D$ to be the greatest k such that $f \in \mathfrak{m}_x^k$. We prove that for any curve D' on X such that D and D' are in general position at x,

$$(DD')_x \geq k, \tag{5}$$

and that there exist curves for which $(DD')_x = k$.

Let f' be a local equation of D'. Write $\overline{\mathcal{O}}$ for $\mathcal{O}_x/\mathfrak{m}_x^k$ and $\overline{f} \in \overline{\mathcal{O}}$ for the image of f'. Since $f \in \mathfrak{m}_x^k$, we have $(DD')_x = \ell(\mathcal{O}_x/(f,f')) \geq \ell(\overline{\mathcal{O}}/(\overline{f}))$. By the theorem on power series expansion (Chap. II, 2.2), $\overline{\mathcal{O}}$ is isomorphic to $k[u,v]/(u,v)^k$. Therefore it is isomorphic as a vector space to the space of polynomials of degree $< k$ in u,v, and has dimension $1 + 2 + \cdots + k = \binom{k+1}{2}$. If $f' \in \mathfrak{m}_x^l \setminus \mathfrak{m}_x^{l+1}$ then elements of the ideal $(\overline{f})$ correspond to polynomials of the form $f'g$ where g runs through all polynomials of degree $\leq k-l$. Hence $\ell((\overline{f})) \leq 1 + \cdots + (k-l) = \binom{k-l+1}{2}$. Since $f' \in \mathfrak{m}$, we have $l \geq 1$, and hence $\ell(\overline{\mathcal{O}}/(\overline{f})) = \ell(\overline{\mathcal{O}}) - \ell((\overline{f})) \geq k$.

Now we prove that equality in (5) can be achieved. Suppose that $f \equiv \varphi(u,v) \bmod \mathfrak{m}_x^{k+1}$, where φ is a form of degree k. Consider a linear form in u,v not dividing φ. At the cost of a linear transformation of u and v we can assume that this is u, with $\varphi(0,v) \neq 0$. Take D' to be the curve with local equation u. Then $(DD')_x = \ell(\mathcal{O}_x/(u,f))$, and, as we have seen in the treatment of Example 3, this number equals k.

1.2. Additivity

Theorem 1. *If $D_1, \ldots, D_{n-1}, D'_n$ and $D_1, \ldots, D_{n-1}, D''_n$ are in general position at x then*

$$(D_1 \cdots D_{n-1}(D'_n + D''_n))_x = (D_1 \cdots D_{n-1} D'_n)_x + (D_1 \cdots D_{n-1} D''_n)_x. \quad (1)$$

Proof. First of all, it is obviously enough to prove Theorem 1 for effective divisors $D_1, \ldots, D_{n-1}, D'_n, D''_n$. From now on we assume that these divisors are effective.

Let $f_1, \ldots, f_{n-1}, f'_n, f''_n$ be local equations of the divisors $D_1, \ldots, D_{n-1}$, D'_n, D''_n. We denote the ring $\mathcal{O}_x/(f_1, \ldots, f_{n-1})$ by $\overline{\mathcal{O}}$, and the images in $\overline{\mathcal{O}}$ of f'_n, f''_n by f, g. Then

$$(D_1 \cdots D_{n-1} D'_n)_x = \ell(\overline{\mathcal{O}}/(f)), \quad (D_1 \cdots D_{n-1} D''_n)_x = \ell(\overline{\mathcal{O}}/(g)),$$
$$\text{and} \quad (D_1 \cdots D_{n-1}(D'_n + D''_n))_x = \ell(\overline{\mathcal{O}}/(fg)),$$

Since the sequence

$$0 \to (g)/(fg) \to \overline{\mathcal{O}}/(fg) \to \overline{\mathcal{O}}/(g) \to 0$$

is exact, it follows that

$$\ell(\overline{\mathcal{O}}/(fg)) = \ell(\overline{\mathcal{O}}/(g)) + \ell((g)/(fg)). \quad (2)$$

If g is a non-zerodivisor of $\overline{\mathcal{O}}$ then multiplication by g defines isomorphisms $\mathcal{O} \cong (g)$ and $(f) \cong (fg)$, hence an isomorphism $\overline{\mathcal{O}}/(f) \cong (g)/(fg)$, and therefore

$$\ell((g)/(fg)) = \ell(\overline{\mathcal{O}}/(f)). \quad (3)$$

Thus (1) follows from (2) and (3), provided that we can prove that g is a non-zerodivisor of $\overline{\mathcal{O}}$.

A sequence $f_1, \ldots, f_n$ of n elements of the local ring $\mathcal{O}_x$ of a nonsingular point of an n-dimensional variety is called a *regular sequence* if each f_i is a non-zerodivisor of $\mathcal{O}_x/(f_1, \ldots, f_{i-1})$ for $i = 1, \ldots, n$.

The arguments just given show that Theorem 1 follows from the next assertion.

Lemma 1. *If the divisors $D_1, \ldots, D_n$ are in general position at a nonsingular point x, then their local equations $f_1, \ldots, f_n$ form a regular sequence.*

In turn, the proof of Lemma 1 requires the following simple auxiliary result, which is a general property of local rings proved in Appendix, §6, Proposition 5.

Lemma 2. *The property that a sequence of elements is a regular sequence is preserved under permuting the elements of the sequence.* $\square$

Proof of Lemma 1. The proof is by induction on the dimension n of X. From the assumptions of the lemma and the theorem on the dimension of intersection (Chap. I, 6.2) it follows that $\dim_x\left(\mathrm{Supp}(D_1) \cap \cdots \cap \mathrm{Supp}(D_{n-1})\right) = 1$. Hence we can find a function u such that $u(x) = 0$, x is a nonsingular point of $V(u)$ and the n divisors $D_1, \ldots, D_{n-1}, \mathrm{div}\, u$ are in general position at x. For this, we need only take u to be the equation of a hyperplane through x not containing $\Theta_{X,x}$ or any component of the curve $\mathrm{Supp}(D_1)\cap\cdots\cap\mathrm{Supp}(D_{n-1})$. Consider the restriction to $V(u)$ of $f_1, \ldots, f_{n-1}$. They obviously satisfy all the assumptions of Lemma 1, hence by induction form a regular sequence on $V(u)$. Since the local ring of x on $V(u)$ is of the form $\mathcal{O}_x/(u)$, we see that $u, f_1, \ldots, f_{n-1}$ is a regular sequence. It then follows from Lemma 2 that $f_1, \ldots, f_{n-1}, u$ is also a regular sequence.

To prove that $f_1, \ldots, f_{n-1}, f_n$ is a regular sequence, we need only prove that f_n is not a zerodivisor of $\mathcal{O}_x/(f_1, \ldots, f_{n-1})$. By the assumption on $f_1, \ldots, f_n$, in some neighbourhood of x, the equations $f_1 = \cdots = f_n = 0$ have no solution other than x. Thus the Nullstellensatz shows that

$$(f_1, \ldots, f_n) \supset \mathfrak{m}_x^k \quad \text{for some } k.$$

In particular $u^k \in (f_1, \ldots, f_n)$, that is, $u^k \equiv a f_n \bmod (f_1, \ldots, f_{n-1})$ for some $a \in \mathcal{O}_x$.

Now if f_n were a zerodivisor of $\mathcal{O}_x/(f_1, \ldots, f_{n-1})$, it would follow that u^k, hence also u, is a zerodivisor of $\mathcal{O}_x/(f_1, \ldots, f_{n-1})$. But this contradicts the fact just proved that $f_1, \ldots, f_{n-1}, u$ is a regular sequence.

Lemma 1 is proved, and with it Theorem 1.

1.3. Invariance Under Linear Equivalence

We come now to the proof of the basic property of intersection numbers, which is the cornerstone of all their applications.

Theorem 2. *Let X be a nonsingular projective variety and $D_1, \ldots, D_n, D_n'$ divisors such that both $D_1, \ldots, D_{n-1}, D_n$ and $D_1, \ldots, D_{n-1}, D_n'$ are in general position, and suppose that D_n and D_n' are linearly equivalent. Then*

$$D_1 \cdots D_{n-1} D_n = D_1 \cdots D_{n-1} D_n'. \tag{1}$$

By the assumption of the theorem $D_n - D_n' = \mathrm{div}\, f$, and (1) is equivalent to

$$D_1 \cdots D_{n-1} \mathrm{div}\, f = 0, \tag{2}$$

when $D_1, \ldots, D_{n-1}$ and $\mathrm{div}\, f$ are in general position.

Representing D_i for $i = 1, \ldots, n-1$ as a difference of effective divisors, we see that it is enough to prove (2) when $D_i > 0$ for $i = 1, \ldots, n-1$. We assume this from now on.

The proof of Theorem 2 uses a notion of intersection number more general than that used so far. Namely, suppose given $k \leq n$ effective divisors $D_1, \ldots, D_k$ on an n-dimensional nonsingular variety X. We say that these are in *general position* if $\dim \bigcap_{i=1}^{k} \operatorname{Supp} D_i = n - k$ or $\bigcap_{i=1}^{k} \operatorname{Supp} D_i = \emptyset$. Suppose that this property is satisfied, and that

$$\bigcap_{i=1}^{k} \operatorname{Supp} D_i = \bigcup C_j, \tag{3}$$

where the C_j are irreducible $(n - k)$-dimensional varieties.

Under these conditions, we can assign a number to each component C_j, called the *intersection multiplicity* of $D_1, \ldots, D_k$ along C_j; this coincides with the intersection multiplicity at a point if $k = n$, when each C_j is just a point. The definition of intersection multiplicity along C_j uses a general notion that we now introduce.

Definition 1. A module M over a ring A is *of finite length* if it has a finite chain of A-submodules

$$M = M_0 \supset M_1 \supset \cdots \supset M_n = 0 \qquad \text{with } M_i \neq M_{i+1}, \tag{4}$$

such that each quotient M_i / M_{i+1} is a simple A-module, that is, does not contain a submodule other than 0 and the module itself. It follows from the Jordan–Hölder theorem that all such chains are made up of the same number n of modules; this common length n is called the *length* of M, and denoted by $\ell(M)$, or $\ell_A(M)$.

If A is a field, the length of a module becomes simply the dimension of a vector space. If M has finite length then so do all its submodules and quotient modules. If a module M has a chain (4) such that each quotient M_i / M_{i+1} has finite length then also M has finite length, and $\ell(M) = \sum \ell(M_i / M_{i+1})$.

The definition of intersection multiplicity along C_j mimics exactly that of intersection multiplicity at a point. Let C be one of the components C_j in (3). We choose a point $x \in C$ and local equations f_i of the D_i in a neighbourhood of x. Then $f_i \in \mathcal{O}_C$ (here $\mathcal{O}_C = \mathcal{O}_{X,C}$ is the local ring of X along C, see Chap. II, 1.1), and the ideal $\mathfrak{a} = (f_1, \ldots, f_k) \subset \mathcal{O}_C$ does not depend on the choice of the local equations f_i or of the point x. Indeed, if $g_1, \ldots, g_k$ are other local equations in a neighbourhood of another point of C then the f_i and g_i are both local equations of D_i in a whole open set that intersects C. It follows that the $f_i g_i^{-1}$ are units of $\mathcal{O}_C$, and hence $(f_1, \ldots, f_k) = (g_1, \ldots, g_k)$.

Lemma 1. $\mathcal{O}_C / \mathfrak{a}$ *is a module of finite length over* $\mathcal{O}_C$.

Indeed, since C is an irreducible component of the subvariety defined by equations $f_1 = \cdots = f_k = 0$, there exists an affine open set $U \subset X$ intersecting C in which these equations define C. Then by the Nullstellensatz,

$(f_1, \ldots, f_k) \supset \mathfrak{a}_C^r$ for some $r > 0$. Here $\mathfrak{a}_C \subset k[U]$ is the ideal of the affine coordinate ring of U defining $C \cap U$. Now set $A = k[U]$ and $\mathfrak{p} = \mathfrak{a}_C$, and consider the local ring $A_{\mathfrak{p}}$ and the natural homomorphism $\varphi \colon A \to A_{\mathfrak{p}}$ as in Chap. II, 1.1. Then $A_{\mathfrak{p}} = \mathcal{O}_C$, $\varphi(\mathfrak{a}_C) = \mathfrak{m}_C$ and $\varphi((f_1, \ldots, f_k)) = \mathfrak{a}$. Hence in $\mathcal{O}_C$, we have $\mathfrak{a} \supset \mathfrak{m}_C^r$.

The lemma now follows from the following general property of local rings: if $\mathfrak{a}$ is an ideal of a Noetherian local ring $\mathcal{O}$ with maximal ideal $\mathfrak{m}$ and $\mathfrak{a} \supset \mathfrak{m}^r$ for some $r > 0$ then $\ell_{\mathcal{O}}(\mathcal{O}/\mathfrak{a}) < \infty$. See Appendix, §9, Proposition 1. The lemma is proved.

Definition 2. The number $\ell_{\mathcal{O}_C}(\mathcal{O}_C/\mathfrak{a})$ is called the *intersection multiplicity* of $D_1, \ldots, D_k$ along C, and denoted by $(D_1 \cdots D_k)_C$.

From now on, we consider the case $k = n - 1$, so that the components C_i of the intersection $D_1 \cap \cdots \cap D_{n-1}$ are curves. Write $\overline{\mathcal{O}}$ for the quotient ring $\mathcal{O}_x/\mathfrak{a}$, where $\mathfrak{a} = (f_1, \ldots, f_{n-1})$; this is obviously a local ring, with maximal ideal $\overline{\mathfrak{m}}$ the image of the maximal ideal $\mathfrak{m} \subset \mathcal{O}_x$ under the quotient homomorphism $\mathcal{O}_x \to \overline{\mathcal{O}}$.

We first need to determine the prime ideals of $\overline{\mathcal{O}}$. Write $\mathfrak{p}_i$ for the set of functions of $\mathcal{O}_x$ that vanish identically on the curve C_i, and $\overline{\mathfrak{p}}_i$ for its image in $\overline{\mathcal{O}}$. Obviously $\overline{\mathcal{O}}/\overline{\mathfrak{p}}_i = \mathcal{O}_x/\mathfrak{p}_i = \mathcal{O}_{C_i, x}$ is the local ring of x on C_i.

Lemma 2. For a fixed point $x \in X$, suppose that $C_1, \ldots, C_r$ are the components of the intersection $D_1 \cap \cdots \cap D_{n-1}$ through x. Then $\overline{\mathfrak{p}}_1, \ldots, \overline{\mathfrak{p}}_r$ and $\overline{\mathfrak{m}}$ are all the prime ideals of $\overline{\mathcal{O}}$.

Proof. This is equivalent to saying that $\mathfrak{p}_1, \ldots, \mathfrak{p}_r$ and $\mathfrak{m}_x$ are all the prime ideals of $\mathcal{O}_x$ containing $\mathfrak{a}$. Let $\mathfrak{p}$ be a prime ideal with $\mathfrak{a} \subset \mathfrak{p} \subset \mathcal{O}_x$. Consider an affine neighbourhood U of x such that $f_1, \ldots, f_{n-1}$ are regular in U, and set $A = k[U]$ and $P = A \cap \mathfrak{p}$. Obviously P is a prime ideal. Let V be the subvariety of U defined by P; because $\mathfrak{p} \supset \mathfrak{a}$, clearly $V \subset C_1 \cup \cdots \cup C_r$, and V is irreducible since P is prime. Hence V is either equal to one of the components C_i, and then $P = A \cap \mathfrak{p}_i$, or is a point $y \in U$ (recall that the C_i are 1-dimensional). In the latter case, if $y \neq x$ then P, hence also $\mathfrak{p}$, contains a function that is nonzero at x. This gives $\mathfrak{p} = \mathcal{O}_x$ in the local ring $\mathcal{O}_x$, but $\mathcal{O}_x$ does not count as a prime ideal. Thus the unique remaining possibility is $P = A \cap \mathfrak{m}_x$. Since $\mathfrak{p} = P\mathcal{O}_x$ it follows at once that $\mathfrak{p} = \mathfrak{p}_i$ for $i = 1, \ldots, r$ or $\mathfrak{p} = \mathfrak{m}_x$, as asserted in the lemma. The lemma is proved

The ideals $\overline{\mathfrak{p}}_i$ are obviously minimal prime ideals of $\overline{\mathcal{O}}$. A local ring in which every prime ideal except for the maximal ideal is minimal is said to be 1-*dimensional*. Thus $\overline{\mathcal{O}}$ is a 1-dimensional local ring.

If $f \in \overline{\mathcal{O}}$ is an element of a 1-dimensional local ring which is a non-zerodivisor then the length $\ell(\overline{\mathcal{O}}/(f))$ can be expressed in terms of invariants connected with the localisation of $\overline{\mathcal{O}}$ at minimal prime ideals:

$$\ell_{\overline{\mathcal{O}}}(\overline{\mathcal{O}}/(f)) = \sum_{\mathfrak{p}_i} \ell_{\overline{\mathcal{O}}_{\mathfrak{p}_i}}(\overline{\mathcal{O}}_{\mathfrak{p}_i}) \times \ell_{\overline{\mathcal{O}}}(\overline{\mathcal{O}}/(\mathfrak{p}_i + f\overline{\mathcal{O}})). \tag{6}$$

This is a general property of 1-dimensional local rings. The proof is given in Appendix, §9, Proposition 2. In our case $f = f_n$, so that $\overline{\mathcal{O}}/(f) = \mathcal{O}_x/(f_1, \ldots, f_n)$, and therefore the left-hand side is $\ell(\overline{\mathcal{O}}/(f)) = (D_1 \cdots D_n)_x$.

For the right-hand side, it is easy to check that $\overline{\mathcal{O}}_{\mathfrak{p}_i} \cong \mathcal{O}_{\mathfrak{p}_i}/\varphi_{\mathfrak{p}_i}(\mathfrak{a})$, so that $\ell_{\overline{\mathcal{O}}_{\mathfrak{p}_i}}(\overline{\mathcal{O}}_{\mathfrak{p}_i}) = \ell(\mathcal{O}_{C_i}/\mathfrak{a}) = (D_1 \cdots D_{n-1})_{C_i}$. Finally

$$\overline{\mathcal{O}}/(\mathfrak{p}_i + f\overline{\mathcal{O}}) = (\overline{\mathcal{O}}/\mathfrak{p}_i)/(f) = \mathcal{O}_{C_i,x}/(f),$$

and therefore $\ell_{\overline{\mathcal{O}}}(\overline{\mathcal{O}}/(\mathfrak{p}_i + f\overline{\mathcal{O}})) = \ell(\mathcal{O}_{C_i,x}/(f_n)) = (\rho_{C_i}(D_n))_x$, where $\rho_{C_i}(D_n)$ is the restriction of the divisor D_n to the irreducible curve C_i (see Chap. III, 1.2). Thus (6) can be rewritten

$$(D_1 \cdots D_n)_x = \sum_{i=1}^{r} (D_1 \cdots D_{n-1})_{C_i} \times (\rho_{C_i}(D_n))_x. \tag{7}$$

We now prove that the multiplicity $(D)_x$ at a point x of a locally principal divisor D on a curve C is given by the formula

$$(D)_x = \sum_{\nu(y)=x} (\nu^*(D))_y, \tag{8}$$

where $\nu \colon C^\nu \to C$ is the normalisation. Indeed, let f be the local equation of a divisor D in a neighbourhood of a point $x \in C$. Then (8) can be rewritten

$$\ell(\mathcal{O}_x/(f)) = \sum_{\nu(y)=x} \ell(\mathcal{O}_y/(f)), \tag{9}$$

where $\mathcal{O}_x$ and $\mathcal{O}_y$ are the local rings of points $x \in C$ and $y \in C^\nu$.

Write $\tilde{\mathcal{O}} = \bigcap_{\nu(y)=x} \mathcal{O}_y$. Since $\tilde{\mathcal{O}}$ is contained in the field of fractions of $\mathcal{O}_x$, for every $u \in \tilde{\mathcal{O}}$ there exists $v \in \mathcal{O}_x$ such that $uv \in \mathcal{O}_x$. According to Chap. III, 2.1, Lemma, $\tilde{\mathcal{O}}$ is a finite $\mathcal{O}_x$-module. Suppose that $\tilde{\mathcal{O}} = \mathcal{O}_x u_1 + \cdots + \mathcal{O}_x u_r$, and for each i, let $v_i \in \mathcal{O}_x$ be such that $u_i v_i \in \mathcal{O}_x$; set $v = v_1 \cdots v_r$. Then $v\tilde{\mathcal{O}} \subset \mathcal{O}_x$. It follows in particular that $\ell(\tilde{\mathcal{O}}/\mathcal{O}_x) \le \ell(\tilde{\mathcal{O}}/v\tilde{\mathcal{O}})$, and by Chap. III, 2.1, Theorem 2, $\ell(\tilde{\mathcal{O}}/v\tilde{\mathcal{O}}) = \sum_{\nu(y)=x} v_y(v) < \infty$, and hence $\ell(\tilde{\mathcal{O}}/\mathcal{O}_x) < \infty$.

From the diagram

$$\begin{array}{ccc} f\tilde{\mathcal{O}} & \subset & \tilde{\mathcal{O}} \\ \cup & & \cup \\ f\mathcal{O}_x & \subset & \mathcal{O}_x \end{array}$$

it follows that $\ell(\tilde{\mathcal{O}}/(f)) + \ell(f\tilde{\mathcal{O}}/f\mathcal{O}_x) = \ell(\tilde{\mathcal{O}}/\mathcal{O}_x) + \ell(\mathcal{O}_x/(f))$. Since $\tilde{\mathcal{O}}$ has no zerodivisors, $\tilde{\mathcal{O}}/\mathcal{O}_x \cong f\tilde{\mathcal{O}}/f\mathcal{O}_x$ and $\ell(\tilde{\mathcal{O}}/\mathcal{O}_x) = \ell(f\tilde{\mathcal{O}}/f\mathcal{O}_x)$, hence $\ell(\mathcal{O}_x/(f)) = \ell(\tilde{\mathcal{O}}/(f))$. By Chap. III, 2.1, Theorem 2, $\ell(\tilde{\mathcal{O}}/(f)) = \sum_{\nu(y)=x} v_y(f) = \sum_{\nu(y)=x} \ell(\mathcal{O}_y/(f))$. This proves (9) and (8).

The proof of Theorem 2 follows almost at once by combining (7) and (8). We write the intersection number in the form

$$D_1 \cdots D_n = \sum_{x \in X} (D_1 \cdots D_n)_x.$$

By (7),

$$D_1 \cdots D_n = \sum_{j=1}^{r} (D_1 \cdots D_{n-1})_{C_j} \times \sum_{x \in C_j} (\rho_{C_j}(D_n))_x,$$

and by (8),

$$\sum_{x \in C_j} (\rho_{C_j}(D_n))_x = \sum_{y \in C_j^\nu} \left(\nu^*(\rho_{C_j}(D_n))\right)_y.$$

Now if $D_n = \text{div } f$ is a principal divisor, with $f \in k(X)$, then so are the divisors $\nu^*(\rho_{C_j}(D_n))$ on the curves C_j^ν: that is, $\nu^*(\rho_{C_j}(D_n)) = \text{div } g$, where $g = \nu^*(\rho_{C_j}(f)) \in k(C_j)$, and therefore $(\text{div } g)_y = v_y(g)$. Because X is projective, so are the C_j, and so are their normalisations C_j^ν by Chap. II, 5.3, Theorem 7. Now by Chap. III, 2.1, Theorem 1, Corollary, $\sum_{y \in C_j^\nu} v_y(g) = \deg(\text{div } g) = 0$, and it follows from this that $D_1 \cdots D_{n-1} \text{ div } f = 0$. Theorem 2 is proved.

1.4. The General Definition of Intersection Number

Theorem 2, together with Chap. III, 1.3, Theorem 1 on moving the support of a divisor away from a point, enables us to define an intersection number of any n divisors on an n-dimensional nonsingular projective variety without assuming any restriction such as general position. For this we need two lemmas.

Lemma 1. *For any n divisors $D_1, \ldots, D_n$ on an n-dimensional variety X, there exist n divisors $D_1', \ldots, D_n'$ such that $D_i \sim D_i'$ (linear equivalence) for $i = 1, \ldots, n$ and $D_1', \ldots, D_n'$ are in general position.*

Proof. Suppose that we have found divisors $D_1', \ldots, D_k'$ such that $D_i \sim D_i'$ for $i = 1, \ldots, k$, and either $\dim(\text{Supp } D_1' \cap \cdots \cap \text{Supp } D_k') = n - k$ or this intersection is empty. Let

$$\text{Supp } D_1' \cap \cdots \cap \text{Supp } D_k' = C_1 \cup \cdots \cup C_r$$

be its decomposition into irreducible components. We choose a point $x_j \in C_j$ on each component, and, using the theorem on moving the support of a divisor, find a divisor D_{k+1}' such that $D_{k+1}' \sim D_{k+1}$ and $x_j \notin \text{Supp } D_{k+1}'$

for $j = 1, \ldots, r$. Then a fortiori $\operatorname{Supp} D'_{k+1}$ does not contain any of the components C_j, and by the theorem on dimension of intersections

$$\dim\big(\operatorname{Supp} D'_1 \cap \cdots \cap \operatorname{Supp} D'_{k+1}\big) = n - k - 1,$$

if this intersection is nonempty. Proceeding in the same way until $k = n$ we get the required system of n divisors. The lemma is proved.

Lemma 2. *If $D_1, \ldots, D_n$ and $D'_1, \ldots, D'_n$ are two n-tuples of divisors in general position and $D_i \sim D'_i$ for $i = 1, \ldots, n$ then*

$$D_1 \cdots D_n = D'_1 \cdots D'_n. \tag{1}$$

Proof. If $D_i = D'_i$ for $i = 1, \ldots, n-1$ then this is the assertion of Theorem 2. Let us prove that (1) hold if $D_i = D'_i$ for $i = 1, \ldots, n - k$. For $k = n$ we get our assertion.

We use induction on k. Suppose that the assertion holds for smaller values of k. Since both $D_1, \ldots, D_n$ and $D'_1, \ldots, D'_n$ are in general position, both

$$Y = \bigcap_{i \neq n-k+1} \operatorname{Supp} D_i \quad \text{and} \quad Y' = \bigcap_{i \neq n-k+1} \operatorname{Supp} D'_i$$

are 1-dimensional. We choose one point on each component of each of Y and Y', and, by the theorem on moving the support of a divisor, find a divisor D''_{n-k+1} such that $\operatorname{Supp} D''_{n-k+1}$ does not contain any of these points and $D''_{n-k+1} \sim D_{n-k+1}$. Then both $D_1, \ldots, D_{n-k}, D''_{n-k}, \ldots, D_n$ and $D'_1, \ldots, D'_{n-k}, D''_{n-k}, \ldots, D'_n$ are in general position. Then by Theorem 2

$$\begin{aligned} D_1 \cdots D_n &= D_1 \cdots D_{n-k} D''_{n-k+1} \cdots D_n \\ \text{and} \quad D'_1 \cdots D'_n &= D'_1 \cdots D'_{n-k} D''_{n-k+1} \cdots D'_n. \end{aligned} \tag{2}$$

Now the right-hand sides in (2) are equal by induction, since they involve $n - k + 1$ equal factors. This proves Lemma 2.

Using Lemmas 1–2 we define the *intersection number* $D_1 \cdots D_n$ of any n divisors on a nonsingular n-dimensional variety, without requiring them to be in general position. For this, we find any divisor $D'_1, \ldots, D'_n$ satisfying the assumptions of Lemma 1, so that the intersection number $D'_1 \cdots D'_n$ is defined, and define $D_1 \cdots D_n$ by $D_1 \cdots D_n = D'_1 \cdots D'_n$. We need to verify that this definition is independent of the choice of the auxiliary divisors $D'_1, \ldots, D'_n$, but this is exactly what Lemma 2 guarantees.

For example, we can now speak of the selfintersection number CC of a curve C on a surface X. This number is also denoted by C^2. We give some examples of how C^2 is computed.

Example 1. Let $X = \mathbb{P}^2$, and let $C \subset \mathbb{P}^2$ be a line. By definition $C^2 = C'C''$ where $C' \sim C'' \sim C$ and C' and C'' are in general position. We can, for example, take C' and C'' to be two distinct lines. These intersect in a single point x, and since they are transverse at x, we have $C'C'' = (C'C'')_x = 1$. Hence $C^2 = 1$.

Example 2. Let $X \subset \mathbb{P}^N$ be an n-dimensional nonsingular projective variety. Write E for the intersection of X with a hyperplane of $\mathbb{P}^N$. Obviously $E \in \mathrm{Div}\, X$. Our aim is to give an interpretation of the number E^n. (We have seen in Chap. III, 1.4 that all hyperplanes define linearly equivalent divisors, so that this number does not depend on the choice of the hyperplane E.)

By definition $E^n = E^{(1)} \cdots E^{(n)}$, where $E^{(i)}$ for $i = 1, \ldots, n$ are hyperplane sections of X in general position. By Chap. I, 6.2, these always exist. Then the points $x_i \in E^{(1)} \cap \cdots \cap E^{(n)}$ are the points of intersection of X with a $(N - n)$-dimensional projective linear subspace $L = \mathbb{P}^{N-n}$ in general position with X. Since $E^{(1)} \cdots E^{(n)} = \sum_{x_i \in X \cap L} (E^{(1)} \cdots E^{(n)})_{x_i}$ and each $(E^{(1)} \cdots E^{(n)})_{x_i} > 0$, it follows that E^n is $\geq$ the number of points of $X \cap L$. Now if L is transversal to X at every point of intersection then $(E^{(1)} \cdots E^{(n)})_{x_i} = 1$ for each x_i, and E^n is equal to the number of points of $X \cap L$. We check that such a subspace L does exist, which gives us the following interpretation of the number E^n: it is the maximum number of points of intersection of X with a projective linear subspace $\mathbb{P}^{N-n}$ of complimentary dimension in general position with respect to X. This number is called the *degree* of X and denoted by $\deg X$. For the case of a hypersurface see 2.1, Example 1.

The existence of the required subspace L is proved by the traditional method of dimension counting (compare Chap. I, 6.4). Consider the variety of projective linear subspaces $L = \mathbb{P}^{N-n} \subset \mathbb{P}^N$; this is the Grassmannian $G = \mathrm{Grass}(N - n + 1, V)$ (see Chap. I, 4.1, Example 1), where $\mathbb{P}^N = \mathbb{P}(V)$, that is, $\dim V = N + 1$. In the product $X \times G$, consider the subvariety Γ of pairs (x, L) such that the subspace L is not in general position with $\Theta_{X,x}$. This is obviously a closed subspace (for example, the conditions that $x \in L$ and $0 \subsetneq L \cap \Theta_{X,x} \subset \Theta_{\mathrm{Proj}^N, x}$ can be written as the vanishing of minors of matrixes made up by the linear equations of the subspace L). The fibre of the first projection $\Gamma \to X$ above $x \in X$ consists of subspaces $L \in \mathrm{Grass}(N - n, \Theta_{\mathbb{P}^N, x})$ that are in nongeneral position with respect to $\Theta_{X,x}$. Its dimension is at most $\dim \mathrm{Grass}(N - n, \Theta_{\mathbb{P}^N, x}) - 1 = (N - n)n - 1$. Hence $\dim \Gamma \leq (N - n)n - 1 + n$. A fortiori the projection of Γ to G has dimension $\leq (N - n)n - 1 + n$. But $\dim G = (N - n + 1)n$, and hence there exists a point of G not contained in the projection of Γ.

Example 3. Let $X \subset \mathbb{P}^3$ be a nonsingular surface of degree m and $L \subset X$ a line; we calculate L^2.

Consider a plane of $\mathbb{P}^3$ containing L and not tangent to X at at least one point of L, and let E be the hyperplane section of X by this plane. Then L

is contained in E as a component of multiplicity 1

$$E = L + C, \quad \text{with } C = \sum k_i C_i \text{ and } \sum k_i \deg C_i = m - 1.$$

We compute first C^2. For this, we observe that the curve E is singular at a point of intersection of L and C, which means that the plane cutting out E equals the tangent plane to X at this point. Consider another plane through L distinct from the tangent planes at all the points of $L \cap C$. This plane defines a divisor $E' = L + C'$, and the points of $L \cap C$ and $L \cap C'$ are all distinct. This means that $C \cap C' = \emptyset$; hence $C^2 = CC' = 0$.

Now using $EL = 1$ (since $L \subset \mathbb{P}^3$ is a line), we get

$$m = E^2 = E(L + C) = EL + EC = 1 + EC,$$

and hence $EC = m - 1$; since we have just proved that $C^2 = 0$,

$$m - 1 = EC = (L + C)C = LC + C^2 = LC;$$

and finally,

$$1 = EL = L^2 + LC = L^2 + m - 1, \quad \text{therefore} \quad L^2 = 2 - m.$$

Note that $L^2 < 0$ if $m > 2$. Lines can indeed lie on a nonsingular surface of arbitrary degree, for example the line $x_0 = x_1$, $x_2 = x_3$ on the surface $x_0^m - x_1^m + x_2^m - x_3^m = 0$.

Exercises to §1

1. Let X be a surface, $x \in X$ a nonsingular point, u, v local parameters at x, and f a local equation of a curve C in a neighbourhood of x. If $f = (au + bv)(cu + dv) + g$ with $g \in \mathfrak{m}_x^3$, and the linear forms $au + bv$ and $cu + dv$ are not proportional then we say that x is a *double point with distinct tangent directions* or *node* and the lines in Θ_x with equations $au + bv = 0$ and $cu + dv = 0$ are called the *tangent lines* to C at x (compare Chap. II, §3, Ex. 12). Under the stated assumptions, let C' be a nonsingular curve on X passing through x. Prove that $(CC')_x > 2$ if and only if $\Theta_{C',x}$ is one of the tangent directions to C at x.

2. Let $C = V(F)$ and $D = V(G)$ be two plane curves in $\mathbb{A}^2$ and x a nonsingular point on both of them. Let f be the restriction of F to the curve D and $v_x(f)$ the order of zero of f at x. Prove that this number is unchanged if F and G are interchanged.

3. Let Y be a nonsingular irreducible codimension 1 subvariety of an n-dimensional nonsingular variety X. Prove that if $D_1, \ldots, D_{n-1}$ are divisors on X in general position with Y at x then $(D_1 \cdots D_{n-1} Y)_x = \left(\rho_Y(D_1) \cdots \rho_Y(D_{n-1}) \right)_x$, where the right-hand side is the intersection number computed in Y.

4. Find the degree of the surface $v_m(\mathbb{P}^2)$, where v_m is the Veronese embedding.

5. Let $X \subset \mathbb{P}^n$ be a nonsingular projective surface and $L = \mathbb{P}^{n-2} \subset \mathbb{P}^n$ a projective linear subspace of dimension $n-2$. Suppose that L and X intersect in a finite number of points, and that at k of these points, L intersects the tangent plane $\Theta_{X,x}$ in a line. Prove that the number of points of intersection of L and X is at most $\deg X - k$.

6. Let $X \subset \mathbb{P}^n$ be a nonsingular projective surface not contained in any $\mathbb{P}^{n-1}$ and $L = \mathbb{P}^{n-m}$ a $(n-m)$-dimensional projective linear subspace such that $X \cap L$ is finite. Suppose that at k of these points, L intersects the tangent plane $\Theta_{X,x}$ in a line. Prove that the number of points of intersection of L and X is $\leq \deg X - k - m + 2$. [Hint: Find a suitable projective linear subspace passing through L and satisfying the assumptions of Ex. 5. Start from the case $k = 0$.]

7. Prove that if $D_1, \ldots, D_{n-1}$ are effective divisors in general position on a nonsingular n-dimensional variety and $C \subset \operatorname{Supp} D_1 \cap \cdots \cap \operatorname{Supp} D_{n-1}$ is an irreducible component then $(D_1 \cdots D_{n-1})_C = \min(D_1 \cdots D_{n-1}D)_x$, where the minimum is taken over all $x \in C$ and all effective divisors with $x \in \operatorname{Supp} D$.

8. Compute $(D_1 D_2)_C$, where $D_1, D_2 \subset \mathbb{A}^3$ are given by $x = 0$ and $x^2 + y^2 + xz = 0$ respectively, and C is the line $x = y = 0$.

2. Applications of Intersection Numbers

2.1. Bézout's Theorem in Projective and Multiprojective Space

Theorems 1 and 2 of 1.2–3 allow us to compute intersection numbers of any divisors on a variety X provided that we know the divisor class group $\operatorname{Cl} X$ well enough. We illustrate this with two examples.

Example 1. $X = \mathbb{P}^n$. We know that $\operatorname{Cl} X \cong \mathbb{Z}$, and that we can take a hyperplane divisor E as generator. Any effective divisor D is the divisor of a form F, and if $\deg F = m$ then $D \sim mE$. It follows that if $D_i \sim m_i E$ for $i = 1, \ldots, n$ then

$$D_1 \cdots D_n = m_1 \cdots m_n E^n = m_1 \cdots m_n, \tag{1}$$

since obviously $E^n = 1$.

If the divisors D_i are effective, that is, they correspond to forms F_i of degree m_i, and are in general position, then the points of $\bigcap \operatorname{Supp} D_i$ are exactly the nonzero solutions of the system of equations

$$F_1(x_0, \ldots, x_n) = \cdots = F_n(x_0, \ldots, x_n) = 0.$$

Here we only consider nonzero solutions, and consider proportional solutions to be the same. For such a point x (or solution), it is natural to call the local intersection number $(D_1 \cdots D_n)_x$ the *multiplicity* of the solution. Then (1) says that the number of solutions of a system of n homogeneous equations in $n+1$ unknowns is either infinite, or is equal to the product of the degrees of the equations, provided that solutions are counted with their multiplicities. This result is called *Bézout's theorem* in projective space $\mathbb{P}^n$.

In particular, if $D_2, \ldots, D_n$ are hyperplanes then we see that $DE^{n-1} = \deg F$, where $F = 0$ is the equation of D. If D is a nonsingular hypersurface then by definition the intersection number DE^{n-1} in $\mathbb{P}^n$ is equal to the intersection number E^{n-1} on D. Therefore $\deg F = \deg D$ in the sense of the definition in 1.5, Example 2.

Example 2. $X = \mathbb{P}^n \times \mathbb{P}^m$. In this case $\operatorname{Cl} X = \mathbb{Z} \oplus \mathbb{Z}$, since by Chap. I, 5.1, Theorem 1, any effective divisor D is defined by a polynomial F that is homogeneous separately in the two sets of variables $x_0, \ldots, x_n$ and $y_0, \ldots, y_m$ (the coordinates in $\mathbb{P}^n$ and $\mathbb{P}^m$ respectively). If F has degree of homogeneity k and l then $D \mapsto (k, l)$ defines an isomorphism $\operatorname{Cl} X \cong \mathbb{Z} \oplus \mathbb{Z}$. In particular, we can take as generators of $\operatorname{Cl} X$ the divisors E and F defined respectively by linear forms in the x_i and y_i; then $D \sim kE + lF$.

Suppose that $D_i \sim k_i E + l_i F$ for $i = 1, \ldots, n+m$. Then

$$D_1 \cdots D_{n+m} = \prod_{i=1}^{n+m} (k_i E + l_i F) = \sum k_{i_1 \ldots i_r} l_{j_1 \ldots j_s} E^r F^s, \qquad (2)$$

where the sum runs over all permutations $(i_1 \ldots i_r j_1 \ldots j_s)$ of $\{1, \ldots, n+m\}$ with $i_1 < i_2 < \cdots < i_r$ and $j_1 < j_2 < \cdots < j_s$. We now compute the intersection number $E^r F^s$. If $r > n$ then we can find r linear forms $E_1, \ldots, E_r$ having no common zeros in $\mathbb{P}^r$, so that

$$E^r F^s = E_1 \cdots E_r F^s = 0.$$

The same thing happens if $s > m$. Since $r + s = n + m$, the intersection number $E^r F^s$ can only be nonzero for $r = n$ and $s = m$. In this case we can take $E_1, \ldots, E_n$ and $F_1, \ldots, F_m$ to be the divisors defined by the forms $x_1, \ldots, x_n$ and $y_1, \ldots, y_m$. These divisors have a unique common point, $(1 : 0 : \cdots : 0; 1 : 0 : \cdots : 0)$. They intersect transversally there, as one checks easily on passing to the open subset $x_0 \neq 0$, $y_0 \neq 0$, which is isomorphic to $\mathbb{A}^{n+m}$. Thus

$$D_1 \cdots D_{n+m} = \prod_{i=1}^{n+m} (k_i E + l_i F) = \sum k_{i_1 \ldots i_n} l_{j_1 \ldots j_m}, \qquad (3)$$

where the sum runs over all permutations $(i_1 \ldots i_r j_1 \ldots j_s)$ of $\{1, \ldots, n+m\}$ with $i_1 < i_2 < \cdots < i_r$ and $j_1 < j_2 < \cdots < j_s$. This result is called *Bézout's theorem* in $\mathbb{P}^n \times \mathbb{P}^m$.

One common feature of the two examples just treated is that $\operatorname{Cl} X$ is finitely generated. It is natural to ask whether this holds for any nonsingular variety X. This is not so; a counterexample is provided by the nonsingular plane cubic curve, which has a subgroup $\mathrm{Cl}^0 X \subset \mathrm{Cl}\, X$ with $\mathrm{Cl}\, X / \mathrm{Cl}^0 X \cong \mathbb{Z}$, and the elements of $\mathrm{Cl}^0 X$ are in one-to-one correspondence with points of X. Hence, for example, if $k = \mathbb{C}$ then Cl^0 is not even countable.

This big subgroup $\mathrm{Cl}^0 X$ has, however, no effect on intersection numbers, since $\deg D = 0$ for $D \in \mathrm{Cl}^0 X$. The same thing also holds for an arbitrary nonsingular projective variety. Namely, one can prove[14] that if a divisor D is algebraically equivalent to 0 (see Chap. III, 4.4 for the definition), then $D_1 \cdots D_{n-1} D = 0$ for any divisors $D_1, \ldots, D_{n-1}$. Thus intersection numbers depend only on the classes of divisors in $\mathrm{Div}\, X / \mathrm{Div}^a X$ (the Néron–Severi group $\mathrm{NS}\, X$). Chap. III, 4.4, Theorem D asserts that this group is finitely generated. Obviously, if $E_1, \ldots, E_r$ are generators of this group, then in order to know any intersection numbers of divisors on X, it is enough to know the finitely many numbers $E_1^{i_1} \cdots E_r^{i_r}$ with $i_1 + \cdots i_r = \dim X$, by analogy with what we saw in Examples 1–2. In other words, an analogue of Bézout's theorem holds for X.

2.2. Varieties over the Reals

The different versions of Bézout's theorem proved in 2.1 have some pretty applications to algebraic geometry over $\mathbb{R}$.

We return to 2.1, Example 1, and suppose that the equations $F_i = 0$ for $i = 1, \ldots, n$ have real coefficients, and that we are interested in real solutions. If $\deg F_i = m_i$ and the divisors D_i are in general position then $D_1 \cdots D_n = m_1 \cdots m_n$, as proved in 2.1, Example 1. By definition, $D_1 \cdots D_n = \sum (D_1 \cdots D_n)_x$, where the sum runs over solutions x of the system of equations $F_1 = \cdots = F_n = 0$. In this we must of course consider both real and complex solutions x. However, since the F_i have real coefficients, whenever x is a solution then so is the complex conjugate $\bar{x}$. By definition of the intersection number it follows at once that $(D_1 \cdots D_n)_x = (D_1 \cdots D_n)_{\bar{x}}$, and hence $D_1 \cdots D_n \equiv \sum (D_1 \cdots D_n)_y \bmod 2$, where now the sum takes place only over real solutions y. In particular if $D_1 \cdots D_n$ is odd (which holds if and only if all the degrees $\deg F_i = m_i$ are odd), then we deduce that there exists at least one real solution. This assertion is proved under the assumption that the D_i are in general position. But the following simple argument allows us to get rid of this restriction.

The point is that in our case the theorem on moving the support of a divisor can be proved very simply and in a more explicit form. Namely, we can choose a linear form l nonzero at all the points $x_1, \ldots, x_r$ we want to move the support of the divisor away from. If D is defined by a form F of degree m then the divisor D' defined by the form $F_\varepsilon = F + \varepsilon l^m$ will satisfy all the conditions in the conclusion of the theorem if $F(x_i) + \varepsilon l(x_j)^m \neq 0$ for

---

[14]See Fulton [27], Chap. 10 for a proof (in much more advanced terms).

$j = 1, \ldots, r$. These conditions can be satisfied for arbitrarily small values of ε.

We now show how to get rid of the general position restriction in the assertion we proved above on the existence of a real solution of a system of equations of odd degrees. Let

$$F_1 = \cdots = F_n = 0 \qquad (1)$$

be any such system. By what we have said above we can find linear forms l_i and arbitrarily small values of ε such that the divisors defined by the forms $F_{i,\varepsilon} = F_i + \varepsilon l_i^{m_i}$ are in general position. Now we proved above that the system $F_{1,\varepsilon} = \cdots = F_{n,\varepsilon} = 0$ has a real solution x_ε. Because projective space is compact, we can find a sequence of numbers $\varepsilon_m \to 0$ such that the sequence x_{ε_m} converges to a point $x \in \mathbb{P}^n$. Now $F_{j,\varepsilon_m} \to F_j$ as $\varepsilon \to 0$, so that x is a solution of the system (1).

We state the result we have just proved.

Theorem 1. *A system of n homogeneous real equations in $n + 1$ variables has a nonzero real solution if the degree of each equation is odd.* $\square$

Entirely analogous arguments apply to the variety $\mathbb{P}^n \times \mathbb{P}^m$ (see 2.1, Example 2). We get the following result.

Theorem 2. *A system of real equations*

$$F_i(x_0 : \cdots : x_n; y_0 : \cdots : y_m) = 0 \qquad for \ i = 1, \ldots, n + m$$

has a nonzero real solution if the number $\sum k_{i_1} \cdots k_{i_n} l_{j_1} \cdots l_{j_m}$ is odd. Here k_i and l_i are the degrees of homogeneity of F_i in the two sets of variables, and we consider a solution to be zero if either $x_0 = \cdots = x_n = 0$ or $y_0 = \cdots = y_m = 0$. $\square$

Theorem 2 has interesting applications to algebra. One of these is concerned with the question of division algebras over $\mathbb{R}$. If an algebra over $\mathbb{R}$ has rank n then it has a basis $e_1, \ldots, e_n$, and the algebra structure is determined by a multiplication table

$$e_i e_j = \sum_{k=1}^{n} c_{ij}^k e_k \qquad for \ i, j = 1, \ldots, n. \qquad (2)$$

We do not assume that the algebra is associative, so that the structure constants c_{ij}^k can be arbitrary. The algebra is called a *division algebra* if the equation

$$ax = b \qquad (3)$$

has a solution for every $a \ne 0$ and every b. It is easy to see that this is equivalent to the nonexistence of zerodivisors in the algebra. For this it is enough to consider the linear map φ given by $\varphi(x) = ax$ in the real vector space formed by elements of the algebra. The condition that (3) has a solution means that the image of φ is the whole space, and this is equivalent to $\ker \varphi = 0$, as is well known. This condition means just that the algebra has no zerodivisors, that is $xy = 0$ implies either $x = 0$ or $y = 0$. If $x = \sum_{i=1}^{n} x_i e_i$ and $y = \sum_{j=1}^{n} y_j e_j$, then (2) gives

$$xy = \sum_{k=1}^{n} z_k e_k, \quad \text{where} \quad z_k = \sum_{i=1}^{n} \sum_{j=1}^{n} c_{ij}^k x_i y_j \quad \text{for } k = 1, \ldots, n.$$

Thus the algebra is a division algebra if the system of equations

$$F_k(x, y) = \sum_{i=1}^{n} \sum_{j=1}^{n} c_{ij}^k x_i y_j = 0 \quad \text{for } k = 1, \ldots, n \tag{4}$$

has no real solutions with $(x_1, \ldots, x_n), (y_1, \ldots, y_n) \neq (0, \ldots, 0)$. These equations very nearly satisfy the conditions of Theorem 2. The difference is that the F_k define divisors in $\mathbb{P}^{n-1} \times \mathbb{P}^{n-1}$, the number n of which is not equal to the dimension $2n - 2$ of the variety. We therefore choose any integer r with $1 \leq r \leq n - 1$ and set $x_{r+2} = \cdots = x_n = 0$ and $y_{n-r+2} = \cdots = y_n = 0$. The equations $F_k\big((x_1, \ldots, x_{r+1}, 0, \ldots, 0), (y_1, \ldots, y_{n-r+1}, 0, \ldots, 0)\big) = 0$ for $k = 1, \ldots, n$ are now defined in $\mathbb{P}^r \times \mathbb{P}^{n-r}$, and a fortiori have no nonzero real roots. According to Theorem 2 this is only possible if the sum

$$\sum k_{i_1} \cdots k_{i_r} l_{j_1} \cdots l_{j_{n-r}} \tag{5}$$

is even, and this must moreover hold for all $r = 1, \ldots, n - 1$. In our case the forms F_k are bilinear, so that $k_i = l_i = 1$, and the sum (5) equals the number of summands, which is $\binom{r}{n}$. We see that if (4) has no nonzero real solutions then all the integers $\binom{r}{n}$ are even for $r = 1, \ldots, n - 1$. This is only possible if $n = 2^k$. Indeed, our condition on $\binom{r}{n}$ can be expressed as follows: over the field with 2 elements $\mathbb{F}_2$ we have $(T + 1)^n = T^n + 1$. If $n = 2^l m$ with m odd and $m > 1$ then over $\mathbb{F}_2$,

$$(T + 1)^{2^l m} = (T^{2^l} + 1)^m = T^{2^l m} + mT^{2^l(m-1)} + \cdots + 1 \neq T^{2^l m} + 1.$$

We have proved the following result:

Theorem 3. *The rank of a division algebra over* $\mathbb{R}$ *is a power of* 2. $\quad\square$

It can be proved that a division algebra over $\mathbb{R}$ exists only for $n = 1, 2, 4$ and 8. The proof of this fact uses rather delicate topological arguments.

Applying analogous arguments, one can investigate for which values of m and n the system of equations

$$\sum_{k=1}^{n}\sum_{j=1}^{m}c_{ij}^{k}x_{k}y_{j}=0 \qquad \text{for } i=1,\dots,n, \tag{6}$$

does not have nonzero real solutions. Based on the interpretation of the tangent space to $\mathbb{P}(V)$ given in Chap. II, 1.3, one can easily show that under the stated assumption, (6) defines $(m-1)$ linearly independent tangent vectors at each point of $\mathbb{P}^{n-1}$, that is, $(m-1)$ everywhere linearly independent vector fields on $\mathbb{P}^{n-1}$. In this form, the question of the possible values of m and n has been completely answered by topological methods. The question is interesting because it is equivalent to that of knowing whether the system of partial differential equations

$$\sum_{k=1}^{n}\sum_{j=1}^{m}c_{ij}^{k}\frac{\partial u_{j}}{\partial x_{k}}=0 \qquad \text{for } i=1,\dots,m$$

is elliptic.

2.3. The Genus of a Nonsingular Curve on a Surface

The following formula plays an enormous role in the geometry on a nonsingular projective surface X. It is usually called the *adjunction formula* or the *genus formula*, and expresses the genus of a nonsingular curve $C \subset X$ in terms of certain intersection numbers:

$$g_C = \frac{1}{2}C(C+K)+1; \tag{1}$$

here g_C is the genus of C and K the canonical class of X.

This formula can be proved using only the methods we already know. However, a clearer and more transparent geometric proof follows from the elementary properties of vector bundles. This is given in Chap. VI, 1.4, Theorem 4. Here we only discuss a number of applications.

Example 1. The projective plane. If $X = \mathbb{P}^2$ then $\operatorname{Cl} X = \mathbb{Z}$, with generator L, the class containing all the lines of $\mathbb{P}^2$. If $C \subset \mathbb{P}^2$ has degree n then $C \sim nL$. In view of $K = -3L$ and $L^2 = 1$, in this case (1) gives

$$g = \frac{n(n-3)}{2}+1 = \frac{(n-1)(n-2)}{2}.$$

We obtained the same result in Chap. III, 6.4 by a different method.

Example 2. The nonsingular quadric surface. Let $X \subset \mathbb{P}^3$ be a nonsingular quadric surface in $\mathbb{P}^3$. Let's see how to classify nonsingular curves on X in terms of their geometric properties.

The algebraic classification is clear. Since $X \cong \mathbb{P}^1 \times \mathbb{P}^1$, any curve on X is defined by an equation $F(x_0 : x_1; y_0 : y_1) = 0$, where F is a polynomial homogeneous in the two sets of variables x_0, x_1 and y_0, y_1; write m and n for the degrees of homogeneity. F has $(m+1)(n+1)$ coefficients, and hence the curves of bidegree (m, n) correspond to points of $\mathbb{P}^N$, where $N = (m+1)(n+1) - 1$. There exists a nonsingular irreducible curve of any bidegree (m, n) with $m > 0$, $n > 0$, for example the curve given by

$$2x_0^m y_0^n + x_0^m y_1^n + x_1^m y_0^n + x_1^m y_1^n = 0;$$

thus nonsingular irreducible curves correspond to points of a nonempty open set of $\mathbb{P}^N$.

We saw in 2.1 that $\operatorname{Cl} X = \mathbb{Z} \oplus \mathbb{Z}$, and that a curve C given by a polynomial of bidegree (m, n) satisfies

$$C \sim mE + nF, \tag{2}$$

where $E = \mathbb{P}^1 \times x$ and $F = x \times \mathbb{P}^1$. Thus the curves corresponding to the given bidegree (m, n) are the effective divisors of the class $mE + nF$.

The classes E and F correspond to the two families of line generators on X. It is easy to find the intersection number of curves given in the form (2): if

$$C \sim mE + nF \quad \text{and} \quad C' \sim mE' + nF' \tag{3}$$

then

$$CC' = mn' + nm'. \tag{4}$$

In particular

$$m = CF \quad \text{and} \quad n = CE. \tag{5}$$

This shows the geometric meaning of m and n: just as the degree of a plane curve equals the number of points of intersection with a line, so m and n are the two degrees of C with respect to the two families of line generators E and F on X.

Taking account of the embedding $X \subset \mathbb{P}^3$ provides a new geometric invariant of a curve, its degree. We know that the families of curves on X are simply classified by the invariants m and n. Our aim at present is to recover this classification in terms of the invariants $\deg C$ and g_C.

We know that

$$\deg C = CH, \tag{6}$$

where H is a hyperplane section of X. Now note that

$$H \sim E + F \tag{7}$$

which follows at once from (5) and from the fact that H intersects both E and F transversally in one point. Substituting in (6) and using (4) gives

$$\deg C = m + n. \tag{8}$$

Note that, except for the case C a line, any irreducible curve C has $m > 0$, $n > 0$. Indeed, if C is not, say, in the first family of line generators, then taking any point $x \in C$ and the line E of the first family through x, we see that C and E are in general position and $CE = n \geq (CE)_x > 0$.

We proceed to calculate g_C. To apply (1), we need to know the canonical class of X, which we now determine. We use the fact that $X \cong \mathbb{P}^1 \times \mathbb{P}^1$. It is easy to solve the even more general question, to find the canonical class of a surface $X = Y_1 \times Y_2$ which is the product of two nonsingular projective curves Y_1 and Y_2. We write $\pi_1 \colon X \to Y_1$ and $\pi_2 \colon X \to Y_2$ for the two projections, consider arbitrary 1-forms $\omega_1 \in \Omega^1(Y_1)$ and $\omega_2 \in \Omega^1(Y_2)$ and the pullback 1-forms $\pi_1^*(\omega_1)$ and $\pi_2^*(\omega_2)$ on X. Then $\omega = \pi_1^*(\omega_1) \wedge \pi_2^*(\omega_2)$ is a 2-form on X and its divisor $\operatorname{div} \omega$ belongs to the canonical class. We now calculate this divisor.

Let $x = (y_1, y_2) \in X$ where $y_1 \in Y_1$ and $y_2 \in Y_2$, and write t_1, t_2 for local parameters on Y_1 and Y_2 in a neighbourhood of y_1 and y_2. An obvious verification then shows that $\pi_1^*(t_1), \pi_2^*(t_2)$ is a local system of parameters at $x \in X$. Write ω_1 and ω_2 in the form $\omega_1 = u_1 dt_1$ and $\omega_2 = u_2 dt_2$. Then $\operatorname{div}(\omega_1) = \operatorname{div}(u_1)$ and $\operatorname{div}(\omega_2) = \operatorname{div}(u_2)$ in a neighbourhood of y_1 and y_2. Obviously $\omega = \pi_1^*(u_1)\pi_2^*(u_2) \cdot d\pi_1^*(t_1) \wedge d\pi_2^*(t_2)$, and it follows that in some neighbourhood of x,

$$\operatorname{div}(\omega) = \operatorname{div}(\pi_1^*(u_1)) + \operatorname{div}(\pi_2^*(u_2)) = \pi_1^*(\operatorname{div}(\omega_1)) + \pi_2^*(\operatorname{div}(\omega_2)).$$

Since this holds for any point $x \in X$ it follows that $\operatorname{div}(\omega) = \pi_1^*(\operatorname{div}(\omega_1)) + \pi_2^*(\operatorname{div}(\omega_2))$, or in other words,

$$K_X = \pi_1^*(K_{Y_1}) + \pi_2^*(K_{Y_2}). \tag{9}$$

Now return to the case $X = \mathbb{P}^1 \times \mathbb{P}^1$. We know that $K_{\mathbb{P}^1} = -2y$ for a point $y \in \mathbb{P}^1$. Thus in our case, (9) gives $K_X = -2\pi_1^*(y_1) - 2\pi_2^*(y_2)$. Since $\pi_1^*(y_1) = E$ and $\pi_2^*(y_2) = F$, we finally get

$$K_X = -2E - 2F. \tag{10}$$

Now for the genus of a curve $C \sim mE + nF$, we substitute this formula into (1) and use (4). We get

$$g_C = (m-1)(n-1). \tag{11}$$

The numbers m and n are thus determined uniquely up to permutation by the degree and genus of C. We see that curves on X of given degree d form $d+1$ families $M_0, \ldots, M_d$. Curves in M_k have genus $(k-1)(d-k-1)$; curves in M_k and M_l have the same genus if and only if $k = l$ or $k+l = d$, that is, the two families are obtained from one another by the automorphism of $\mathbb{P}^1 \times \mathbb{P}^1$ that interchanges the two factors. The dimension of M_k is $(k+1)(d-k+1)-1$, which in terms of the degree and genus is $g + 2d - 1$.

In his "Vorlesungen über die Entwicklung der Mathematik im 19. Jahrhundert," Chap. VII, p. 319, Felix Klein gives the classification of curves

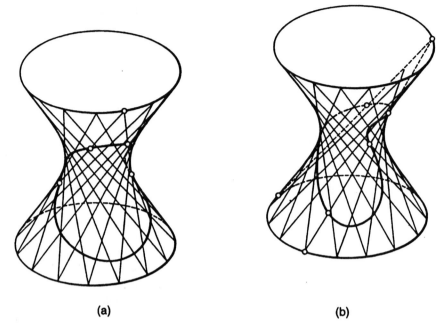

(a)   (b)

Figure 16. Curves on a Quadric Surface

of degree 3 and 4 on the hyperboloid as an example of the application of ideas of birational geometry. We take pictures of curves with $d = 4$ from this reference: Figure 16, (b) has $m = 1$, $n = 3$ and Figure 16, (a) has $m = n = 2$.

Example 3. Curves on the cubic surface. As a further application of (1), we determine the possible negative values for the selfintersection of curves C on a cubic surface of $\mathbb{P}^3$. According to the result of Chap. III, 6.4, in this case $K = -E$, where E is the hyperplane section. Thus (1) takes the form

$$C^2 - \deg C = 2g - 2.$$

Obviously $C^2 < 0$ only if $g = 0$ and $\deg C = 1$, that is, C is a line of the cubic surface. In this case $C^2 = -1$.

2.4. The Riemann–Roch Inequality on a Surface

Recall from Chap. III, 1.5 that we write $\ell(D)$ for the dimension of the vector space associated with a divisor D. Another fundamental relation involving intersection numbers on an irreducible nonsingular projective surface X is the Riemann–Roch inequality:

$$\ell(D) + \ell(K - D) \geq \frac{1}{2}D(D - K) + \chi(\mathcal{O}_X); \tag{1}$$

here D is an arbitrary divisor on X and $\chi(\mathcal{O}_X)$ is an invariant depending only on X and not on D; in the case of a field of characteristic 0 we have $\chi(\mathcal{O}_X) = 1 - h^1(X) + h^2(X)$, where $h^r = \dim \Omega^r[X]$ are as defined in Chap. III, 6.1. Inequality (1) is obtained by omitting one term from the Riemann–Roch equality, which we do not treat here. The Riemann–Roch equalities for curves and surfaces generalise to varieties of arbitrary dimension.

We illustrate the usefulness of the Riemann–Roch inequality by discussing one of its consequences. As mentioned in 2.1, the intersection number of divisors D_1, $D_2 \in \mathrm{Div}\, X$ depends only on their images in $\mathrm{Div}\, X / \mathrm{Div}^a X$, which is a finitely generated group. We can also pass to the quotient by the torsion subgroup, since torsion elements, of course, give zero intersection numbers. As a result we get a group isomorphic to $\mathbb{Z}^m$, and if $u_1, \dots, u_m$ is a basis, intersection numbers define a symmetric integral matrix $(u_i u_j)$, that is, an integral quadratic form. This is an extremely important invariant of a surface.

We now determine the crudest invariant of this quadratic form, its index of inertia. It certainly takes positive values, since $E^2 = \deg X > 0$, where E is a hyperplane section. It turns out that on reducing the quadratic form to a sum of squares, all but one of the nonzero diagonal entries are negative. We prove this result in a form that does not use the result that $\mathrm{Div}\, X / \mathrm{Div}^a X$ is finitely generated.

Hodge Index Theorem. *If D is a divisor on a surface X and $DE = 0$, where E is the hyperplane section, then $D^2 \leq 0$.*

Proof. Suppose that $D^2 > 0$. We prove that for all sufficiently large $n > 0$, either $\ell(nD) > 0$ or $\ell(-nD) > 0$. The theorem will follow from this: if, say, $\ell(nD) > 0$, then nD is linearly equivalent to an effective divisor, that is; $nD \sim D' > 0$; therefore $nDE = D'E > 0$, because every curve intersects a hyperplane. Hence $nDE > 0$ and so also $DE > 0$, which contradicts the assumption.

Using (1), the assumption $D^2 > 0$ implies that

$$\ell(nD) + \ell(K - nD) \geq c(n) \quad \text{and} \quad \ell(-nD) + \ell(K + nD) \geq c(n), \quad (2)$$

where $c(n)$ grows with n without bound. If $\ell(nD) = \ell(-nD) = 0$, we get $\ell(K - nD) \geq c(n)$ and $\ell(K + nD) \geq c(n)$. But now if $\ell(D_1) > 0$, we always have $\ell(D_1 + D_2) \geq \ell(D_2)$; thus we would deduce that $\ell(2K) \geq c(n)$, which is an obvious contradiction. The theorem is proved.

2.5. The Nonsingular Cubic Surface

Let $X \subset \mathbb{P}^3$ be a nonsingular cubic surface. X contains a line L, by Chap. I, 6.4, Theorem 10. Through L, pass two distinct planes E_1 and E_2 with equations $\varphi_1 = 0$ and $\varphi_2 = 0$, and consider the rational map $\varphi \colon X \to \mathbb{P}^1$ given by $\varphi(x) = (\varphi_1(x) : \varphi_2(x))$. The linear system $\lambda_1 \varphi_1 + \lambda_2 \varphi_2$ corresponding to this map has L as a fixed component: if E_{λ_1,λ_2} is the section of X by the plane with equation $\lambda_1 \varphi_1 + \lambda_2 \varphi_2 = 0$ then $E_{\lambda_1,\lambda_2} = L + F_{\lambda_1,\lambda_2}$, where F_{λ_1,λ_2} is a plane conic. The linear system F_{λ_1,λ_2} obviously defines the same map φ. We prove that φ is regular. For this it is enough to prove that $F_{\lambda_1,\lambda_2} \cap F_{\mu_1,\mu_2} = \emptyset$ if $(\lambda_1 : \lambda_2) \neq (\mu_1 : \mu_2)$. Note that F_{λ_1,λ_2} cannot contain L as a component: an equality $E_{\lambda_1,\lambda_2} = 3L$ or $2L + L'$ would contradict the relations $L^2 = -1$, $E_{\lambda_1,\lambda_2} L = 1$ and $LL' \geq 0$ (see 1.4, Example 3). Moreover, F_{λ_1,λ_2} and F_{μ_1,μ_2} cannot have a common component; indeed, this would have to be a line distinct from L, and would determine the plane containing it. Thus F_{λ_1,λ_2} and F_{μ_1,μ_2} are in general position, and it is enough to prove that $F_{\lambda_1,\lambda_2} F_{\mu_1,\mu_2} = 0$, that is, $F^2 = 0$, where $F = E - L$. This follows from $E^2 = 3$, $EL = 1$ and $L^2 = -1$.

If L is the line given by $\xi_0 = \xi_1 = 0$ then the equation of X can be written as

$$
\begin{aligned}
A(\xi_0,\xi_1)\xi_2^2 &+ 2B(\xi_0,\xi_1)\xi_2\xi_3 + C(\xi_0,\xi_1)\xi_3^2 \\
&+ 2D(\xi_0,\xi_1)\xi_2 + 2E(\xi_0,\xi_1)\xi_3 + F(\xi_0,\xi_1) = 0,
\end{aligned}
\tag{1}
$$

where A, B, C, D, E and F are forms in ξ_0, ξ_1 of degrees $\deg A = \deg B = \deg C = 1$, $\deg D = \deg E = 2$ and $\deg F = 3$. We see from this that our map $\varphi \colon X \to \mathbb{P}^1$ represents an open set $V = \varphi^{-1}(\mathbb{A}^1) \subset X$ as a pencil of conics $V \to U$ over the affine line $\mathbb{A}^1 \subset \mathbb{P}^1$; we always choose the coordinates so that the fibre over the point at infinity $\mathbb{P}^1 \setminus \mathbb{A}^1$ is a nondegenerate conic. From Chap. II, 6.4, Example 1, it follows that the degenerate fibres correspond to zeros of the discriminant, each zero has multiplicity 1, and each degenerate fibre is a pair of distinct lines. Then the number of degenerate fibres equals the degree of the discriminant

$$
\Delta = \det \begin{vmatrix} A & C & D \\ C & B & E \\ D & E & F \end{vmatrix},
$$

which is 5. The next result follows from this.

Proposition. *Every line L on a nonsingular projective cubic surface X meets exactly 10 other lines on X, which break up into 5 pairs of intersecting lines.* $\square$

It follows from Chap. I, 6.2, Corollary 6 that a nonsingular cubic surface is rational: Δ is not identically zero, since it has only simple roots. The rationality of X can also be proved otherwise: consider any line L' intersecting L, and apply Proposition 1 to it. Then L' meets 10 lines, of which only L and one further line meet L. Therefore there exists a line M not intersecting L, and the rationality of X follows by Chap. I, 3.3, Example 2.

The line M just found obviously satisfies $MF = 1$, where F is the fibre of the conic bundle, since $ME = 1$, $ML = 0$ and $E \sim L + F$. Hence M intersects F in exactly one point, and, in particular, it intersects exactly one of each pair of lines meeting L. Write L_i' for this one, and L_i for the other, for $i = 1, \ldots, 5$. Then $L_i M = 0$, $L_i' M = 1$. The configuration of lines obtained thus is illustrated in Figure 17.

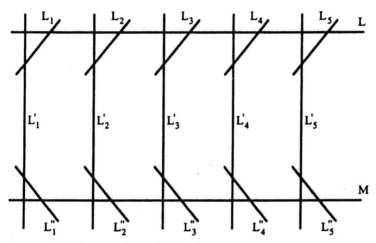

Figure 17. Lines on the Cubic Surface

It follows from Chap. III, 1.6, Theorem 4 that the group $\operatorname{Cl} X$ has generators the classes defined by the divisors $L_1, L_2, L_3, L_4, L_5, F, S$, where S is some section of the conic bundle $X \to \mathbb{P}^1$. We prove that S can be taken to be the line M found above. Indeed, since $M \cap L = \emptyset$, the equations of M can be written $\xi_2 = a\xi_0 + b\xi_1$, $\xi_3 = c\xi_0 + d\xi_1$; that is, passing to inhomogeneous coordinates, $x_2 = \xi_2/\xi_0$ and $x_3 = \xi_3/\xi_0$ can be expressed as rational functions of $x_1 = \xi_1/\xi_0$, the coordinate of $\mathbb{P}^1$, and these expressions satisfy (1).

We thus obtain the following result.

Proposition. $\operatorname{Cl} X$ *is a free group with 7 generators, the classes of the lines* $L_1, L_2, L_3, L_4, L_5, M$ *and* F. $\square$

The intersection numbers of $L_1, L_2, L_3, L_4, L_5, M$ and F are easily determined; they are tabulated as follows:

	L_1	L_2	L_3	L_4	L_5	M	F
L_1	-1	0	0	0	0	0	0
L_2	0	-1	0	0	0	0	0
L_3	0	0	-1	0	0	0	0
L_4	0	0	0	-1	0	0	0
L_5	0	0	0	0	-1	0	0
M	0	0	0	0	0	-1	1
F	0	0	0	0	0	1	0

The group $\operatorname{Cl} X$ to a significant extent determines the geometry of X. In particular, we now show how to use it to find all the lines on X. A line C on X satisfies $C^2 = -1$. We know L and a further 10 lines intersecting it. We now try to find the lines disjoint from L. These satisfy $CL = 0$, and therefore $CF = 1$. Suppose that $C \sim \sum_1^5 x_i L_i + yM + zF$. Then $CF = 1$ implies $y = 1$, and $C^2 = -1$ and $CL = 0$ give

$$-\sum_1^5 x_i^2 + 2z = 0, \qquad \sum_1^5 x_i + 2z = 0; \tag{2}$$

It follows that $\sum_1^5 (x_i^2 + x_i) = 0$, that is, each $x_i = 0$ or -1. Moreover, (2) implies also that the number of i for which $x_i = -1$ is even, so that either (a) all $x_i = 0$; or (b) all $x_i = -1$ except one; or (c) $x_i = x_j = -1$, and the three remaining $x_k = 0$. Of these possibilities, (a) gives the class of the line M, (b) and (c) 5 and 10 cases, that is, 16 classes altogether. Each class contains at most one line: for if C and C' are distinct lines then $CC' = 0$ or 1, whereas if $C \sim C'$ are in the same class then $CC' = C^2 = -1$. Thus it remains to exhibit at least one line in each class. In case (a), this is M.

In case (b), if $x_i = 0$ and $x_j = -1$ for $j \neq i$ we get the class $C_i = -\sum_{j \neq i} L_j + M + 2F$. We note that the lines L_i' and M lie in the same plane, in which there must be a third line L_i'', so that $L_i' + L_i'' + M = E$. Setting $E \sim \sum \alpha_k L_k + \beta M + \gamma F$ and arguing as before, we get that $E \sim -\sum L_k + 2M + 3F$. Substituting this expression for E and $L_i' \sim F - L_i$ for L_i', we find easily that $L_i'' \sim C_i$.

In case (c) we get a class $D_{ij} = -L_i - L_j + M + F$. Note that $L_i'' L_j = C_i L_j = 1$, so that the lines L_i'' and L_j for $i \neq j$ intersect, and hence there is a third line L_{ij} in the plane through them. Arguing exactly as before we show that $L_{ij} \sim D_{ij}$. Thus we have found 1 line in case (a), 5 in case (b) and (10) in case (c), altogether $1+5+10=16$. Together with L and the 10 lines meeting it, this gives 27 lines. This proves the next result.

Theorem. *A nonsingular cubic surface of $\mathbb{P}^3$ contains exactly 27 lines.* $\square$

2.6. The Ring of Cycle Classes

The theory of divisors and their intersection numbers is a particular case of a general theory that deals with subvarieties of any dimension. The notion of divisor is replaced by that of k-cycle. A k-*cycle* is an element of the free Abelian group generated by irreducible k-dimensional subvarieties. Two irreducible subvarieties Y_1 and Y_2 are *in general position*, by definition, if every irreducible component Z_i of the intersection $Y_1 \cap Y_2$ has the same dimension, and
$$\text{codim } Z_i = \text{codim } Y_1 + \text{codim } Y_2.$$

Two k-cycles are in general position if all components of the first are in general position with those of the second.

The foundation of the theory is a method of assigning to each component Z_i of $Y_1 \cap Y_2$ a positive integral multiplicity $n_i(Y_1, Y_2)$. The cycle $Y_1 \cdot Y_2 = \sum n_i(Y_1, Y_2)Z_i$ is called the product of the subvarieties Y_1 and Y_2. The notion extends by additivity to any two cycles in general position. The reader can learn about this theory from Fulton [27].

The basic property of this product is that it is invariant under an equivalence relation that we now describe. It generalises the algebraic equivalence of divisors introduced in Chap. III, 4.4, and is defined in an entirely similar way. Namely, let T be an arbitrary irreducible nonsingular variety and $Z \subset X \times t$ a cycle such that Z and the fibre $X \times T$ are in general position for every $t \in T$. The set of cycles $C_t = Z \cdot (X \times t)$ is called an *algebraic family of cycles*. Two cycles C_1 and C_2 are *algebraically equivalent* if there exists a family of cycles C_t with $t \in T$ such that $C_{t_1} = C_1$ and $C_{t_2} = C_2$ for two points $t_1, t_2 \in T$. The set of cycle classes under algebraic equivalence forms a group.

The product of cycles on a nonsingular projective variety is invariant under algebraic equivalence. There is a theorem on reducing to general position (the so-called *moving lemma*), according to which for any two cycles C_1 and C_2 there exist cycles C_1' and C_2' equivalent to C_1 and C_2 respectively, and in general position. These two results allow us to define the product of any two cycle classes.

Now let X be a nonsingular n-dimensional projective variety, and write $\mathfrak{A}^r$ to denote the group of cycle classes (under algebraic equivalence) of codimension r on X. The group
$$\mathfrak{A} = \bigoplus_{r=0}^{n} \mathfrak{A}^r$$

is a ring, where the product is defined on the individual summands as described above, and on arbitrary elements by additivity. This ring is commutative and associative. By the formula for the dimension of intersections (Chap. I, 6.2, (4)),
$$\mathfrak{A}^r \cdot \mathfrak{A}^s \subset \mathfrak{A}^{r+s},$$

where we set $\mathfrak{A}^m = 0$ for $m > n$. That is, $\mathfrak{A}$ is a graded ring. It is easy to prove that all points of X, viewed as 0-cycles, are algebraically equivalent, and the 0-cycle x (for $x \in X$) is not algebraically equivalent to 0. Therefore $\mathfrak{A}_n = \mathbb{Z}u$, with the standard generator u, the class of a point $x \in X$. The classes of divisors under algebraic equivalence form a group $\mathfrak{A}^1$. For n elements $\alpha_1, \ldots, \alpha_n \in \mathfrak{A}^1$ the product $\alpha_1 \cdots \alpha_n \in \mathfrak{A}^n = \mathbb{Z}u$, that is,

$$\alpha_1 \cdots \alpha_n = ku \qquad \text{with } k \in \mathbb{Z}.$$

The number k equals the intersection number $\alpha_1 \cdots \alpha_n$ defined in §1.

The ring $\mathfrak{A}$ is a very interesting invariant of X, and is still not well studied. $\mathfrak{A}^0$ is isomorphic to $\mathbb{Z}$, with generator X itself. As already pointed out, $\mathfrak{A}^n \cong \mathbb{Z}$. The group $\mathfrak{A}^1$ is finitely generated, as asserted in Chapter III, 4.4, Theorem D. However, already $\mathfrak{A}^2$ may have an infinite number of generators. The structure of these groups is quite mysterious.

Exercises to §2

1. Determine $\deg v_m(\mathbb{P}^n)$ where v_m is the Veronese embedding (Chap. I, 4.4).

2. Suppose that a nonsingular plane curve C of degree r lies on a nonsingular surface of degree m in $\mathbb{P}^3$. Determine C^2. (This generalises 1.5, Example 3.)

3. Suppose that a form of degree l on a nonsingular projective surface of degree m in $\mathbb{P}^3$ has divisor consisting of one component of multiplicity 1 that is a nonsingular curve. Find its genus.

4. Consider k sets of variables $x_0^{(i)}, \ldots, x_{n_i}^{(i)}$ and a system of $\sum_{i=1}^{k} n_i$ simultaneous equations

$$f_i(x_0^{(1)}, \ldots, x_{n_1}^{(1)}; \ldots; x_0^{(k)}, \ldots, x_{n_k}^{(k)}) = 0,$$

that are linear in each set of variables $x_0^{(i)}, \ldots, x_{n_i}^{(i)}$. Prove that the number of solutions in $\mathbb{P}^{n_1} \times \cdots \times \mathbb{P}^{n_k}$ of the system equals the multinomial coefficient $\binom{n_1 + \cdots + n_k}{n_1, \ldots, n_k} = (\sum n_i)! / \prod n_i!$. Here the number of solutions is taken as usual in the sense of the corresponding intersection number.

5. Let $X \subset \mathbb{P}^3$ be a nonsingular surface of degree 4 and $C \subset X$ a nonsingular curve. Prove that if $C^2 < 0$ then $C^2 = -2$.

6. Prove that the selfintersection number of a nonsingular curve on a nonsingular surface in $\mathbb{P}^3$ of even degree is always even.

7. Let X be a nonsingular curve and D the diagonal of $X \times X$ (that is, the set of points of the form (x, x)). Prove that $D^2 = -\deg K_X$. [Hint: Use the fact that D and X are isomorphic.]

8. Generalise the result of Ex. 7 to the case that $D \subset C_1 \times C_2$ is the graph of a map $\varphi: C_1 \to C_2$ of degree d.

9. If $D \subset C_1 \times C_2$ is a divisor, prove the inequality

$$D^2 \leq 2(C_1 \times c_2)D \cdot (C_2 \times c_1)D$$

for $c_1 \in C_1$ and $c_2 \in C_2$. [Hint: Cook up α and β such that $D' = D - \alpha(C_1 \times c_2) - \beta(C_2 \times c_1)$ satisfies $(C_1 \times c_2)D' = (C_2 \times c_1)D' = 0$, then apply the Hodge index theorem to D'.]

10. In the notation of Ex. 8–9, suppose that $C_1 = C_2 = C$ is a curve of genus g. Let $\varphi: C \to C$ be a map of degree d with graph $\Gamma_\varphi \subset C \times C$, and write $\Delta \subset C \times C$ for the diagonal. Prove that

$$\left| \Gamma_\varphi \Delta - d - 1 \right| \leq 2g\sqrt{d}.$$

[Hint: Set $D = m\Delta + n\Gamma_\varphi$ and view $D^2 - 2(C \times c)D \cdot (C \times c)D$ as a quadratic form in m and n; then write out the condition for this to be negative definite.] Here $\Gamma_\varphi \Delta$ is the number of fixed points of φ. This inequality, applied to the case that φ is the Frobenius map, generalises the inequality Chap. III, 3.4, Example 2, (3) to curves of arbitrary genus.

3. Birational Maps of Surfaces

This section treats an application of intersection numbers to the proof of some basic properties of birational maps of algebraic surfaces. We start by deriving the elementary properties of blowups of algebraic surfaces.

3.1. Blowups of Surfaces

Let X be an algebraic surface, $\xi \in X$ a nonsingular point, x, y local parameters at ξ and $\sigma: Y \to X$ the blowup centred at ξ. By Chap. II, 4.3, Theorem 1 there exists a neighbourhood U of ξ in X such that $V = \sigma^{-1}(U)$ is the subvariety of $U \times \mathbb{P}^1$ defined by $t_0 y = t_1 x$, where $(t_0 : t_1)$ are coordinates on $\mathbb{P}^1$. In the open set where $t_0 \neq 0$ the blowup is given by the simple equations

$$x = u \quad \text{and} \quad y = uv, \quad \text{where } v = t_1/t_0. \tag{1}$$

Set $L = \sigma^{-1}(\xi)$. A local system of parameters at any point $\eta \in L$ is given by u and $v - v(\eta)$. The local equation of L is obviously $u = 0$.

Let C be an irreducible curve on X passing through ξ. By analogy with Chap. II, 4.3, Theorem 1, the inverse image $\sigma^{-1}(C)$ consists of two components: the exceptional curve L and a curve C' that can be defined as the closure in Y of $\sigma^{-1}(C \setminus \xi)$. The curve C' is called the *birational*[15] *transform*

---

[15]The terms *strict transform* and *proper transform* are also widely used in the literature.

of C. We denote it by $C' = \sigma'(C)$. Now consider C as an irreducible divisor on X. Then

$$\sigma^*(C) = \sigma'(C) + kL, \tag{2}$$

where $\sigma'(C)$ appears with coefficient 1, because σ is an isomorphism of $Y \setminus L$ and $X \setminus \xi$. We now determine the coefficient k in (2). Suppose for this that C has ξ as an r-fold point. This means that the local equation f of C in a neighbourhood of ξ satisfies $f \in \mathfrak{m}_\xi^r \setminus \mathfrak{m}_\xi^{r+1}$. Then $\sigma^*(C)$ has local equation $\sigma^*(f)$ in a neighbourhood of any point $\eta \in L$. Set

$$f = \varphi(x,y) + \psi \qquad \text{with } \varphi \text{ a form of degree } r \text{ and } \psi \in \mathfrak{m}_\xi^{r+1}. \tag{3}$$

Substituting the formulas (1) for σ in (3), we see that $\big(\sigma^*(f)\big)(u,v) = \varphi(u, uv) + \sigma^*(\psi)$. Since $\psi \in \mathfrak{m}_\xi^{r+1}$, it follows that we can write $\psi = F(x,y)$ with F a form of degree $r+1$ in x,y with coefficients in $\mathcal{O}_\xi$. Therefore $\sigma^*(\psi) = \big(\sigma^*(F)\big)(u, uv)$, and finally

$$\big(\sigma^*(f)\big)(u,v) = u^r\Big(\varphi(1,v) + u\big(\sigma^*(F)\big)(1,v)\Big); \tag{4}$$

since $\varphi(1,v)$ is not divisible by u it follows that $k = r$ in (2). We state the result we have proved.

Theorem 1. *If C is a prime divisor on X passing through the centre ξ of a blowup σ then the inverse image $\sigma^*(C)$ of C is given by $\sigma^*(C) = \sigma'(C) + kL$, where $\sigma'(C) \subset Y$ is a prime divisor, $L = \sigma^{-1}(\xi)$ and k is the multiplicity of C at ξ.* $\square$

3.2. Some Intersection Numbers

We start from general properties of a birational regular map $f : Y \to X$ between nonsingular projective surfaces.

Theorem 2. *(i) If D_1 and D_2 are divisors on X then*

$$f^*(D_1)f^*(D_2) = D_1 D_2. \tag{1}$$

(ii) If $\overline{D}$ is a divisor on Y all of whose components are exceptional curves of f and D is any divisor on X then

$$f^*(D)\overline{D} = 0. \tag{2}$$

Proof. We write $S \subset X$ for the finite set of points at which the inverse map f^{-1} is not regular, and set $T = f^{-1}(S)$ for the set-theoretic inverse image. Then f defines an isomorphism

$$Y \setminus T \xrightarrow{\sim} X \setminus S. \tag{3}$$

If $\operatorname{Supp} D_1 \cap S = \operatorname{Supp} D_2 \cap S = \emptyset$ and D_1 and D_2 are in general position then (1) is obvious from (3). Otherwise we use Chap. III, 1.3, Theorem 1 on moving the support of a divisor away from points. Suppose that $D_1' \sim D_1$ and $D_2' \sim D_2$ are divisors with $\operatorname{Supp} D_1' \cap S = \operatorname{Supp} D_2' \cap S = \emptyset$ and D_1' and D_2' in general position. Then $D_1 D_2 = D_1' D_2'$, by 1.4, Lemma 2, and by what we said above $D_1' D_2' = f^*(D_1') f^*(D_2')$. Since $f^*(D_i') \sim f^*(D_i)$, the required equality (1) holds.

Equality (2) is likewise obvious if $\operatorname{Supp} D \cap S = \emptyset$. The general case reduces to this by an entirely similar argument. The theorem is proved.

We now give a corollary that relates directly to blowups. We use the notation of 3.1.

Corollary 1.
$$L^2 = -1. \tag{4}$$

Proof. Consider the curve $C \subset X$ with local equation y. By Theorem 1 $\sigma^*(C) = \sigma'(C) + L$, and moreover, it is clear from 3.1, (1) that the local equation of $\sigma'(C)$ is v. Since u is the local equation for L, it follows that $\sigma'(C)L = 1$ and (4) follows from (2):
$$0 = \sigma^*(C)L = (\sigma'(C) + L)L = 1 + L^2.$$
The corollary is proved.

Corollary 2. *If $C \subset X$ is a curve with multiplicity k at ξ then $\sigma'(C)L = k$.*

This follows at once from (2) and (4) and from 3.1, (2).

Corollary 3.
$$\sigma'(C_1)\sigma'(C_2) = C_1 C_2 - k_1 k_2,$$
where k_1, k_2 are the multiplicities of C_1, C_2 at ξ.

Proof.
$$\begin{aligned}
C_1 C_2 &= \sigma^*(C_1)\sigma^*(C_2) = (\sigma'(C_1) + k_1 L)\sigma^*(C_2) \\
&= \sigma'(C_1)\sigma^*(C_2) = \sigma'(C_1)(\sigma'(C_2) + k_2 L) \\
&= \sigma'(C_1)\sigma'(C_2) + k_1 k_2;
\end{aligned}$$
here the 3rd equality comes from (2) and the final one from Corollary 2. This gives (5). The corollary is proved.

3.3. Resolution of Indeterminacy

We can now prove an important property of rational maps from algebraic surfaces.

Theorem 3. *Let X be a nonsingular projective surface and $\varphi\colon X \to \mathbb{P}^n$ a rational map. Then there exists a chain of blowups $X_m \to \cdots \to X_1 \to X$ such that the composite rational map $\psi = \varphi \circ \sigma_1 \circ \cdots \circ \sigma_m\colon X_m \to \mathbb{P}^n$ is regular; in other words, there is a commutative diagram*

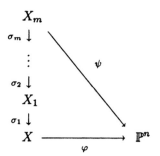

in which the vertical column is a chain of blowups, and the diagonal arrow ψ is a regular map.

Remark. The assumption that X is projective is used in the proof of Theorem 3 in order to be able to use intersection numbers $D_1 D_2$. However, the result itself holds for noncomplete surfaces or for complete nonprojective surfaces, and can be proved without difficulty by reducing to the statement of Theorem 3.

Proof. By Chap. II, 3.1, Theorem 3, we know that φ only fails to be regular at a finite number of points; Chap. III, 1.4, Theorem 2 gives a more precise description of this set, which we now recall. Suppose that $\varphi = (f_0 : \cdots : f_n)$, and set

$$\overline{D} = \mathrm{hcd}\Big(\mathrm{div}(f_0), \ldots, \mathrm{div}(f_n)\Big) \quad \text{and} \quad D_i = \mathrm{div}(f_i) - \overline{D}.$$

Then the set of points of irregularity of φ is exactly $\bigcap_{i=0}^m \mathrm{Supp}\, D_i$.

We introduce an invariant $d(\varphi)$ of a rational map φ as follows. All the divisors D_i are obviously linearly equivalent. Hence we can set

$$d(\varphi) = D_i^2.$$

Let us prove that $d(\varphi) \geq 0$. For this, let $\lambda = (\lambda_0, \ldots, \lambda_n) \in k^{n+1}$, and define $D_\lambda = \mathrm{div}\left(\sum_{i=0}^n \lambda_i f_i\right) - \overline{D}$; obviously D_λ is an effective divisor linearly equivalent to the D_i. It is enough to prove that there exists λ such that D_0 and D_λ have no common components, since then $d(\varphi) = D_0 D_\lambda \geq 0$.

By construction no curve is a common component of all the D_i. Hence for every irreducible component $C \subset D_0$ there exists some D_i such that $v_C(D_i) = 0$. If g_i are local equations for the D_i in a neighbourhood of some point $c \in C$ then

$$v_C(D_\lambda) > 0 \iff \sum \lambda_i \left(g_i|_C \right) = 0.$$

Therefore the set of λ such that $v_C(D_\lambda) > 0$ is a strict vector subspace of k^{n+1}. A vector space (over an infinite field) is not the union of finitely many strict vector subspaces, and hence there exists λ such that $v_C(D_\lambda) = 0$ for every component $C \subset D_0$. Then D_0 and D_λ have no common components for this λ.

If $x_0 \in \bigcap \operatorname{Supp} D_i$ then $x_0 \in \operatorname{Supp} D_\lambda$ for every λ. Hence $d(\varphi) > 0$ if $\operatorname{Supp} D_i \neq \emptyset$, that is, if φ is not regular. If this happens, write $\sigma \colon X' \to X$ for the blowup centred at a point $x_0 \in \bigcap \operatorname{Supp} D_i$, and set $\varphi' = \varphi \circ \sigma \colon X' \to \mathbb{P}^n$. We prove that $d(\varphi') < d(\varphi)$. Theorem 3 of course follows from this.

We define the *multiplicity* of any divisor $D = \sum l_i C_i$ at a point ξ to be $k = \sum l_i k_i$, where k_i are the multiplicities of the C_i at ξ. Obviously if $D \geq 0$ then $k \geq 0$, with $k = 0$ if and only if $\xi \notin \operatorname{Supp} D$. In the same way, we define the birational transform of D by $\sigma'(D) = \sum l_i \sigma'(C_i)$; then Theorem 1 remains true for any divisor D, that is, $\sigma^* D = \sigma'(D) + kL$.

We now write ν_i for the multiplicity of D_i at x_0 and set $\nu = \min \nu_i$. The map φ' is given by the functions $f_i' = \sigma^*(f_i)$, and

$$\operatorname{div}(f_i') = \sigma^* \left(\operatorname{div}(f_i) \right) = \sigma'(D_i) + (\nu_i - \nu)L + \nu L + \sigma^*(\overline{D}),$$

where the divisors $D_i' = \sigma'(D_i) + (\nu_i - \nu)L$ for $i = 0, \ldots, n$ have no common components. Choose some i such that $\nu_i = \nu$. Then by definition

$$d(\varphi') = (D_i')^2 = \left(\sigma'(D_i) \right)^2.$$

Now applying Theorem 2 to the equality $\sigma^*(D_i) = \sigma'(D_i) + \nu L$ gives

$$\left(\sigma'(D_i) \right)^2 = \left(\sigma^*(D_i) - \nu L \right)^2 = \left(\sigma^*(D_i) \right)^2 - \nu^2 = D_i^2 - \nu^2,$$

and hence $d(\varphi') = d(\varphi) - \nu^2$. This proves Theorem 3.

The simplest example of Theorem 3 is the map (occurring in the definition of the projective line) $f \colon \mathbb{A}^2 \to \mathbb{P}^1$ given by $f(x, y) = (x : y)$, which is not regular at $\xi = (0, 0)$. Substituting 3.1, (1) we see that at points of $\sigma^{-1}(\xi)$ with $t_0 \neq 0$ we have $f(x, y) = (1 : v)$, and hence $f \circ \sigma$ is regular.

3.4. Factorisation as a Chain of Blowups

We now have all we need for the proof of the main result on birational maps of surfaces.

Theorem 4. *Let $\varphi\colon X \to Y$ be a birational map of nonsingular projective surfaces. Then there exist a surface Z, surfaces X_i and Y_j with $X_0 = X$, $Y_0 = Y$, $X_k = Y_l = Z$, and maps*

$$\sigma_i\colon X_i \to X_{i-1} \text{ for } i = 1,\dots,k \quad \text{and} \quad \tau_j\colon Y_j \to Y_{j-1} \text{ for } j = 1,\dots,l$$

such that each σ_i and τ_j is a blowup, and $\varphi \circ \sigma_1 \circ \cdots \circ \sigma_k = \tau_1 \circ \cdots \circ \tau_i$. In other words, there is a commutative diagram

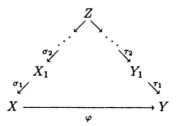

in which the diagonal arrows σ_i and τ_j are blowups.

Theorem 4 is an obvious corollary of Theorem 3 together with the next result.

Theorem 5. *Let $\varphi\colon X \to Y$ be a regular map between nonsingular projective surfaces which is birational. Then there exists a chain of surfaces and blowups $\sigma_i\colon Y_i \to Y_{i-1}$ for $i = 1,\dots,k$ such that $Y_0 = Y$, $Y_k = X$ and*

$$\varphi = \sigma_1 \circ \cdots \circ \sigma_k.$$

We precede the proof of Theorem 5 with some general remarks on birational maps of surfaces.

First of all, for any rational map $\varphi\colon X \to Y$ from a nonsingular surface X to a projective variety Y, it makes sense to talk of the image $\varphi(C)$ of a curve $C \subset X$. Indeed, φ is regular at all points of C except possibly a finite set S. Thus by $\varphi(C)$ we understand the closure in Y of $\varphi(C \setminus S)$.

Moreover, Chap II, 4.4, Theorem 2 on the existence of exceptional subvarieties remains valid in this setup.

Lemma. *Let $\varphi\colon X \to Y$ be a birational map of nonsingular projective surfaces, and suppose that φ^{-1} is not regular at some point $y \in Y$. Then there exists a curve $C \subset X$ such that $\varphi(C) = y$.*

Proof. Consider open sets $U \subset X$ and $V \subset Y$ such that $\varphi: U \to V$ is an isomorphism, and let Z be the closure in $X \times Y$ of the graph of the isomorphism $\varphi: U \to V$. The projections to X and Y define regular birational maps $p: Z \to X$ and $q: Z \to Y$. Obviously $\varphi^{-1} = p \circ q^{-1}$, so that, because we are assuming that φ^{-1} is irregular at y, the same is true of q^{-1}.

We can now apply Chap II, 4.4, Theorem 2 on the existence of exceptional subvarieties to the regular map $q: Z \to Y$. This theorem shows that there exists a curve $D \subset Z$ such that $q(D) = y$. We set $p(D) = C$ and verify that C satisfies the conclusion of the lemma. We really only need prove that $\dim C = 1$, that is, $\dim C = \dim D$. Now otherwise, $p(D)$ would be a point $x \in X$, so that $p(D) = x$ and $q(D) = y$ would imply $D \subset (x, y) \in X \times Y$, which contradicts that D is a curve. The lemma is proved.

We now proceed to the proof of Theorem 5. Suppose that φ is not an isomorphism, that is, φ^{-1} is not regular at some point $y \in Y$. Consider the blowup $\sigma: Y' \to Y$ with centre in y and define $\varphi': X \to Y'$ by $\varphi' = \varphi \circ \sigma^{-1}$, so that the diagram below is commutative:

$$
\begin{array}{c}
X \\
\varphi \downarrow \quad \searrow \varphi' \\
Y \underset{\sigma}{\leftarrow} Y'.
\end{array}
\tag{1}
$$

Auxiliary notation introduced in the course of the proof is summarised in the following diagram:

$$
\begin{array}{ccc}
X \supset Z \ni x \\
\varphi \downarrow \quad \downarrow \quad \searrow \psi \\
Y \ni y \xleftarrow{\;\;\sigma\;\;} L \subset Y'.
\end{array}
$$

The theorem will be proved if we prove that φ' is a regular map. Indeed, from the commutative diagram (1) it then follows that φ' maps the subvariety $\varphi^{-1}(y)$ to $\sigma^{-1}(y) = L \cong \mathbb{P}^1$. Since φ' maps X onto the whole of Y', it follows that it maps X onto the whole of L. Thus not every component of $\varphi^{-1}(y)$ maps to a point. Therefore, for any $y' \in L$ the number of components of $(\varphi')^{-1}(y')$ is less than the number of components of $\varphi^{-1}(y)$. Hence after a finite number of blowups we arrange that X does not contain any exceptional subvarieties, that is, our regular map becomes an isomorphism.

It remains to prove that φ' is regular. Suppose otherwise. Then by the lemma, $\psi = (\varphi')^{-1}$ maps some curve on Y' to a point $x \in X$. It follows from the commutative diagram (1) that this curve can only be L, hence $\psi(L) = x$. Now according to Chap. II, 3.1, Theorem 3, there exists a finite subset $E \subset L$ such that ψ is regular at all point $y \in L \setminus E$. Since $\sigma(y') = y$, it follows from the commutativity of (1) that $\varphi(x) = y$.

We now prove that

$$
d_x\varphi: \Theta_{X,x} \to \Theta_{Y,y}
\tag{2}
$$

is an isomorphism. For this, it is enough to prove that it is onto. Suppose that $d_x\varphi(\Theta_{X,x}) \subset l \subset \Theta_{Y,y}$ for some line l in the plane $\Theta_{Y,y}$. Then from the commutativity of the diagram (1) it follows that also

$$d_{y'}\sigma(\Theta_{Y',y'}) \subset l \tag{3}$$

for every point $y' \in L \setminus E$. However, this contradicts the most elementary property of blowups. Indeed, suppose that C is a nonsingular curve on Y with $y \in C$ and $\Theta_{C,y} \neq l$, for example, the curve given by $\alpha u + \beta v = 0$ where u and v are local parameters at y. Then by 3.1, (2) we have $\sigma(\sigma'(C)) = C$, where $\sigma'(C)$ intersects L in one point y' which has coordinates $(-\beta : \alpha)$ on L, and $\sigma'(C)$ is nonsingular with $\sigma: \sigma'(C) \to C$ an isomorphism. We can choose α and β so that $y' \notin E$ and then already $d_{y'}\sigma(\Theta_{\sigma'(C),y'}) \not\subset l$.

The fact that (2) is an isomorphism contradicts the assumption that φ^{-1} is irregular at y. Indeed, using Chap II, 4.4, Theorem 2 on the existence of exceptional subvarieties, we find a curve $Z \subset X$ with $Z \ni x$ such that $\varphi(Z) = y$. Then $\Theta_{Z,x} \subset \Theta_{X,x}$ (recall that the tangent space is defined even if x is a singular point of Z). Since $\varphi(Z) = y$, we have $d_x\varphi(\Theta_{Z,x}) = 0$, and hence (2) has a kernel. This contradiction proves Theorem 5.

3.5. Remarks and Examples

Consider a regular birational map $f: X \to Y$ between nonsingular projective surfaces. Suppose that f^{-1} fails to be regular at only one point $\eta \in Y$, and that the curve $C = f^{-1}(\eta)$ is irreducible. By Theorem 5, f is a composite $f = \sigma_1 \circ \cdots \circ \sigma_k$ of blowups, and since every blowup gives rise to its own exceptional curve, C is irreducible only if $k = 1$, that is, f is itself a blowup. Then C is the curve L, concerning which we proved in 3.1–2 that

$$L \cong \mathbb{P}^1 \quad \text{and} \quad L^2 = -1. \tag{1}$$

Such a curve is called[16] a -1-curve.

The converse statement is also true: if a nonsingular projective surface X contains a -1-curve C, then there exists a regular birational map $f: X \to Y$ such that Y is nonsingular, $f(C) = \eta \in Y$, and f coincides with the blowup of $\eta \in Y$. Thus the conditions (1) are necessary and sufficient for the curve C to be contracted to a point in the sense just described. This result was proved by Castelnuovo, and is known as *Castelnuovo's contractibility criterion*. We will not give the proof, which can be found, for example, in Shafarevich [67], Chap. II or Hartshorne [35], Chap. V, §5.

Example The standard quadratic transformation. We conclude this section with a construction in a simple example of a factorisation of a birational

---

[16] -1-curves are called *exceptional curves of the first kind* in the older literature.

map into blowups, as in the conclusion of Theorem 4. This example is the birational map f from $\mathbb{P}^2$ to itself given by

$$f(x_0 : x_1 : x_2) = (y_0 : y_1 : y_2),$$
$$\text{where } y_0 = x_1 x_2, \, y_1 = x_0 x_2, \, y_2 = x_0 x_1; \tag{2}$$

it is called the *standard quadratic transformation*.

We consider f as a birational map between two copies $\mathbb{P}^2$ and $\overline{\mathbb{P}^2}$ of the projective plane, the first with homogeneous coordinates $(x_0 : x_1 : x_2)$, and the second $(y_0 : y_1 : y_2)$. Obviously f fails to be regular at the 3 points

$$\xi_0 = (1,0,0), \quad \xi_1 = (0,1,0), \quad \xi_2 = (0,0,1).$$

According to Theorem 3, we must start by performing the blowups σ_0, σ_1, σ_2 in the three points ξ_0, ξ_1, ξ_2. We arrive at a surface X with a regular map $\varphi = \sigma_2 \circ \sigma_1 \circ \sigma_0 \colon X \to \mathbb{P}^2$. We now prove that already the map $\psi = f \circ \varphi \colon X \to \overline{\mathbb{P}^2}$ is regular. Indeed, ψ is regular at a point z if $\varphi(z) \neq \xi_i$. At points $\zeta \in \sigma_1^{-1}(\xi_0)$ the map $f \circ \sigma_1$ is already regular. To verify this, it is enough to set $x = x_1/x_0$, $y = x_2/x_0$ and substitute 3.1, (1) in (2). We see that

$$f(1,x,y) = (xy : x : y) = (u^2 v : uv : u) = (uv : v : 1). \tag{3}$$

Since σ_1 and σ_2 both induce isomorphisms in a neighbourhood of ζ, also ψ is regular at points z for which $\varphi(z) = \xi_0$. The same holds for ξ_1 and ξ_2.

By Theorem 4, ψ is a composite of blowups $\psi = \tau_1 \circ \cdots \circ \tau_k$. We now determine which curves $C \subset X$ can map to points under ψ. Obviously any such curve is either one of the $M_i' = \sigma_i^{-1}(\xi_i)$ for $i = 0, 1$ or 2, or the birational transform in X of a curve $L \subset \mathbb{P}^2$ mapped to a point by f. It is easy to see that f defines an isomorphism of $\mathbb{P}^2 \setminus (L_0 \cup L_1 \cup L_2)$ and $\overline{\mathbb{P}^2} \setminus (M_0 \cup M_1 \cup M_2)$ where L_i is the line $x_i = 0$ in $\mathbb{P}^2$ and M_i the line $y_i = 0$ in $\overline{\mathbb{P}^2}$. Hence the only curves ψ can contract to a point are M_0', M_1', M_2', L_0', L_1', L_2', where the L_i' are the birational transform in X of the lines L_i. But we see from (3) that M_0', given by the local equation $u = 0$, maps onto the whole curve M_0 given by $y_0 = 0$. In the same way, M_i' maps onto M_i for $i = 1, 2$. Thus the only curves ψ can contract are the L_i'. Moreover, ψ^{-1} is not regular at the points $\eta_0 = (1 : 0 : 0)$, $\eta_1 = (0 : 1 : 0)$, $\eta_2 = (0 : 0 : 1)$, since otherwise f^{-1} would be regular at one of them, and f^{-1} is given by the same formulas as f, as one sees from (2). Thus on the one hand, a factorisation $\psi = \tau_0 \circ \cdots \circ \tau_k$ of ψ can have at most 3 blowups as factors, and on the other hand, it must include the blowups at η_0, η_1 and η_2. We deduce that

$$f = \tau_2 \circ \tau_1 \circ \tau_0 \circ \sigma_0^{-1} \circ \sigma_1^{-1} \circ \sigma_2^{-1}.$$

It is easy to visualise the configuration of the curves M_0', M_1', M_2', L_0', L_1', L_2' on the surface X; see Figure 18, where the arrows indicate which curves contract to which points.

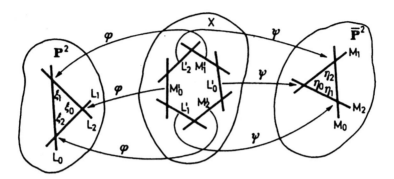

Figure 18. The Standard Quadratic Transformation

Of course, the standard quadratic transformation depends on the choice of the coordinate system in $\mathbb{P}^2$, or, what is the same thing, the choice of ξ_0, ξ_1, ξ_2. Composing different standard quadratic transformations gives different birational maps of the plane to itself. M. Noether proved the theorem that any birational maps of the plane to itself can be written as a composite of quadratic transformations and projective linear transformations. We will not give the proof of this theorem, which is very delicate; it can be found in Shafarevich [67], Chap. V. A description of the relations between these generators has been obtained comparatively recently: see Gizatullin [29] and Iskovskikh [40].

Exercises to §3

1. For every integer k (positive, negative or 0), construct a nonsingular projective surface X and a curve C on it with $C^2 = k$. [Hint: Construct X by blowing up a number of points on $\mathbb{P}^2$.]

2. Let X be a nonsingular projective surface, and C_1, C_2 two curves on X. Suppose that $x \in C_1 \cap C_2$ is a nonsingular point of C_1 and C_2. Let $\sigma \colon Y \to X$ be the blowup of x and C_1', C_2' the birational transforms of C_1, C_2. Prove that C_1' and C_2' intersect at a point $y \in \sigma^{-1}(x)$ if and only if C_1 and C_2 are tangent at x. Moreover, then $\sigma^{-1}(x) \cap C_1' \cap C_2' = y$ is a single point, and the order of tangency of C_1' and C_2' and y is 1 less than that of C_1 and C_2 at x.

3. Suppose that $f \colon \mathbb{P}^2 \to \mathbb{P}^1$ is given by

$$f(x_0 : x_1 : x_2) = \Big(P(x_0, x_1, x_2) : Q(x_0, x_1, x_2) \Big),$$

where P and Q are forms of degree n. How many blowups does one have to perform to get a surface $\varphi \colon X \to \mathbb{P}^2$ such that $f \circ \varphi \colon X \to \mathbb{P}^1$ is regular?

4. Let $X \subset \mathbb{P}^3$ be a nonsingular surface of degree 2 and $f \colon X \to \mathbb{P}^2$ the birational map consisting of projection from a point $x \in X$. Factor f as a composite of blowups.

5. Let $f \colon \mathbb{P}^2 \to \mathbb{P}^2$ be the birational map given in inhomogeneous coordinates by $x' = x$, $y' = y + x^2$. Factor f as a composite of blowups.

6. Let $L \subset \mathbb{P}^2$ be a line, and x, y two points of L. Write $X \to \mathbb{P}^2$ for the composite of the blowups at x and y, and L' for the birational transform of L. Prove that $(L')^2 = -1$. According to Castelnuovo's contractibility theorem stated in 3.5, there is a regular map $f \colon X \to Y$ that is birational and contracts L to a point. Construct f in the given special case. [Hint: Try to find it among the preceding exercises.]

7. Let $f \colon X \to Y$ be a birational regular map of nonsingular n-dimensional projective varieties. Prove that $f^*(D_1) \cdots f^*(D_n) = D_1 \cdots D_n$ for $D_1, \ldots, D_n \in \operatorname{Div} Y$.

8. Let $\sigma \colon X \to Y$ be a blowup with centre in a point $y \in Y$ and $\Gamma = \sigma^{-1}(y)$. For $D_1 \in \operatorname{Div}(Y)$, and $D_2, \ldots, D_{n-1} \in \operatorname{Div}(X)$, prove that $\sigma^*(D_1) D_2 \cdots D_{n-1} \Gamma = 0$.

9. In the notation of Ex. 7–8, calculate Γ^n for any $n > 1$.

10. Prove that if a curve of degree n passes through k of the points ξ_0, ξ_1, ξ_2 (for $k = 0$, 1 or 2) defining the standard quadratic transformation f (see 3.5, Example), and is not singular there, then its image under f has degree $2n - k$.

11. Let φ be the transformation of inversion with respect to a circle with centre O and radius 1, that is, $\varphi(P) = Q$, where P, Q and O are collinear and $|OP| \cdot |OQ| = 1$. Taking coordinates with O as the origin, write out the formulas for φ in coordinates x, y and $u = x + iy$, $v = x - iy$. Prove that after composing with the reflection $(u, v) \mapsto (u, -v)$, φ becomes the standard quadratic transformations defined by O and the two circular points at infinity. Deduce from this that under inversion circles through O transform to lines, and other circles to circles.

4. Singularities

4.1. Singular Points of a Curve

Theorem 1. *Let C be an irreducible curve on a nonsingular surface X; then there exists a surface Y and a regular map $f \colon Y \to X$, such that f is a composite of blowups $Y \to X_1 \to \cdots \to X_n \to X$ and the birational transform C' of C on Y is nonsingular.*

Proof. We can consider separately each singularity of C. Indeed, if we can construct a map $f \colon Y \to X$ for one point $x \in C$, with f a composite of blowups above x and the birational transform C' of C on Y nonsingular at all point of $f^{-1}(x)$, then we can subsequently apply the same argument to the remaining singular points of C'; the number of these equals the number of singularities of C outside x.

Thus let $x \in C$ be a singular point. We blow up x; if some points of the inverse image of x are singular points of the birational transform of C, we blow these up too, and so on. We have to prove that this process stops after finitely many steps.

Write $\mu_x(C)$ for the multiplicity of a singular point $x \in C$; let $\sigma \colon X' \to X$ be the blowup, C' the birational transform of C, and $L = \sigma^{-1}(x)$. By 3.2, Corollary 2, $\mu_x(C) = C'L$. On the other hand, $C'L = \sum_{\sigma(x')=x}(C'L)_{x'}$, where the sum takes place over all points $x' \in C'$ with $\sigma(x') = x$. Since $(C'L)_{x'} \geq \mu_{x'}(C')$, we get

$$\mu_x(C) \geq \sum_{\sigma(x')=x} \mu_{x'}(C').$$

Therefore, if there is more than one point x', each of them must have multiplicity $\mu_{x'}(C') < \mu_x(C)$, so our process must stop after finitely many steps. It remains to consider the case when C' has only one singular point x' with $\sigma(x') = x$, and the same continues to hold after each blowup.

Write $\mathcal{O}_x$ for the local ring of $x \in C$, and $\overline{\mathcal{O}}_x$ for its integral closure in the field $k(C)$. Then $\overline{\mathcal{O}}_x$ is a finite module over $\mathcal{O}_x$; this follows because by Chap. II, 5.2, Theorem 4, x has an affine neighbourhood U for which the normalisation $k[U]^\nu$ of the coordinate ring $k[U]$ is a finite module over $k[U]$. Suppose that $k[U]^\nu = \alpha_1 k[U] + \cdots + \alpha_m k[U]$. Then $\overline{\mathcal{O}}_x = \alpha_1 \mathcal{O}_x + \cdots + \alpha_m \mathcal{O}_x$; in fact if $f \in \overline{\mathcal{O}}_x$ then $f^n + a_1 f^{n-1} + \cdots + a_n = 0$ with $a_i \in \mathcal{O}_x$, that is, $a_i = b_i/c$ with $b_i, c \in k[U]$ and $c(x) \neq 0$. Then cf is integral over $k[U]$, hence $cf = \alpha_1 r_1 + \cdots + \alpha_m r_m$ with $r_i \in k[U]$ and $f = \alpha_1 r_1/c + \cdots + \alpha_m r_m/c$.

Since α_i is in the field of fractions of $\mathcal{O}_x$ (or even in that of the smaller ring $k[U]$), there exists a nonzero element $d \in \mathcal{O}_x$ such that $d\alpha_i \in \mathcal{O}_x$ for each i, and hence $d\overline{\mathcal{O}}_x \subset \mathcal{O}_x$. It follows from this that $\overline{\mathcal{O}}_x/\mathcal{O}_x$ is a finite dimensional vector space. In fact its dimension is at most the dimension of the vector space $\overline{\mathcal{O}}_x/d\overline{\mathcal{O}}_x$, which is generated by the m subspaces $\alpha_i(\mathcal{O}_x/d\mathcal{O}_x)$; but $\mathcal{O}_x/d\mathcal{O}_x$ is finite dimensional, since C is a curve, and so for any function $d \neq 0$ there exists k such that $\mathfrak{m}_x^k \subset d\mathcal{O}_x$.

Obviously $\mathcal{O}_x \subset \mathcal{O}_{x'}$ after one blowup. Moreover, we now prove that $\mathcal{O}_{x'} \subset \overline{\mathcal{O}}_x$. Indeed, let $\nu' \colon C^\nu \to C'$ be the normalisation and $(\nu')^{-1}(x') = \{y_i\}$. Then $\nu = \sigma \circ \nu' \colon C^\nu \to C$ coincides with the normalisation of C and $\nu^{-1}(x) = \{y_i\}$. Obviously $\mathcal{O}_{x'} \subset \bigcap \mathcal{O}_{y_i}$, so that we will be home if we check that $\bigcap \mathcal{O}_{y_i} = \overline{\mathcal{O}}_x$. Again, obviously $\overline{\mathcal{O}}_x \subset \bigcap \mathcal{O}_{y_i}$. Since ν is a finite map, we can assume that C and C^ν are affine. If $u \in \bigcap \mathcal{O}_{y_i}$ then all the poles of u on C^ν are distinct from the y_i, and it follows that there exists a function $v \in k[C]$ such that $v(x) \neq 0$ and $uv \in k[C^\nu]$ (for this, it is sufficient that $\nu^*(v)$ has zeros of sufficiently high degree at each pole of u). Then uv is integral over $k[C]$, and it follows easily that u is integral over $\mathcal{O}_x$, that is, $u \in \overline{\mathcal{O}}_x$.

Hence $\ell(\overline{\mathcal{O}}_x/\mathcal{O}_{x'}) \leq \ell(\overline{\mathcal{O}}_x/\mathcal{O}_x)$. If $\ell(\overline{\mathcal{O}}_x/\mathcal{O}_{x'}) = 0$ then $\overline{\mathcal{O}}_x = \mathcal{O}_{x'}$, so that $\mathcal{O}_{x'}$ is integrally closed, and then x' is nonsingular and our process has stopped. It now only remains to prove that $\ell(\overline{\mathcal{O}}_x/\mathcal{O}_{x'}) < \ell(\overline{\mathcal{O}}_x/\mathcal{O}_x)$, since then our process must stop after at most $\ell(\overline{\mathcal{O}}_x/\mathcal{O}_x)$ steps. But $\ell(\overline{\mathcal{O}}_x/\mathcal{O}_{x'}) = \ell(\overline{\mathcal{O}}_x/\mathcal{O}_x)$ implies that $\mathcal{O}_{x'} = \mathcal{O}_x$. Let u, v be local parameters at $x \in X$, so that the local parameters at $x' \in X'$ are u and $t = v/u$. Since t restricted to C is an element of $\mathcal{O}_{x'}$, from $\mathcal{O}_{x'} = \mathcal{O}_x$ it follows that $t \in \mathcal{O}_x$ and $\mathfrak{m}_x = (u, v) = (u, ut) = (u)$. It follows from this that $\mathfrak{m}_x/\mathfrak{m}_x^2 = (u)/(u^2) \cong \mathcal{O}_x/(u) \cong k$, that is, $x \in C$ was already nonsingular. The theorem is proved.

Theorem 1 enables us to define an important characteristic of a singular point of a curve on a nonsingular surface, its tree of infinitely near points. This is the diagram consisting of the singular point, the singular points arising out of it after one blowup, the singular points arising out of these after blowing them up again, and so on. All these points are said to be *infinitely near* to the original point of the curve. We write the multiplicity of each point. Once we get to a point of multiplicity 1, we don't carry out any further blowups there. Some examples are illustrated in Figure 19.

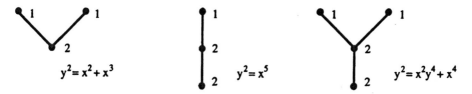

Figure 19. Resolutions of Some Curve Singularities

The genus of the normalisation of a singular curve lying on a nonsingular projective surface is expressed in terms of these invariants: for this, by 2.3, (1), we have to determine how the expression $C(C+K)$ changes on replacing C by $\sigma'(C)$ and K_X by $K_{X'}$, where $\sigma: X' \to X$ is the blowup of a point $x \in C$ of multiplicity k. According to 3.1, Theorem 1, $\sigma'(C) = \sigma^*(C) - kL$. To compute $K_{X'}$, consider a differential form $\omega \in \Omega^2(X)$ such that $x \notin \text{Supp}(\text{div}\,\omega)$; this exists by Chap. III, 1.3, Theorem 1 (on moving the support of a divisor away from a point). Then since $\sigma: X' \setminus L \to X \setminus x$ is an isomorphism, obviously $\text{div}(\sigma^*\omega) = \sigma^*(\text{div}\,\omega)$ over $X' \setminus L$. If x, y are local parameters at x then $\omega = f dx \wedge dy$, where $f \in \mathcal{O}_x$ and $f(x) \neq 0$. If $x = u$, $y = uv$ are as in 3.1, (1) then $\sigma^*(\omega) = \sigma^*(f)v du \wedge dv$ on X, and since $\sigma^*(f) \neq 0$ on L, we get $\text{div}(\sigma^*(\omega)) = \sigma^*(\text{div}\,\omega) + L$, that is, $K_{X'} = \sigma^*(K_X) + L$. Substituting in 2.3, (1) gives

$$\sigma'(C)(\sigma'(C) + K_{X'}) = (\sigma^*(C) - kL)(\sigma^*(C) + \sigma^*(K_X) - (k-1)L)$$
$$= C(C + K_X) - k(k-1).$$

Now using 4.1, Theorem 1, we get

$$\overline{C}(\overline{C} + K_Y) = C(C + K_X) - \sum k_i(k_i - 1),$$

for the nonsingular curve $\overline{C} \subset Y$, where k_i are the multiplicities of all the infinitely near points. It follows from 2.3, (1) that

$$g(\overline{C}) = \frac{1}{2}C(C + K_X) + 1 - \sum \frac{k_i(k_i - 1)}{2}. \tag{1}$$

In particular if $X = \mathbb{P}^2$ and C is a curve of degree n then

$$g = \frac{(n-1)(n-2)}{2} - \sum \frac{k_i(k_i - 1)}{2}.$$

A corollary of (1) that is often used is that since $g(\overline{C}) \geq 0$,

$$C(C + K_X) \geq -2, \tag{2}$$

and equality holds if and only if C is nonsingular (that is, all $k_i = 1$), and $g(\overline{C}) = g(C) = 0$, so that $C \cong \mathbb{P}^1$.

4.2. Surface Singularities

The theorem on resolution of singularities has been proved for algebraic surfaces over a field of arbitrary characteristic; we can suppose that X is normal, so has only finitely many singular points. Resolution of singularities asserts that there exists a nonsingular projective surface Y birational to X. Using the theorem on the resolution of indeterminacies (3.3, Theorem 3), we can assume given a birational regular map $f: Y \to X$. It is often convenient to consider the situation locally, dropping the assumption that X and Y are projective, thus replacing them by open subsets $U \subset X$ and $f^{-1}(U) \subset Y$. Then the map $f: Y \to X$ will be proper (see the remark after Chap. I, 5.2, Theorem 3). It can be shown that Chap. II, 4.4, Theorem 2 remains true in this case, and f contracts a bunch of projective curves $C_1, \ldots, C_r \subset Y$ to each singular point $x \in X$. Moreover, using Castelnuovo's contractibility criterion discussed in 3.5, one can prove that Y can be chosen so that there are no -1-curves among the C_i. In this case Y is a *minimal resolution of singularities* of X. We will not prove all these assertions, and will not make use of them: they only serve as motivation for the questions that we now discuss.

If $x \in X$ is a surface singularity, the bunch of curves $C_1, \ldots, C_r \subset Y$ on the nonsingular surface Y that are contracted by $f: Y \to X$ is an important geometric characteristic of the singularity, and it is interesting to see what can be said in general about such a bunch of curves.

Theorem 2. *Let $f: Y \to X$ be a regular map of algebraic surfaces, with Y nonsingular and $C_1, \ldots, C_r \subset Y$ projective curves that are contracted to $x \in X$; suppose that $f: Y \setminus (C_1 \cup \cdots \cup C_r) \xrightarrow{\sim} X \setminus x$ is an isomorphism. Then the matrix of intersection numbers $\{C_i C_j\}$ is negative definite.*

Proof. Consider a curve E on Y distinct from all the C_i but intersecting each of them (for example, a hyperplane section of Y); set $f(E) = H \subset X$ and choose a function $u \in \mathcal{O}_x$ vanishing along H. Set $g = f^*(u)$. Then $\operatorname{div} g = \sum m_i C_i + F$, where all the $m_i > 0$ and $FC_i > 0$ for $i = 1, \ldots, r$. Since $\operatorname{div} g$ is a principal divisor, if we set $D = \sum m_i C_i$ then the restriction of D to each of the C_j satisfies $D_{|C_j} \sim -F_{|C_j}$; hence $DC_j < 0$ for $j = 1, \ldots, r$.

The theorem now follows from the following result of linear algebra, the proof of which can be found in the Appendix, §1, Proposition 2.

Proposition. *Let M be a $\mathbb{Z}$-module with a scalar product $ab \in \mathbb{Z}$ defined for $a, b \in M$, and $e_1, \ldots, e_r$ a set of generators of M with $e_i e_j \geq 0$ for $i \neq j$; suppose that there exists an element $d = \sum m_i e_i$ with $m_i > 0$ such that $de_i < 0$ for $i = 1, \ldots, r$. Then every nonzero $m \in M$ satisfies $m^2 < 0$ and $e_1, \ldots, e_r$ is a free basis of M.* $\square$

The theorem is proved.

It is interesting to note the analogy between Theorem 2 and the Hodge index theorem in Chap. IV, 2.4.

If $x \in X$ is a surface singularity, the bunch of curves $C_1, \ldots, C_r$ contracted to x under the minimal resolution can be drawn as a graph: each curve C_i is represented by a node, and intersecting curves C_i and C_j are joined by an edge, marked by the intersection number $C_i C_j$ if $C_i C_j \neq 1$, and left unmarked if $C_i C_j = 1$ (that is, when C_i and C_j intersect transversally at one point); the node corresponding to C_i is marked with C_i^2.

Interesting examples of singularities are provided by the quotients $\mathbf{A}^2/G$ of the plane by a finite group G of linear transformations. Recall that these are normal varieties (Chap. I, 2.3, Example 11 and 5.3, Example 1), and points which are images of $x \in \mathbf{A}^2$ for which $g(x) \neq x$ for all $g \neq G$ are nonsingular (Chap. II, 2.1, Example).

Suppose for example that $G = \langle g \rangle$ is a cyclic group of order n generated by $g(x, y) = (\varepsilon x, \varepsilon^q y)$, where ε is a primitive nth root of 1 and q is coprime to n. It can be shown that after excluding certain uninteresting cases, every action of a cyclic group reduces to this form. In this case G acts freely on $\mathbf{A}^2 \setminus (0, 0)$, and hence $\mathbf{A}^2/G$ has a single singularity, the image of $(0, 0) \in \mathbf{A}^2$. This is called a singularity of type (n, q).

For example, if $q = -1$, the ring of invariants $k[x, y]^G$ is generated by $u = x^n$, $v = y^n$ and $w = xy$, with the single relation

$$uv = w^n. \tag{1}$$

This is the equation of the surface $\mathbf{A}^2/G$.

For $q = 1$ the generators of $k[x, y]^G$ are $u_i = x^i y^{n-i}$ for $i = 0, \ldots, n$. The relations holding between these are the same as those between the coordinates of the Veronese curve in Chap. I, 5.4. Thus in this case $\mathbf{A}^2/G$ is the cone over the Veronese curve.

It is not hard to verify that the resolution graph corresponding to an arbitrary singularity of type (n, q) has the form of a chain

$$
\begin{array}{cccc}
-e_1 & -e_2 & -e_{n-1} & -e_n \\
\bullet\!\!-\!\!-\!\!-\!\!\bullet & \cdots & \bullet\!\!-\!\!-\!\!-\!\!\bullet
\end{array}
$$

The curves C_i and C_{i+1} intersect transversally, and $C_i^2 = -e_i$, where the $e_i \geq 2$, and are defined by an expansion which is very close to a continued fraction expansion of n/q:

$$
\frac{n}{q} = e_1 - \cfrac{1}{e_2 - \cfrac{1}{e_3 - \dots}} .
$$

See for example de la Harpe and Siegfried [22] for the proof.

4.3. Du Val Singularities

An extremely important class of singularities is defined by the following condition.

Definition. A point $x \in X$ of a normal surface is called *a Du Val singularity* if[17] there exists a minimal resolution $f : Y \to X$ contracting curves $C_1, \dots, C_r$ to x, such that $K_Y C_i = 0$ for all i, where K_Y is the canonical class of Y.

The meaning of the Du Val singularities, as formulated by Du Val himself, is that they "do not affect the canonical class". For example, it is easy to see that for a surface $X \subset \mathbb{P}^3$ of degree n with only Du Val singularities, the invariant $h^2 = \dim \Omega^2[Y]$ of its minimal resolution Y is the same as that of a nonsingular surface of degree n. This is a sharp contrast between surfaces and curves, for which, according to 4.1, (1), any singularity decreases the genus of the normalisation of the curve.

The types of resolution graphs corresponding to Du Val singularities $x \in X$ can be completely determined. Indeed, if C_i is one of the irreducible projective curves contracted to x by $f : Y \to X$ then $K_Y C_i = 0$, and according to the inequality 4.1, (2),

$$
C_i^2 \geq -2.
$$

Because $C_i^2 < 0$, and $C_i^2 \neq -1$ by minimality of the resolution, it follows that $C_i^2 = -2$ and $C_i \cong \mathbb{P}^1$. From the fact that $(C_i + C_j)^2 < 0$ for $i \neq j$ it now follows that $C_i C_j \leq 1$, that is, C_i and C_j either do not intersect, or intersect transversally in one point.

---

[17]There are many alternative names in the literature: Kleinian singularities, rational double points, simple singularities, etc.

The purely algebraic question of classifying $\mathbf{Z}$-modules $\mathbf{Z}e_1 + \cdots + \mathbf{Z}e_r$ having a negative definite scalar product satisfying $e_i e_j > 0$ and $e_i^2 = -2$ for $i, j = 1, \ldots, r$ occurs in a number of problems. It first appeared in connection with the classification of simple Lie algebras in the theory of root systems (see Bourbaki [17]). The answer is as follows: the basis $e_1, \ldots, e_r$ breaks up into disjoint "connected components" with $e_i e_j = 0$ for e_i and e_j belonging to different components, and the module decomposes as a direct sum of submodules corresponding to the different components. Thus the problem reduces to describing the "connected" modules, which can have only the graphs of Figure 20 (each vertex is marked with -2):

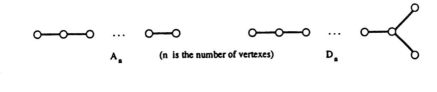

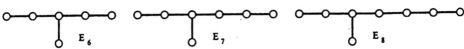

Figure 20. The Dynkin Diagrams A_n, D_n, E_6, E_7 and E_8

It can be shown that the set of curves that appear on resolving a singularity is always connected. For $k = \mathbf{C}$ we will prove this in Chapter VII, 2.5, Theorem 2. Thus the Du Val singularities correspond only to graphs of type A_n, D_n, E_6, E_7 and E_8. It can be proved that a Du Val singularity is determined up to formal analytic equivalence by its graph. They can be given by equations

$$
\begin{array}{ll}
A_n: & x^2 + y^2 + z^{n+1} = 0 \quad \text{for } n \geq 1, \\
D_n: & x^2 + y^2 z + z^{n-1} = 0 \quad \text{for } n \geq 4, \\
E_6: & x^2 + y^3 + z^4 = 0, \\
E_7: & x^2 + y^3 + yz^3 = 0, \\
E_8: & x^2 + y^3 + z^5 = 0.
\end{array}
$$

One of the realisations of these singularities is as follows:

Theorem 3. *Suppose that* char $k = 0$ *and that* G *is a finite group of linear transformations of the plane* $\mathbf{A}^2$, *with* $\det g = 1$ *for all* $g \in G$. *Then the image of the origin* $0 \in \mathbf{A}^2$ *is a Du Val singularity* $y_0 \in \mathbf{A}^2/G$.

The proof uses the following construction, which we discuss in complete generality in Chapter V, 4.1. Let X, Y and S be three varieties, and $f: X \to S$

and $h\colon Y \to S$ regular maps. The *fibre product* of X and Y over S is the closed subset in $X \times Y$ consisting of pairs (x, y) for which $f(x) = h(y)$; it is denoted by $X \times_S Y$. The maps $f\colon X \to S$ and $h\colon Y \to S$ define a map $X \times_S Y \to S$, and the projections $X \times Y \to X$ and $X \times Y \to Y$ define projections $X \times_S Y \to X$ and $X \times_S Y \to Y$.

Let $h\colon X \to \mathbf{A}^2/G = S$ be the minimal resolution of the singularity $y_0 \in \mathbf{A}^2/G$. Consider the fibre product $Z = \mathbf{A}^2 \times_S X$ and its normalisation Z^ν. (Here we are using the existence of the normalisation, proved in Chap. II, 5.2, Theorem 4 and 5.3, Theorem 6 only for affine varieties and curves; in Chap. VI, 1.1, the normalisation will be constructed in sufficient generality for our present purposes.) We have the diagram of maps

$$
\begin{array}{ccc}
Z^\nu & \overset{q}{\to} & X \\
p\downarrow & & \downarrow h \\
\mathbf{A}^2 & \underset{f}{\to} & \mathbf{A}^2/G.
\end{array}
$$

Consider the differential form $\omega = \mathrm{d}x \wedge \mathrm{d}y$ on $\mathbf{A}^2$. From the condition $\det g = 1$ for all $g \in G$ it follows that $g^*(\omega) = \omega$. Write ω in the form $h\,\mathrm{d}s \wedge \mathrm{d}t$ with $s, t \in k(\mathbf{A}^2/G)$ and $h \in k(\mathbf{A}^2)$. Then from the fact that $g^*(\omega) = \omega$ it follows that $g^*(h) = h$; writing h in the form P/Q with $P, Q \in k[\mathbf{A}^2]$ we see that

$$
h = \left(P \prod_{g \neq e} g^*(Q) \right) \Big/ \left(\prod_g g^*(Q) \right),
$$

and it follows that $h \in k(\mathbf{A}^2/G)$. Thus $\omega = f^*(\omega_0)$ with $\omega_0 \in \Omega^2(\mathbf{A}^2/G)$. Write $\omega_1 = h^*(\omega_0)$ and $\overline{\omega} = q^*(\omega_1) = p^*(\omega)$. From the fact that $\overline{\omega} = p^*(\omega)$ it follows that $\overline{\omega}$ is regular on the set of nonsingular points of the surface Z^ν. On the other hand, for any maps $f\colon X \to S$ and $h\colon Y \to S$ it is easy to check that if f is finite then so is $X \times_S Y \to X$. Thus $Z \to X$ is finite, and hence also $Z^\nu \to X$. We use the following fact:

Lemma. *If $\varphi\colon U \to V$ is a finite map of nonsingular surfaces and ω_1 a rational differential 2-form on V such that $\varphi^*(\omega_1)$ is regular, then ω_1 is also regular.*

Proof of Theorem 3. We leave the proof of the lemma until after that of Theorem 3. It follows from the lemma that ω_1 is regular outside the image of the finite set of singular points of Z^ν, and hence is regular on the whole of X. Let us determine the divisor $\mathrm{div}(\omega_1)$ on X. At any point $\alpha \in \mathbf{A}^2$ with $\alpha \neq (0, 0)$ we can find local parameters of the form $f^*(u), f^*(v)$ (see Chap. II, 2.1, Example), and it follows that ω_0 is regular and nonzero at all points $y \neq y_0 \in \mathbf{A}^2/G$, and these points are nonsingular. In exactly the same way, h is an isomorphism on $X \setminus f^{-1}(y_0)$, and ω_1 is nonzero on $X \setminus f^{-1}(y_0)$. Thus $D = \mathrm{div}(\omega_1) = \sum r_i C_i$ with $r_i \geq 0$; obviously $D \sim K_X$. From the inequality 4.1, (2), and the minimality of the resolution, we get $DC_i = K_X C_i \geq 0$. But then $D^2 = \sum r_i DC_i \geq 0$, which by 4.1, Theorem 1 is only possible if $DC_i = 0$ for all i. Thus $K_X C_i = 0$, that is, y_0 is a Du Val singularity. The theorem is proved.

Proof of the lemma. It is enough to prove that if $v_C(\mathrm{div}(\omega_1)) < 0$ for any irreducible curve $C \subset V$ then also $v_{C'}(\varphi^* \mathrm{div}(\omega_1)) < 0$ for any component C' of the inverse image of C. This can be checked on any open subset $V' \subset V$ meeting C.

What makes the problem nontrivial is that $\varphi^*(\mathrm{div}(\omega_1)) \neq \mathrm{div}(\varphi^*(\omega_1))$ in general. However, if the differential $d_\alpha\varphi \colon \Theta_{U,\alpha} \xrightarrow{\sim} \Theta_{V,\varphi(\alpha)}$ of φ at a point $\alpha \in U$ is an isomorphism on the tangent spaces then the inverse images $\varphi^*(v_1)$, $\varphi^*(v_2)$ of local parameters v_1, v_2 at $\varphi(\alpha)$ are local parameters at α. Thus if $\omega_1 = f dv_1 \wedge dv_2$ then $\varphi^*(\omega_1) = \varphi^*(f) d\varphi^*(v_1) \wedge d\varphi^*(v_2)$, and in an neighbourhood of α this has divisor $\mathrm{div}(\varphi^*(f)) = \varphi^*(\mathrm{div}(f)) = \varphi^*(\mathrm{div}(\omega_1))$. Thus we need only consider curves $C' \subset U$ such that $d_\alpha\varphi$ is degenerate at every point $\alpha \in C'$. Set $\varphi(C') = C$. Since the map $\varphi \colon C' \to C$ is an isomorphism of the tangent spaces over an open set, we can assume that at α the local parameter along C' is $\varphi^*(v_1)$, where v_1, v_2 are local parameters at $\varphi(\alpha)$, where $v_2 = 0$ is a local equation of C and v_1 restricts to a local parameter along C. Set $w = v_1$ and let (w, t) be local parameters at α, with t a local equation of C'. Suppose $\varphi^*(v_2) = t^e h$ where $v_{C'}(h) = 0$, and set $w_1 = f dv_1 \wedge dv_2$. Then

$$\varphi^*(\omega_1) = \varphi^*(f) dw \wedge d(\varphi^*(v_2)) = \varphi^*(f)\left(et^{e-1}h dw \wedge dt + t^e dw \wedge dh\right),$$

and it follows that $v_{C'}\big(\varphi^*(\mathrm{div}(\omega_1))\big) = v_{C'}\big(\varphi^*(\mathrm{div}(f))\big) + e - 1$. But if C is in the divisor of poles of ω_1 then $v_C(f) = -l$ with $l > 0$, and then $v_{C'}\big(\varphi^*(\mathrm{div}(\omega_1))\big) = -le + e - 1$, which is also < 0. In other words, an effective term gets added on to the divisor $\varphi^*(\mathrm{div}(\omega_1))$, but not enough to compensate for the pole that arises. The lemma is proved.

The groups G appearing in Theorem 3 are well known. Write $\mathrm{SL}(2, k)$ for the group of linear transformations with determinant 1, and consider the homomorphism $\pi \colon \mathrm{SL}(2, k) \to \mathrm{PSL}(2, k)$ to the group of projective transformations of $\mathbb{P}^1$; the kernel of π is ± 1. Then the finite subgroups $G \subset \mathrm{SL}(2, k)$ are the following: either the *cyclic group* of order n consisting of transformations $(x, y) \mapsto (\varepsilon x, \varepsilon^{-1} y)$ for $\varepsilon^n = 1$, or the *binary dihedral* group of order $4n$, generated by $(x, y) \mapsto (\varepsilon x, \varepsilon^{-1} y)$ for $\varepsilon^{2n} = 1$, and $(x, y) \mapsto (-y, x)$, or the binary tetrahedral, binary octahedral or binary icosahedral groups, that is, the inverse image under π of the subgroups of $\mathrm{PSL}(2, k)$ isomorphism to the tetrahedral, octahedral or icosahedral groups. These groups have order n, $4n$, 24, 48 and 120 respectively (see for example Springer [72]). It is not hard to find the corresponding Du Val singularities, which turn out to be A_{n-1} for the cyclic group of order n, D_{n+2} for the binary dihedral group of order $4n$, and E_6, E_7 and E_8 for the binary tetrahedral, binary octahedral or binary icosahedral groups (see de la Harpe and Siegfried [22]).

4.4. Degeneration of Curves

Let X be a nonsingular projective irreducible surface and $f: X \to S$ a regular map to a curve S; fix some point $s_0 \in S$, and assume that for every $s \in S$ with $s \neq s_0$, the fibre $f^{-1}(s)$ is a nonsingular irreducible curve. We can consider $\{f^{-1}(s) \mid s \in S \setminus s_0\}$ as a family of nonsingular curves, and $f^{-1}(s_0)$ as a degeneration of this. By the Castelnuovo contractibility criterion mentioned in 3.5, any -1-curve among the components of $f^{-1}(s_0)$ can be contracted to a point without affecting the nonsingularity of X. Hence we assume in what follows that there are no such components. Moreover, it can be proved that $f^{-1}(s_0)$ is connected, that is, cannot be written as a union of two closed disjoint curves. In the case $k = \mathbb{C}$ this will be proved in Chap. VII, 2.4, Theorem 2.

Theorem 4. *Under the above assumptions, consider the divisor s_0 on S consisting of one point, and suppose that its inverse image $f^*(s_0)$ decomposes as $f^*(s_0) = \sum r_i C_i$, where the C_i are irreducible components and $r_i > 0$. Then any divisor $D = \sum l_i C_i$ satisfies $D^2 \le 0$, with $D^2 = 0$ if and only if D is proportional to $\sum r_i C_i$.*

Proof. Obviously $\sum r_i C_i = f^*(s_0) \sim f^*(\Delta)$, where Δ is a divisor on C not containing s_0. Hence the restriction to any component C_i of these divisors satisfies $f^*(s_0)_{|C_i} \sim f^*(\Delta)_{|C_i} = 0$. It follows from this that $f^*(s_0)C_i = 0$, that is $(\sum r_i C_i)B = 0$ for every $B = \sum l_i C_i$. In particular $(f^*(s_0))^2 = 0$. Theorem 4 now follows from the following result of linear algebra proved in Appendix, §1, Proposition 3:

Proposition. *Let M be a free $\mathbb{Z}$-module with a scalar product $ab \in \mathbb{Z}$ for a, $b \in M$. Suppose that M has a basis $e_1, \ldots, e_r$ satisfying $e_i e_j \ge 0$ for each $i \neq j$, and that the $\{e_i\}$ cannot be split up into two components with $e_i e_j = 0$ for e_i, e_j in different components; assume that there exists an element $d = \sum l_i e_i$ with $l_i > 0$ such that $d e_i = 0$ for $i = 1, \ldots, r$. Then every $m \in M$ satisfies $m^2 \le 0$, with equality only if m is proportional to d.* $\square$

Theorem 4 is proved.

It is interesting to note that Theorem 4 occupies an intermediate position between the Hodge index theorem of 2.4, and 4.2, Theorem 2: the curves considered in Theorem 4 are contained in a fibre of a map $f: X \to C$ to a curve C, those of Theorem 2 in a fibre of a map $f: X \to Y$ to a surface Y, and those of the Hodge index theorem in a fibre of a map $f: X \to z$ to a point z.

We now study the simplest examples of the situation described in Theorem 4. If the genus of the curves $f^*(s)$ for $s \neq s_0$ is 0, that is, if $f^*(s) \cong \mathbb{P}^1$

then under the above assumption one can show that there is no degeneration, that is, the curve $f^{-1}(s_0)$ is also nonsingular and isomorphic to $\mathbb{P}^1$. See, for example, Shafarevich [67], Chap. V or Griffiths and Harris [31], IV.5.

Next in order of difficulty is the case when the curves $f^{-1}(s)$ for $s \neq s_0$ have genus 1, that is, they are isomorphic to nonsingular plane cubic curves. Consider the pencil X of elliptic curves of Chap. II, 6.4, Example 2 given by the equation

$$\xi_2^2 \xi_0 = \xi_1^3 + a(t)\xi_1\xi_o^2 + b(t)\xi_0^3$$

in $\mathbb{P}^2 \times \mathbb{A}^1$. In the affine part $\mathbb{A}^2 \times \mathbb{A}^1$ this equation is $y^2 = x^3 + a(t)x + b(t)$. The fibre of $f : X \to \mathbb{A}^1$ over a point $c \in \mathbb{A}^1$ will be nonsingular when $\Delta(c) \neq 0$, where $\Delta = 4a^3 + 27b^2$. We suppose that Δ does not vanish identically on $\mathbb{A}^1$, but that $\Delta(0) = 0$, and study the fibre $f^{-1}(0)$. In order to work with a projective surface we consider the closure of X in $\mathbb{P}^2 \times \mathbb{P}^1 \supset \mathbb{P}^2 \times \mathbb{A}^1$. The surface we obtain is in general singular: points of a fibre $f^{-1}(c)$ will be nonsingular if $\Delta(c) \neq 0$, but when $\Delta(c) = 0$ this will only happen if c is a simple root of Δ (see Chap. II, 6.4, Example 2). We consider the minimal resolution $\varphi : Y \to X$, which maps to $\mathbb{P}^1$ by $g = f \circ \varphi : Y \to \mathbb{P}^1$; here at a point such that $\Delta(c) \neq 0$, the fibre of g is the same as that of the original pencil f.

On Y, consider the differential 2-form $\omega = y^{-1}dx \wedge dt$. One sees easily that above points $c \in \mathbb{A}^1$ where $\Delta(c) \neq 0$, this form is regular and nowhere vanishing; this comes from the fact that the 1-form $y^{-1}dx$ is regular and nowhere vanishing on the curve $f^{-1}(c)$. It follows from this that the canonical class K_Y contains a divisor consisting only of components of fibres. Suppose that $g^*(0) = \sum r_i C_i$, where the C_i are components of the fibre $g^{-1}(0)$ and $r_i > 0$. We write K_Y in the form $K_Y = \sum n_i C_i + D$, where D consists of components of fibres other than $g^{-1}(0)$. Since $g^*(0) \sim g^*(c)$ for $c \neq 0$, we can add in a multiple of $g^*(0) - g^*(c)$ to arrange that all $n_i > 0$. Since all fibres $g^*(c)$ are linearly equivalent, $g^*(c)K_Y = 0$. We consider two cases.

Case A. $g^{-1}(0)$ is an irreducible curve C. Since in this case $C^2 = 0$ and $K_Y C = 0$, from 4.1, (1) and the fact that $g(C) \geq 0$ we get that

$$\delta = \sum \frac{k_i(k_i - 1)}{2} \leq 1,$$

that is, δ equals 0 or 1. If $\delta = 0$ then C is a nonsingular curve. If $\delta = 1$ then C has just one singularity of multiplicity 2 which is resolved by one blowup. It follows that the singular point is formally analytically equivalent to the singularity $y^2 = x^2$ or $y^2 = x^3$ (see Chap. II, §3, Ex. 12). These are exactly the types of singularities that can appear on irreducible plane cubics.

Case B. $g^{-1}(0)$ is reducible. Then Theorem 4 implies that any component C_i of the fibre satisfies $C_i^2 < 0$. We write K_Y in the form $K_Y = \sum n_j C_j + D$ with all the $n_j > 0$ and $\operatorname{Supp} D$ disjoint from $g^{-1}(0)$. Then if $(\sum n_j C_j)^2 < 0$ we would have $K_Y C_i < 0$ for at least one component C_i. But then, by the same argument as in 4.5, the inequality 4.1, (1) implies that $C_i^2 = -1$, $g(C_i) = 0$ and $k_i = 0$, so that C_i is a -1-curve, and we are assuming that there are no such components in fibres. Hence $(\sum n_j C_j)^2 = 0$, and so it follows from Theorem 4 that $\sum n_j C_j$ is proportional to the fibre $f^*(0)$, and so $K_Y C_i = 0$. Now 4.1, (1) gives that $C_i^2 = -2$, $g(C_i) = 0$, $k_i = 0$, in other words, all the components of fibres are isomorphic to $\mathbb{P}^1$ and have $C_i^2 = -2$.

If the fibre has just two components C_1 and C_2 and $g^*(0) = n_1 C_1 + n_2 C_2$ then $(C_1 + C_2)^2 \leq 0$ gives $C_1 C_2 \leq 2$, and $(n_1 C_1 + n_2 C_2)^2 = 0$ implies that $n_1^2 + n_2^2 = n_1 n_2 C_1 C_2$, which is only possible if $C_1 C_2 = 2$. The two components can intersect transversally in 2 points or have a point of tangency.

If the fibre has more than two components, then $(C_i + C_j)^2 < 0$ implies that $C_i C_j = 0$ or 1. Thus the curves C_i and C_j are either disjoint, or they meet transversally. We draw the system of curves $C_1, \ldots, C_r$ as a graph with the same conventions as for the resolution of isolated singularities.

We have seen that these define a basis of the $\mathbb{Z}$-module $\oplus \mathbb{Z} C_i$ satisfying the assumptions of Theorem 4, and the additions condition $C_i^2 = -2$. All such $\mathbb{Z}$-modules have been found in connection with root systems (see Bourbaki [17]); their graphs are as follows:

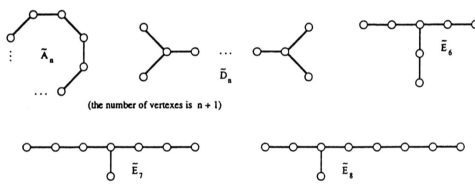

Figure 21. The Extended Dynkin Diagrams $\tilde{A}_n$, $\tilde{D}_n$, $\tilde{E}_6$, $\tilde{E}_7$ and $\tilde{E}_8$

The relation with the theory of Du Val singularities is as follows. Suppose that the elliptic pencil is given by the equation

$$y^2 = x^3 + a(t)x + b(t), \tag{1}$$

where $a(t)$ and $b(t)$ are polynomials. We will assume that a and b are not simultaneously divisible by a 4th and 6th power of any polynomial $c(t)$, since in that case one could get rid of the factors by means of the birational transformation $y = y_1 c^3$, $x = x_1 c^2$. Then the surface given by (1) has a Du Val singularity on every fibre $f^{-1}(c)$ with $\Delta(c) = 0$, and the fibre of the nonsingular surface consists of the curves appearing in the minimal resolution of this singularity, together with the birational transform of the fibre. A singularity of type A_n gives a fibre of type $\tilde{A}_n$, D_n a fibre of type $\tilde{D}_n$, and E_6, E_7, E_8 a fibre of type $\tilde{E}_6$, $\tilde{E}_7$, $\tilde{E}_8$.

Exercises to §4

1. Find the graph of infinitely near points for the curve singularity $y^2 = x^n$.

2. Generalise the notion of *class* (see Chap. III, §6, Ex. 15) to singular plane curves. Prove that the class of a plane projective curve of degree n with d ordinary double points is $n(n-1) - d$.

3. What are the singularities of a curve lying on a nonsingular surface that can be resolved by a single blowup? Give a characterisation of these in terms of the local equation of the curve, or more precisely, in terms of all its terms of degree r and $r + 1$, where r is the multiplicity of the singularity.

4. Prove that an irreducible plane curve C of degree n satisfies $\sum r_i(r_i - 1) \le (n-1)(n-2)$, where r_i are the multiplicities of the singular points. What happens in case of equality?

5. Find the resolution graph corresponding to the Du Val singularities $\mathbf{A}^2/G$ of type $(n, -1)$; for this, embed $\mathbf{A}^2/G$ in affine 3-space $\mathbf{A}^3$, and use a sequence of blowups of $(0,0,0)$ and in the singularities of the variety arising after the blowup.

Do the same for the quotient singularity $\mathbf{A}^2/G$ where $G \subset \mathrm{SL}(2, \mathbb{C})$ is the binary dihedral group of order $4n$, generated by

$$\alpha = \begin{pmatrix} \varepsilon & 0 \\ 0 & \varepsilon^{-1} \end{pmatrix} \quad \text{and} \quad \beta = \begin{pmatrix} 0 & 1 \\ -1 & 0 \end{pmatrix},$$

where $\varepsilon = \exp(2\pi i/2n)$.

6. Suppose that X is a nonsingular projective surface such that, for some $n > 0$, the rational map φ corresponding to the class nK_X is regular and a birational embedding with normal image. Prove that $\varphi(X)$ has only Du Val singularities.

7. Find all the types of degenerate fibres of a pencil of elliptic curves in Weierstrass normal form for which $\Delta = 4a^3 + 27b^2$ has a double root at the point of degeneration.

8. Resolve the singularity of the surface $y^2 = x^3 + \alpha t^2 x + \beta t^3$, where α, $\beta \in k$ and $4\alpha^3 + 27\beta^2 \ne 0$. For this, blow up the ambient space at $(0,0,0)$, then again at the new singular points, and so on. Verify that the singularity is a Du Val singularity of type D_4, and that the singular fibre of the pencil of elliptic curves arising after resolving is of type $\tilde{D}_4$.

Algebraic Appendix

1. Linear and Bilinear Algebra

Recall that a *scalar product* on an Abelian group M with values in an Abelian group B is a function (a, b) for $a,\ b \in M$ with values in B satisfying the conditions

$$(b, a) = (a, b), \tag{1}$$

$$(a_1 + a_2, b) = (a_1, b) + (a_2, b). \tag{2}$$

Proposition 1. *Let M be an arbitrary Abelian group and B an Abelian group in which division by 2 is possible and unique. A function $f(a)$ on M with values in B can be expressed as $f(a) = (a, a)$ for some scalar product (a, b) if and only if*

$$f(a + b) + f(a - b) = 2\Big(f(a) + f(b)\Big). \tag{3}$$

Proof. If $f(a) = (a, a)$ then (3) follows at once from (1) and (2). Assume (3), and set

$$(a, b) = \frac{1}{2}\Big(f(a + b) - f(a) - f(b)\Big). \tag{4}$$

Then (1) is obvious and (2) is equivalent to

$$(a + b, c) - (a, c) - (b, c) = 0. \tag{5}$$

We write $\psi(a, b, c)$ for the left-hand side of (5). It follows from (4) that

$$2\psi(a, b, c) = f(a + b + c) - f(a + b) - f(a + c) - f(b + c) + f(a) + f(b) + f(c),$$

so that $\psi(a, b, c)$ is a symmetric expression in a, b and c. Applying (3) with $a = b = 0$ implies that $f(0) = 0$, and with $a = 0$ that $f(-b) = f(b)$. Now from (3) and (4) we deduce that $(a, -b) = -(a, b)$, and so in view of (1), $(-a, b) = -(a, b)$. Putting this all together gives $\psi(a, b, -c) = -\psi(a, b, c)$, and the same equality for a and b by symmetry. But (5) also gives $\psi(-a, -b, c) =$

$-\psi(a,b,c)$, whereas we have just proved that $\psi(-a,-b,c) = \psi(a,b,c)$. Hence $\psi(a,b,c) = 0$. The proposition is proved.

Proposition 2. *Let M be a $\mathbb{Z}$-module having a scalar product $(a,b) \in \mathbb{Z}$ for $a,\,b \in M$, and suppose that $e_1,\ldots,e_r$ is a set of generators of M satisfying $(e_i,e_j) \geq 0$ for $i \neq j$. Assume that there exists an element $d = \sum_{i=1}^{r} m_i e_i$ with $m_i > 0$ and $(d,e_i) < 0$ for $i = 1,\ldots,r$. Then $(m,m) < 0$ for all nonzero $m \in M$, and $e_1,\ldots,e_r$ are linearly independent in M.*

Proof. Write A for the symmetric $r \times r$ matrix with entries $a_{ij} = (e_i,e_j)$, and define a scalar product φ on $\mathbb{R}^r$ by $\varphi(x,y) = \sum a_{ij}x_iy_j$. It is enough to prove that $\varphi(x,x) < 0$ for all nonzero $x \in \mathbb{R}^r$; because then the map $\mathbb{Z}^r \to M$ taking the basis element $f_i = (0,\ldots,1,\ldots,0)$ to e_i is an isometry, hence an isomorphism, and so $e_1,\ldots,e_r$ are linearly independent in M.

The function $\varphi(x,x)/|x|^2$ on $\mathbb{R}^r \setminus 0$ achieves it supremum λ because the unit sphere S^{r-1} is compact; moreover, it is easy to see that

$$\lambda = \sup_{0 \neq u \in \mathbb{R}^r}\left\{\frac{\varphi(u,u)}{|u|^2}\right\} = \frac{\varphi(x,x)}{|x|^2}$$

holds if and only if x is a nonzero eigenvector of A belonging to the maximum eigenvalue λ, so that $Ax = \lambda x$. Now because $a_{ij} \geq 0$ for $i \neq j$, we can assume that the coordinates x_i of $x = (x_1,\ldots,x_r)$ are all ≥ 0; for $\varphi(x,x) = \sum a_{ij}x_ix_j \leq \sum a_{ij}|x_i||x_j| = \varphi(y,y)$, where $y = (|x_1|,\ldots,|x_r|)$.

Since x is an eigenvector of A, we have

$$\lambda \sum_{ij} x_i m_j = \sum_{ij} a_{ij}x_i m_j = \sum_{ij} \varphi(e_i,e_j)x_i m_j \qquad (*)$$

for any $m = (m_1,\ldots,m_r) \in \mathbb{R}^r$. We apply this to $m = (m_1,\ldots,m_r) \in \mathbb{Z}^r$, where $d = \sum m_i e_i$ is the element given in the assumption. Then the right-hand side of $(*)$ equals $\sum x_i \varphi(e_i,d)$, which is negative since $x_i \geq 0$ and $\varphi(e_i,d) < 0$. Finally, $\sum_{ij} x_i m_j > 0$ on the left-hand side of $(*)$, and therefore $\lambda < 0$; thus the matrix A has maximum eigenvalue $\lambda < 0$, and it follows that it is negative definite.

Proposition 3. *Let M be a $\mathbb{Z}$-module having a scalar product $(a,b) \in \mathbb{Z}$ for $a,\,b \in M$, and suppose that $e_1,\ldots,e_r$ is a set of generators of M satisfying $(e_i,e_j) \geq 0$ for $i \neq j$. Assume that there exists an element $d = \sum_{i=1}^{r} l_i e_i$ with $l_i > 0$ and $(d,e_i) = 0$ for $i = 1,\ldots,r$. Then $(m,m) \leq 0$ for all $m \in M$. If in addition the elements $e_1,\ldots,e_r$ cannot be partitioned into two components in such a way that $(e_i,e_j) = 0$ for e_i and e_j in different components then $(m,m) = 0$ only for m proportional to d.*

Proof. The proof is almost the same as for Proposition 2. Arguing as there, we find that the matrix A has maximum eigenvalue $\lambda = 0$, which proves that

$\varphi(x, x) \leq 0$ for all $x \in \mathbf{R}^r$; and moreover, $\varphi(x, x) = 0$ for $x \in \mathbf{R}^r$ only if $Ax = 0$. Suppose that there exist two linearly independent vectors $x^{(1)}$ and $x^{(2)} \in \mathbf{R}^r$ with $\varphi(x^{(i)}, x^{(i)}) = 0$, hence $Ax^{(i)} = 0$. Then there is a nonzero linear combination x of $x^{(1)}$ and $x^{(2)}$ which has some zero coefficients.

As in the proof of Proposition 2, passing to the vector $y = (|x_1|, \ldots, |x_r|)$, we can again assume that all the coefficients of x are ≥ 0, some $= 0$; and x still satisfies $Ax = 0$, that is, $\varphi(x, f_i) = 0$ for $i = 1, \ldots, r$, where $f_i = (0, \ldots, 1, \ldots, 0) \in \mathbf{R}^r$ is the standard basis. Suppose that $x = \sum_{i=1}^{s} x_i f_i$ with $x_i > 0$ and $s < r$. Then for $j > s$ we have $0 = \varphi(x, f_j) = \sum_{i=1}^{s} x_i \varphi(f_i, f_j)$, and since $\varphi(f_i, f_j) \geq 0$ for all i and j and $x_i > 0$ for $i \leq s$ it follows that $(e_i, e_j) = \varphi(f_i, f_j) = 0$ for $i \leq s$ and $j > s$. This partitions the set of vectors $\{e_1, \ldots, e_r\}$ into two components $\{e_1, \ldots, e_s\}$ and $\{e_{s+1}, \ldots, e_r\}$ consisting of pairwise orthogonal vectors. The proposition is proved.

2. Polynomials

Proposition 1. *Let $a_n \in \mathbf{Q}$ be a sequence of numbers, and suppose that there exists a polynomial $g(T) \in \mathbf{Q}[T]$ such that $a_{n+1} - a_n = g(n)$ for all sufficiently large n. Then there exists a polynomial $f(T) \in \mathbf{Q}[T]$ such that $a_n = f(n)$ for all sufficiently large n.*

Proof. For any $g(T) \in \mathbf{Q}[T]$, there exists a polynomial $h(T) \in \mathbf{Q}[T]$ such that $h(T + 1) - h(T) = g(T)$. This assertion can be proved by induction on $n = \deg g$; for if g has leading term equal to aT^n then setting $h_0(T) = a/(n+1)T^{n+1}$, we find that $h_0(T+1) - h_0(T) - g(T)$ has degree $< n$, and then we can use induction. Note that h is determined up to an additive constant.

For any choice of the polynomial h we get

$$a_{n+1} - a_n = h(n + 1) - h(n), \quad \text{that is,} \quad h(n + 1) - a_{n+1} = h(n) - a_n$$

for all sufficiently large n, that is, $h(n) - a_n = c$. The polynomial $f = h - c$ satisfies the requirements of the proposition. The proposition is proved.

3. Quasilinear Maps

Let L be a vector space over a field K and $\varphi: L \to L$ a map. We say that φ is *quasilinear* if $\varphi(x + y) = \varphi(x) + \varphi(y)$ for x, $y \in L$, and there exists an automorphism g of K such that

$$\varphi(\alpha x) = g(\alpha)\varphi(x) \qquad \text{for all } \alpha \in K \text{ and } x \in L;$$

then we say that g is the *automorphism* of K *associated with* φ.

Proposition 1. *Let L be a finite dimensional vector space over a field K, and G a finite group of quasilinear maps of L. Assume that every element*

$e \neq \varphi \in G$ has associated automorphism $g \neq \mathrm{id}_K$. Then L has a basis consisting of elements invariant under G.

Remark. Obviously, in this basis, each map in G has the identity matrix; this does not mean, of course, that it is the identity map: it acts on coordinates by the corresponding automorphism of K.

We start with a well-known lemma.

Lemma. *Any set $\{g_1, \ldots, g_n\}$ of distinct field homomorphism $g_i : K \to K$ is linearly independent over K; that is, there does not exist any nontrivial relation of the form*

$$\sum_{i=1}^{n} \lambda_i g_i(\xi) = 0 \qquad \text{for all } \xi \in K \tag{1}$$

with $\lambda_i \in K$.

In other words, there exist $\alpha_1, \ldots, \alpha_n \in K$ such that

$$\det \big| g_i(\alpha_j) \big| \neq 0. \tag{2}$$

Proof of the lemma. Among all relations of the form (1), choose one with the minimal number of nonzero coefficients λ_i. There are obviously at least two such nonzero coefficients, say $\lambda_j \neq 0$ and $\lambda_k \neq 0$ with $j \neq k$. Since by assumption $g_j \neq g_k$ there exists $\alpha \in K$ such that $g_j(\alpha) \neq g_k(\alpha)$. Substituting $\alpha\xi$ for ξ in (1) gives

$$\sum_{i=1}^{n} \lambda_i g_i(\alpha) g_i(\xi) = 0 \qquad \text{for all } \xi \in K. \tag{3}$$

Subtracting $g_j(\alpha)$ times (1) from (3) gives a relation

$$\sum_{i=1}^{n} \lambda_i (g_i(\alpha) - g_j(\alpha)) g_i(\xi) = 0, \tag{4}$$

in which $g_j(\xi)$ has coefficient 0, but the coefficient $g_k(\xi)$ is

$$\lambda_k (g_k(\alpha) - g_j(\alpha)) \neq 0.$$

This contradicts the minimality of the choice of relation (1). The lemma is proved.

Proof of Proposition. By assumption the different maps in G have different associated homomorphisms. Thus we can index the elements of G by their associated homomorphisms. Write $A_g \in G$ for the map with associated homomorphism g.

Let L^G be the set of vectors $x \in L$ invariant under all $A_g \in G$; let us prove that L^G generates L. For this, set $S(x) = \sum A_g(x)$; obviously $S(x) \in L^G$ for any $x \in L$. We prove that already the vectors $S(x)$ with $x \in L$ generate L over K.

For this, note that the space spanned by the $S(x)$ contains the elements $S(\alpha x) = \sum g(\alpha) A_g(x)$ for all $\alpha \in K$. By the lemma, we can choose elements $\alpha_1, \ldots, \alpha_n$ for which $\det |g_i(\alpha_j)| \neq 0$. We see that the $A_g(x)$ can be expressed as linear combinations of the $S(\alpha_i x)$. In particular for $g = e$ we get an expression for x itself, which was what we wanted.

Now it is enough to choose vectors $y_1, \ldots, y_r \in L^G$ that generate L, and a maximal linearly independent subset among these. This will be the required basis. The proposition is proved.

4. Invariants

Proposition 1. *Let A be a finitely generated algebra over k and G a finite group of automorphisms of A. Assume that the order n of G is not divisible by char k. Write A^G for the subalgebra of elements $a \in A$ such that $g(a) = a$ for all $g \in G$. Then A^G is finitely generated as an algebra over k.*

Proof. We write S for the averaging operator

$$S(a) = \frac{1}{n} \sum_{g \in G} g(a).$$

For any $a \in A$, the coefficients of the polynomial

$$P_a(T) = \prod_{g \in G} (T - g(a)) = T^n + \sigma_1 T^{n-1} + \cdots + \sigma_n$$

belong to A^G. The coefficients σ_i are the *elementary symmetric functions* in $g(a)$, that can be expressed in terms of the Newton sums $S(a^i)$ for $i = 1, \ldots, n$. Let $u_1, \ldots, u_m$ be a set of generators of A. Write B for the subalgebra of A^G generated by the elements $S(u_i^j)$ for $i = 1, \ldots, m$ and $j = 1, \ldots, n$. Then $P_{u_i}(u_i) = 0$, and hence the u_i^n can be expressed as linear combinations of $1, u_i, \ldots, u_i^{n-1}$ with coefficients in B. Therefore it follows by induction that any monomial $u_1^{a_1} \cdots u_m^{a_m}$ can be expressed as a linear combination of monomials of the same kind with $a_1, \ldots, a_m < n$. Thus any element $a \in A$ has an analogous expression

$$a = \sum_{a_i < n} \varphi_{a_1 \ldots a_m} u_1^{a_1} \cdots u_m^{a_m} \qquad \text{with } \varphi_{a_1 \ldots a_m} \in B.$$

In particular, let $a \in A^G$. Applying the operator S to this, we get

$$a = S(a) = \sum \varphi_{a_1 \ldots a_m} S(u_1^{a_1} \cdots u_m^{a_m}).$$

It follows that A^G is generated by elements $S(u_1^{a_1} \cdots u_m^{a_m})$ with $a_i < n$ and $S(u_i^n)$. The proposition is proved.

5. Fields

Proposition 1. *Let k be an algebraically closed field and $k \subset K$ a finitely generated extension. Then there exist elements $z_1, \ldots, z_{d+1} \in K$ with $K = k(z_1, \ldots, z_{d+1})$, and such that $z_1, \ldots, z_d$ are algebraically independent over k and z_{d+1} is separable over $k(z_1, \ldots, z_d)$.*

Proof. Suppose that K is generated over k by a finite number of elements $t_1, \ldots, t_n$, and let d be the maximal number of algebraically independent elements among the t_i. Suppose that $t_1, \ldots, t_d$ are algebraically independent. Then any element $y \in K$ is algebraically dependent on $t_1, \ldots, t_d$, and moreover, there exists a relation $f(t_1, \ldots, t_d, y) = 0$ with $f(T_1, \ldots, T_d, T_{d+1})$ irreducible over k.

Let $f(T_1, \ldots, T_{d+1})$ be such a polynomial for $t_1, \ldots, t_{d+1}$. We assert that the partial derivative $f'_{T_i}(T_1, \ldots, T_{d+1}) \neq 0$ for at least one $i = 1, \ldots, d+1$. Indeed, if not, then each T_i only occurs in f in powers that are multiples of the characteristic p of k; that is, f is of the form $f = \sum a_{i_1 \ldots i_{d+1}} T_1^{p i_1} \cdots T_{d+1}^{p i_{d+1}}$. Set $a_{i_1 \ldots i_{d+1}} = b_{i_1 \ldots i_{d+1}}^p$ and $g = \sum b_{i_1 \ldots i_{d+1}} T_1^{i_1} \cdots T_{d+1}^{i_{d+1}}$; then we get $f = g^p$, which contradicts the irreducibility of f.

If $f'_{T_i} \neq 0$, the d elements $t_1, \ldots, t_{i-1}, t_{i+1}, \ldots, t_{d+1}$ are algebraically independent over k. Indeed, t_i is algebraic over $k(t_1, \ldots, t_{i-1}, t_{i+1}, \ldots, t_{d+1})$ because $f'_{T_i} \neq 0$, so that T_i occurs in f. Thus if $t_1, \ldots, t_{i-1}, t_{i+1}, \ldots, t_{d+1}$ were algebraically dependent, the transcendence degree of $k(t_1, \ldots, t_{d+1})$ would be less than d, which contradicts the algebraic independence of $t_1, \ldots, t_d$.

Thus we can always renumber the t_i, so that $t_1, \ldots, t_d$ are algebraically independent over k, and $f'_{T_{d+1}} \neq 0$. This shows that t_{d+1} is separable over $k(t_1, \ldots, t_d)$. Since t_{d+2} is algebraic over $k(t_1, \ldots, t_d)$, by the primitive element theorem (see van der Waerden, [73], §46), we can find an element y such that $k(t_1, \ldots, t_{d+2}) = k(t_1, \ldots, t_d, y)$. Repeating the process of adjoining elements $t_{d+1}, \ldots, t_n$, we express K as $k(z_1, \ldots, z_{d+1})$, where $z_1, \ldots, z_d$ are algebraically independent over k and

$$f(z_1, \ldots, z_d, z_{d+1}) = 0,$$

with f an irreducible polynomial over k with $f'_{T_{d+1}} \neq 0$. Proposition 1 is proved.

Proposition 2. *Let k be an algebraically closed field of characteristic p, and K a finitely generated field extension of k, having transcendence degree 1 over k. Let $K^{(p)}$ be the subfield consisting of pth powers α^p, with $\alpha \in K$. Then $[K : K^{(p)}] = p$. If $L \subset K$ is a subfield such that K is an inseparable extension of L then $L \subset K^{(p)}$.*

Proof. Recall that $\alpha \mapsto \alpha^p$ defines a field homomorphism $K \to K$, whose image is the subfield $K^{(p)}$. Let $t \in K$ be transcendental over k. The first assertion follows from the diagram

$$K^{(p)} \subset K$$
$$\cup \qquad \cup$$
$$k(t)^{(p)} \subset k(t).$$

Indeed, this implies that

$$[K : k(t)^{(p)}] = [K : k(t)][k(t) : k(t)^{(p)}] = [K : K^{(p)}][K^{(p)} : k(t)^{(p)}].$$

Since $\alpha \mapsto \alpha^p$ defines an isomorphic inclusion, it follows that $[K : k(t)] = [K^{(p)} : k(t)^{(p)}]$, and therefore $[K : K^{(p)}] = [k(t) : k(t)^{(p)}]$. Finally, it is obvious that $k(t)^{(p)} = k(t^p)$, and hence

$$[k(t) : k(t)^{(p)}] = p \quad \text{and} \quad [K : K^{(p)}] = p.$$

To prove the second assertion, write L' for the set of all elements of K that are separable over L. It is very easy to prove that this is a subfield. We can obviously replace L by L', and thus assume that any element of K that is separable over L belongs to L. Let $\alpha \in K$ and suppose that its minimal polynomial is of the form $P(T) = a_0 T^{p^m r} + a_1 T^{p^m (r-1)} + \cdots + a_{r-1} T^{p^m} + a_r$ where $Q(T) = a_0 T^r + a_1 T^{(r-1)} + \cdots + a_{r-1} T + a_r$ is a separable polynomial, that is $Q'(T) \neq 0$. Then $\beta = \alpha^{p^m}$ satisfies $Q(\beta) = 0$, that is, β is separable over L, and therefore belongs to L.

It follows that K can be obtained from L by successively adjoining pth roots; that is, there is a chain $L = K_1 \subset \cdots \subset K_m = K$, with $K_i = K_{i-1}(\sqrt[p]{\alpha_i})$, for some $\alpha_i \in K_{i-1}$. Set $K' = K_{m-1}$ and $\alpha = \alpha_{m-1}$, so that $K = K'(\sqrt[p]{\alpha})$. We prove that $K' = K^{(p)}$, and it is at this point that we use that K has transcendence degree 1 over k.

Any element $\beta \in K$ has an expression $\beta = a_0 + a_1 \sqrt[p]{\alpha} + \cdots + a_{p-1}(\sqrt[p]{\alpha})^{p-1}$ with $a_i \in K'$, and hence $\beta^p = a_0^p + a_1^p \alpha + \cdots + a_{p-1}^p \alpha^{p-1}$, that is $K^{(p)} \subset K'$. But $[K : K'] = p$, and we proved that $[K : K^{(p)}] = p$ in the first part of the proof. Therefore $K' = K^{(p)}$ and $L \subset K^{(p)}$. Proposition 2 is proved.

6. Commutative Rings

Proposition 1 (The Hilbert Nullstellensatz). *Let k be an algebraically closed field and $F_1, \ldots, F_m \in k[T_1, \ldots, T_n]$. If the ideal $(F_1, \ldots, F_m) \neq (1)$ then the system of equations $F_1 = \cdots = F_m = 0$ has a solution in k.*

Lemma. *If a system of equations $F_1 = \cdots = F_m = 0$ with $F_i \in k[T_1, \ldots, T_n]$ has a solution in some finitely generated extension field K of k, then it has a solution in k.*

Proof. By §5, Proposition 1, K is of the form $k(x_1,\ldots,x_r,\theta)$ where $x_1,\ldots,x_r$ are algebraically independent over k, and θ is a root of a polynomial

$$P(\underline{X},U) = p_0(\underline{X})U^d + \cdots + p_d(\underline{X}) \in k(\underline{X})[U],$$

with $P(\underline{X},U)$ irreducible over $k(\underline{X}) = k(x_1,\ldots,x_r)$; here we write $\underline{X} = (x_1,\ldots,x_r)$. Suppose that $F_j(\xi_1,\ldots,\xi_n) = 0$ with $\xi_i \in K$. Write the ξ_i in the form $\xi_i = C_i(\underline{X},\theta)$, with $C_i(\underline{X},U) \in k(\underline{X})[U]$. The relation $F_j(\xi_1,\ldots,\xi_n) = 0$ gives the identity

$$F_j\big(C_1(\underline{X},U),\ldots,C_n(\underline{X},U)\big) = P(\underline{X},U)Q_j(\underline{X},U) \tag{1}$$

in $\underline{X} = (x_1,\ldots,x_r)$ and U, where $Q_j(\underline{X},U) \in k(\underline{X})[U]$. Choose values $x_i = \alpha_i \in k$ for $i = 1,\ldots,n$ such that $(\alpha_1,\ldots,\alpha_n)$ is not a zero of the denominators of any coefficient of $P, Q_i, C_1,\ldots,C_n \in k(\underline{X})[U]$, nor a zero of the leading coefficient of P. Now choose $U = \tau \in k$ to be one of the roots of $P(\alpha_1,\ldots,\alpha_n,\tau) = 0$, and set $C_j(\alpha_1,\ldots,\alpha_n,\tau) = \lambda_j$ for $j = 1,\ldots,m$. Then it follows from (1) that $F_j(\lambda_1,\ldots,\lambda_n) = 0$, that is, $(\lambda_1,\ldots,\lambda_n)$ is a solution of the system $F_1 = \cdots = F_m = 0$. The lemma is proved.

Proof of Proposition 1. If the ideal $(F_1,\ldots,F_m) \neq (1)$ then it is contained in some maximal ideal $M \subset k[T_1,\ldots,T_n]$, and $K = k[T_1,\ldots,T_n]/M$ is a field. Write ξ_i for the image of T_i in K. Obviously $K = k(\xi_1,\ldots,\xi_n)$ and $(\xi_1,\ldots,\xi_n)$ is a solution in K of the system $F_1 = \cdots = F_m = 0$. We get a solution in k by applying the lemma. This proves Proposition 1.

Corollary. *If $G, F_1,\ldots,F_m \in k[T_1,\ldots,T_n]$ and G is 0 at all solutions of the system $F_1 = \cdots = F_m = 0$, then $G^N \in (F_1,\ldots,F_m)$ for some $N \geq 0$.*

Proof. It is enough to consider the case $G \neq 0$. We introduce a new variable U, and consider the polynomials

$$F_1,\ldots,F_m \text{ and } UG - 1 \in k[T_1,\ldots,T_n,U].$$

By assumption these have no common solutions in k, and therefore by Proposition 1 there exist polynomials $P_1,\ldots,P_m, Q \in k[T_1,\ldots,T_n,U]$ such that

$$P_1F_1 + \cdots + P_mF_m + Q(UG - 1) = 1.$$

This identity is preserved if we set $U = 1/G$. Clearing denominators we get

$$G^N \equiv 0 \mod (F_1,\ldots,F_m).$$

The corollary is proved.

Proposition 2. *Let A be a commutative ring with a 1. An element $a \in A$ is nilpotent (that is, $a^n = 0$ for some $n > 0$) if and only if a belongs to every prime ideal of A.*

Proof. A nilpotent element is obviously contained in every prime ideal. Conversely, suppose that a is not nilpotent. We construct a prime ideal not containing a. Consider the ideals $I \subset A$ not containing any power of a. By assumption, $I = (0)$ has this property. Let $\mathfrak{a}$ be a maximal element of this set of ideals, which exists by Zorn's lemma. We prove that $\mathfrak{a}$ is a prime ideal; then since $a \notin \mathfrak{a}$, this will prove the proposition.

For this, set $B = A/\mathfrak{a}$ and write b for the image of a; we prove that B is an integral domain. By assumption, any nonzero ideal $\mathfrak{b} \subset B$ contains some power of b, but b itself is not nilpotent. Suppose $b_1, b_2 \in B$ with $b_1, b_2 \neq 0$. Then by assumption, $b^{n_1} \in (b_1)$ and $b^{n_2} \in (b_2)$ for some $n_1, n_2 > 0$. Hence $b^{n_1+n_2} \in (b_1 b_2)$, and therefore $b_1 b_2 \neq 0$. Proposition 2 is proved.

Proposition 3 (Nakayama's Lemma). *Let M be a finite module over a ring A and $\mathfrak{a} \subset A$ an ideal. Suppose that for any element $a \in 1 + \mathfrak{a}$, $aM = 0$ implies $M = 0$. Then $\mathfrak{a}M = M$ implies that $M = 0$.*

Proof. Suppose that $M = (\mu_1, \ldots, \mu_n)$. The assumption $\mathfrak{a}M = M$ implies that there are equalities

$$\mu_i = \sum_{j=1}^{n} \alpha_{ij}\mu_j \qquad \text{with } \alpha_{ij} \in \mathfrak{a}.$$

Thus $\sum_{j=1}^{n}(\alpha_{ij} - \delta_{ij})\mu_j = 0$ for $i = 1, \ldots, n$, and by Cramér's rule $d\mu_i = 0$ for $i = 1, \ldots, n$, where $d = \det(\alpha_{ij} - \delta_{ij})$; therefore $dM = 0$. Since $d \in 1 + \mathfrak{a}$, it follows by assumption that $M = 0$. The proposition is proved.

Corollary 1. *If $A \subset B$ are rings with B a finite A-module and $\mathfrak{a} \subset A$ an ideal, then $\mathfrak{a} \neq A$ implies $\mathfrak{a}B \neq B$.*

Proof. Since B contains the unit element of A, $\mathfrak{a}B = 0$ only if $\mathfrak{a} = 0$, and if $\mathfrak{a} \neq (1)$ then $0 \notin 1 + \mathfrak{a}$. This verifies the assumptions of Proposition 3, and so $\mathfrak{a}B \neq B$. The corollary is proved.

Corollary 2. *If $\mathfrak{a} \subset A$ is an ideal such that every element of $1 + \mathfrak{a}$ is invertible, M a finite A-module and $M' \subset M$ any submodule, then $M' + \mathfrak{a}M = M$ implies that $M' = M$.*

Proof. Apply Proposition 3 to the module M/M'. The corollary is proved.

Remark. It is easy to see that the assumption on the ideal $\mathfrak{a}$ in Corollary 2 holds if $A/\mathfrak{a}$ is a local ring.

Corollary 3. *Under the assumptions of Corollary 2, elements $\mu_1, \ldots, \mu_n \in M$ generate M if and only if their images generate $M/\mathfrak{a}M$.*

Proof. Apply Corollary 2 to the submodule $M' = (\mu_1, \ldots, \mu_n)$. The corollary is proved.

Proposition 4. *Let A be a Noetherian ring, and $\mathfrak{a} \subset A$ an ideal such that every element of $1 + \mathfrak{a}$ is invertible in A. Then $\bigcap_{n>0}(\mathfrak{b} + \mathfrak{a}^n) = \mathfrak{b}$ for any ideal $\mathfrak{b} \subset A$.*

(1) The case $\mathfrak{b} = 0$. Apply Proposition 3 to $M = \bigcap \mathfrak{a}^n$.

(2) The general case. Set $B = A/\mathfrak{b}$ and let $\bar{\mathfrak{a}} = (\mathfrak{a} + \mathfrak{b})/\mathfrak{b}$ be the image of $\mathfrak{a}$ in B. Then $(\bar{\mathfrak{a}})^n = (\mathfrak{a} + \mathfrak{b})^n/\mathfrak{b} = (\mathfrak{a}^n + \mathfrak{b})/\mathfrak{b}$ is the image of $\mathfrak{a}^n$ in B. By the case (1), $\bigcap_{n>0}(\bar{\mathfrak{a}})^n = 0$, and hence $\bigcap_{n>0}(\mathfrak{b} + \mathfrak{a}^n) = \mathfrak{b}$. The proposition is proved.

Proposition 5. *Suppose that $\mathfrak{a}$ is an ideal of a Noetherian ring A such that every element of $1 + \mathfrak{a}$ is invertible in A. Then the property that a sequence of elements $f_1, \ldots, f_m \in \mathfrak{a}$ is a regular sequence (see Chap. IV, 1.2) is preserved under permutations of the f_i.*

Proof. It is enough to prove that permuting two adjacent elements f_i, f_{i+1} of a regular sequence again gives a regular sequence. Set $(f_1, \ldots, f_{i-1}) = \mathfrak{b}$ and $A/\mathfrak{b} = B$, and write a, b for the images in B of f_i, f_{i+1}. Everything reduces to the proof of Proposition 5 for a regular sequence a, b of B. We need to prove (1) that b is not a zerodivisor in B, and (2) that a is not a zerodivisor modulo b.

(1) Suppose that $xb = 0$. We prove then that

$$x \in (a^k) \qquad \text{for all } k. \tag{1}$$

Since A is Noetherian, it follows by Proposition 4 that $x = 0$. We verify (1) by induction. If $x = x_1 a^k$ then $x_1 a^k b = 0$. Since a, b is a regular sequence, a is a non-zerodivisor, and hence $x_1 b = 0$. Again because a, b is a regular sequence, it follows that $x_1 \in (a)$, hence $x \in (a^{k+1})$.

(2) Suppose that $xa = yb$. Because a, b is a regular sequence, it follows that $y = az$ with $z \in A$, and hence $x = zb$. The proposition is proved.

7. Unique Factorisation

Proposition 1. *Suppose that a Noetherian local ring A is contained in a local ring $\widehat{A}$ which is a UFD. Suppose that the maximal ideals $\mathfrak{m} \subset A$ and $\widehat{\mathfrak{m}} \subset \widehat{A}$ satisfy the following conditions:*

(a) $\mathfrak{m}\widehat{A} = \widehat{\mathfrak{m}}$;

(b) $(\mathfrak{m}^n \widehat{A}) \cap A = \mathfrak{m}^n$ for $n > 0$;

(c) for any $\alpha \in \widehat{A}$ and any integer $n > 0$ there exists $a_n \in A$ such that $\alpha - a_n \in \mathfrak{m}^n \widehat{A}$.

Then A is also a UFD.

Proof (taken from Mumford [58], §1C). The usual method of proving unique factorisation into prime factors deduces it from the statement that if a divides bc, and a and b have no common factors then a divides c. We need to establish this result in A, knowing that it holds in $\widehat{A}$. For this it is enough to prove the following two assertions:

(1) for $a, b \in A$,
$$a \mid b \text{ in } \widehat{A} \implies a \mid b \text{ in } A;$$

(2) if a and b have no commons factors in A, then they have no common factors in $\widehat{A}$.

Both these assertions are based on the following lemma.

Lemma. $(a\widehat{A}) \cap A = a$ *for any ideal* $a \subset A$.

Proof. It is enough to prove that $(a\widehat{A}) \cap A \subset a$. Suppose that $a = (a_1, \ldots, a_n)$, and let $x \in (a\widehat{A}) \cap A$. Then $x = \sum a_i \alpha_i$ with $\alpha_i \in \widehat{A}$. By assumption (c), there exist elements $a_i^{(n)} \in A$ such that $\alpha_i = a_i^{(n)} + \xi_i^{(n)}$ with $\xi_i^{(n)} \in \widehat{m}^n$. Then $x = \sum a_i^{(n)} a_i + \sum \xi_i^{(n)} a_i = a + \xi$ with $a \in a$ and $\xi \in \widehat{m}^n$. Hence $\xi = x - a \in A \cap \widehat{m}^n = m^n$. Therefore $x \in a + m^n$ for all $n > 0$ and so $x \in a$ by §6, Proposition 4. The lemma is proved.

Proof of (1). If a divides b in $\widehat{A}$ then $b \in A \cap (a)\widehat{A}$, which by the lemma is equal to (a). This just means a divides b in A.

Proof of (2). If a and b have a common factor in $\widehat{A}$ then they can be written $a = \gamma\alpha$, $b = \gamma\beta$ where $\alpha, \beta \in \widehat{A}$ are proper divisors of a and b with no common factors. Then $a\beta - b\alpha = 0$. By assumption (c), there exist x_n, $y_n \in A$ and $u_n, v_n \in m^n$ such that $\alpha = x_n + u_n$, $\beta = y_n + v_n$. Hence $ay_n - bx_n \in (a, b)\widehat{m}^n = (a, b)m^n\widehat{A}$. By the lemma, $ay_n - bx_n \in (a, b)m^n$, that is, $ay_n - bx_n = at_n + bs_n$ with $s_n, t_n \in m^n$. Hence $a(y_n - t_n) = b(x_n + s_n)$ and so $\alpha(y_n - t_n) = \beta(x_n + s_n)$. From the assumption that α and β have no common factors in $\widehat{A}$ it follows that $x_n + s_n$ is divisible by α, that is, $x_n + s_n = \alpha\lambda$. Since $\bigcap \widehat{m}^k = 0$, for sufficiently large n we have $\alpha, \beta \notin \widehat{m}^{n-1}$. Then also $x_n + s_n \notin m^{n-1}$, and hence $\lambda \notin m$, that is, λ is invertible in $\widehat{A}$. Hence $\widehat{A}(x_n + s_n) = (\alpha)\widehat{A}$, and $x_n + s_n$ divides α, and so divides a in $\widehat{A}$. By (1) it also divides a in A, that is, $a = (x_n + s_n)h$. But $a(y_n - t_n) = b(x_n + s_n)$, and hence $b = (y_n - t_n)h$. Since a and b have no common factors in A it follows that h is invertible in A, that is, $(a) = (x_n + s_n) = (\alpha)$, and this contradicts the assumption that α is a proper divisor of a in $\widehat{A}$.

The proposition is proved.

8. Integral Elements

Proposition 1. *Let $B = k[T_1, \ldots, T_n]$ be the polynomial ring and $L = k(T_1, \ldots, T_n)$ its field of fractions, and suppose that $L \subset K$ is a finite field extension. Write A for the integral closure of B in L. Then A is a finite B-module.*

Proof. The proof in the case that the extension $L \subset K$ is separable, which is very simple, is given in Atiyah and Macdonald [7], Proposition 5.17. We do not reproduce it here, but show how to reduce everything to the case of a separable extension.

Suppose that $K = L(\alpha_1, \ldots, \alpha_s)$. If α_1 is not separable over L then its minimal polynomial is of the form $\alpha_1^{p^l m} + a_1 \alpha_1^{p^l(m-1)} + \cdots + a_m = 0$, where $a_i \in k(T_1, \ldots, T_r)$ and $\alpha_1^{p^l}$ is separable over L. Write $B' = k[T_1^{1/p^l}, \ldots, T_r^{1/p^l}]$, $L' = k(T_1^{1/p^l}, \ldots, T_r^{1/p^l})$ and $K' = K(T_1^{1/p^l}, \ldots, T_r^{1/p^l})$, and let A' be the integral closure of B' in K'. Now set $a_i = b_i^{p^l}$, with $b_i \in L'$. Then $K' = L'(\alpha_1, \ldots, \alpha_s)$ and $\alpha_1^m + b_1 \alpha_1^{(m-1)} + \cdots + b_m = 0$, so that α_1 is separable over L'. On the other hand $A \subset A'$, and if the proposition is proved for A' then A' is a finite B'-module. But B' is itself a finite B-module: it has a basis consisting of monomials $T_1^{i_1/p^l} \cdots T_r^{i_r/p^l}$ with $0 \leq i_r, \ldots, i_r < p^l$. Therefore A' is a finite B-module, and hence so is its submodule A.

We see that the proof of the proposition reduces to the case that α_1 is separable. By the primitive element theorem, then $L(\alpha_1, \ldots, \alpha_s) = L(\alpha_2', \alpha_3, \ldots, \alpha_s)$. Applying the same argument s times we reduce the proof to the case of a separable extension. The proposition is proved.

9. Length of a Module

Definition. A module M over a ring A has *finite length* if there exists a chain of A-submodules

$$M = M_0 \supset M_1 \supset \cdots \supset M_n = 0 \qquad \text{with } M_i \neq M_{i+1}, \tag{1}$$

such that each quotient M_i/M_{i+1} is a simple A-module, that is, does not contain any proper submodule. By the Jordan-Hölder theorem, all such chains have the same length n; this common length n is called the *length* of M, and denoted by $\ell(M)$, or $\ell_A(M)$ to stress the role of the ring A.

Obviously, the quotient modules M_i/M_{i+1} in (1) are isomorphic to $A/\mathfrak{m}$ where $\mathfrak{m}$ are maximal ideals of A. If M has finite length then the same holds for all its submodules and quotient modules. If a module M has a chain (1) such that each quotient M_i/M_{i+1} has finite length then M has finite length, and

$$\ell(M) = \sum \ell(M_i/M_{i+1}).$$

Proposition 1. *Let $\mathcal{O}$ be a Noetherian local ring, with maximal ideal $\mathfrak{m}$, and suppose that $\mathfrak{a} \subset \mathcal{O}$ is an ideal such that $\mathfrak{a} \supset \mathfrak{m}^k$ for some $k > 0$; then the module $\mathcal{O}/\mathfrak{a}$ has finite length.*

Proof. It is enough to prove that $\mathcal{O}/\mathfrak{m}^k$ is a module of finite length. By considering the chain of submodules $M_i = \mathfrak{m}^i/\mathfrak{m}^k$ for $i = 0, \dots, k$, we see that it is enough to check that each module $\mathfrak{m}^i/\mathfrak{m}^{i+1}$ has finite length. But in the $\mathcal{O}$-module structure of $\mathfrak{m}^i/\mathfrak{m}^{i+1}$, multiplication by $\mathfrak{m}$ kills every element. Therefore $\mathcal{O}/\mathfrak{m} = k$ acts on $\mathfrak{m}^i/\mathfrak{m}^{i+1}$, so that it is a vector space over the field k, and its length equals the dimension of this vector space over k. Since $\mathcal{O}$ is a Noetherian ring, $\mathfrak{m}^i/\mathfrak{m}^{i+1}$ has a finite number of generators, that is, it is a finite dimensional vector space. This proves the proposition.

If M is an A-module and $\mathfrak{p}$ a prime ideal of A then we write $M_\mathfrak{p}$ for the localisation of M at $\mathfrak{p}$, that is the module $M \otimes_A A_\mathfrak{p}$, where $A_\mathfrak{p}$ is the local ring of $\mathfrak{p}$.

Example. If $M = A/\mathfrak{p}$ then $M_\mathfrak{q} = 0$ if $\mathfrak{q} \not\supset \mathfrak{p}$. If $\mathfrak{q} \supset \mathfrak{p}$ then $M_\mathfrak{q} = (A/\mathfrak{p})_{\overline{\mathfrak{q}}}$ where $\overline{\mathfrak{q}} = \mathfrak{q}/\mathfrak{p}$ is the image of $\mathfrak{q}$ in $A/\mathfrak{p}$.

Lemma. *A finite module M over a Noetherian ring A has a chain* (1) *of submodules such that $M_i/M_{i+1} \cong A/\mathfrak{p}_i$, where $\mathfrak{p}_i \subset A$ is a prime ideal.*

Proof. For an element $m \in M$ with $m \neq 0$, write $\operatorname{Ann} m$ for the ideal of elements of a such that $am = 0$. Because A is Noetherian, a chain of ideals of the form $\operatorname{Ann}(m_1) \subset \operatorname{Ann}(m_2) \subset \cdots$ must terminate. Hence we can choose $m \in M$ with the following property: $\operatorname{Ann} m \subset \operatorname{Ann}(m')$ with $m' \neq 0$ implies that $\operatorname{Ann} m = \operatorname{Ann}(m')$. We prove that $\operatorname{Ann} m$ is then a prime ideal. Let $ab \in \operatorname{Ann} m$ with $b \notin \operatorname{Ann} m$. Then $\operatorname{Ann} m \subset \operatorname{Ann}(bm)$ and $bm \neq 0$, but then by assumption $\operatorname{Ann} m = \operatorname{Ann}(bm)$. But $a \in \operatorname{Ann}(bm)$, hence $a \in \operatorname{Ann} m$.

Set $\mathfrak{p} = \operatorname{Ann} m$. Then the submodule $Am \subset M$ is isomorphic to $A/\mathfrak{p}$. In $M' = M/Am$ we can again find a submodule isomorphic to $A/\mathfrak{p}'$ where $\mathfrak{p}'$ is a prime ideal of A. In this way we construct a chain $M^{(1)} \subset M^{(2)} \subset \cdots$ such that $M^{(i-1)}/M^{(i)} \cong A/\mathfrak{p}_i$. By the assumption that M is Noetherian, this chain terminates. The lemma is proved.

Definition. A local ring A with maximal ideal $\mathfrak{m}$ is 1-*dimensional* if there exists a prime ideal $\mathfrak{p} \subsetneq \mathfrak{m}$, and every such prime ideal $\mathfrak{p}$ is minimal, that is, does not contain any strictly smaller prime ideal.

Proposition 2. *Let $\mathcal{O}$ be a 1-dimensional local ring having a finite number of minimal prime ideals $\mathfrak{p}_1, \dots, \mathfrak{p}_n$, and $a \in A$ a non-zerodivisor of A not contained in any of the $\mathfrak{p}_i$. Then*

$$\ell(\mathcal{O}/(a)) = \sum_{i=1}^{n} \ell_{\mathcal{O}_{\mathfrak{p}_i}}(\mathcal{O}_{\mathfrak{p}_i}) \times \ell_{\mathcal{O}}\big(\mathcal{O}/(\mathfrak{p}_i + a\mathcal{O})\big). \tag{2}$$

Proof (taken from Fulton [27], A.1–3). At the same time as (2) we prove a generalisation to an arbitrary finite $\mathcal{O}$-module M. For this, we set $e(M, a) = \ell_A(M/aM) - \ell_A(\text{Ann}_M(a))$, where $\text{Ann}_M(a)$ denotes the A-module $\{m \in M \mid am = 0\}$. The generalisation of (2) is the following:

$$e(M, a) = \sum_{i=1}^{n} \ell_{\mathcal{O}_{\mathfrak{p}_i}}(M_{\mathfrak{p}_i}) \times \ell_{\mathcal{O}}(\mathcal{O}/(\mathfrak{p}_i + a\mathcal{O})). \tag{3}$$

The advantage of the invariant $e(M, a)$ is that it is additive: if $a \in \mathcal{O}$ and $0 \to M' \to M \to M'' \to 0$ is an exact sequence, then

$$e(M, a) = e(M', a) + e(M'', a),$$

and the left-hand side is finite if both terms on the right-hand side are. This follows at once from the following exact sequence

$$0 \to \text{Ann}_{M'}(a) \to \text{Ann}_{M}(a) \to \text{Ann}_{M''}(a)$$
$$\to M'/aM' \to M/aM \to M''/aM'' \to 0,$$

which is trivial to verify. By induction we get that for any chain (1),

$$e(M, a) = \sum e(M_i/M_{i+1}, a).$$

It follows from these considerations and from the lemma that we need only prove (3) for modules M isomorphic to $\mathcal{O}/\mathfrak{p}$, where $\mathfrak{p}$ is a prime ideal of $\mathcal{O}$. If $\mathfrak{p} = \mathfrak{m}$ is the maximal ideal then $M \cong k$ (as an $\mathcal{O}$-module), so that $e(M, a) = 0$ and $M_{\mathfrak{p}_i} = 0$. If $\mathfrak{p}$ is a minimal prime ideal $\mathfrak{p} = \mathfrak{p}_i$ then $M_{\mathfrak{p}_j} \neq 0$ for $j \neq i$, and $M_{\mathfrak{p}_i}$ is the field of fraction of the quotient ring, so that $\ell_{\mathcal{O}_{\mathfrak{p}_i}}(M_{\mathfrak{p}_i}) = 1$. Hence in either case (3) is obvious.

Finally to deduce (2) from (3), we must set $M = \mathcal{O}$. Indeed, under the assumptions of the proposition,

$$e(\mathcal{O}, a) = \ell(\mathcal{O}/(a)) \quad \text{and} \quad e(\mathcal{O}/\mathfrak{p}_i, a) = \ell_{\mathcal{O}}(\mathcal{O}/(\mathfrak{p}_i + a\mathcal{O})),$$

so that (3) implies (2). The proposition is proved.

References

1. Abhyankar, S. S.: Local Analytic Geometry, Academic Press, New York London 1964; MR **31**-173.
2. Abraham, R., Robbin, J.: Transversal mappings and flows, Benjamin, New York Amsterdam 1967.
3. Ahlfors, L.: The complex analytic structure of the space of closed Riemann surfaces, in book Analytic functions, Princeton Univ. Press, Princeton N.J. 1960, 45–66.
4. Aleksandrov, P. S., Efimov, V. A.: Combinatorial topology, Vol. 1, Graylock, Rochester, N.Y. 1956.
5. Altman, A. B., Kleiman, S. L.: Compactifying the Picard scheme, I, Adv. in Math. **35** (1980), 50–112; and II, Amer. J. Math. **101** (1979), 10–41; MR 81f:14025a–b.
6. Artin, M., Mumford, D.: Some elementary examples of unirational varieties which are not rational, Proc. London Math. Soc. **25** (1972), 75–95; MR **48** #299.
7. Atiyah, M. F., Macdonald, I. G.: Introduction to commutative algebra, Addison-Wesley, Reading, Mass. 1969; MR **39**-4129.
8. Barth, W., Peters, C., Van de Ven, A. D. M.: Compact complex surfaces, Springer-Verlag, Berlin Heidelberg New York 1984.
9. Bers, L.: Spaces of Riemann surfaces, in Proc. Int. Congr. Math. (Edinburgh 1958), 349–361.
10. Bôcher, M.: Introduction to higher algebra, Dover, New York 1964.
11. Bogomolov, F. A.: Brauer groups of quotient varieties, Izv. Akad. Nauk SSSR, Ser. Mat. **51** (1987), 485–516; English translation: Math. USSR Izvestiya **30** (1988), 455–485.
12. Bombieri, E., Husemoller, D.: Classification and embeddings of surfaces, in Proc. Symp. in Pure Math. **29** (1975), AMS, Providence, R.I. 1975, 329–420; MR **58** #22085.
13. Borevich, Z. I., Shafarevich, I. R.: Number theory, Nauka, Moscow, 1985 (Second edition); English translation, Academic Press, New York London 1966
14. Bourbaki, N.: Élements de Mathématiques, Topologie générale, Hermann, Paris; English translation: General topology, I–II, Addison-Wesley, Reading, Mass. 1966; reprint Springer-Verlag, Berlin Heidelberg New York 1989
15. Bourbaki, N.: Élements de Mathématiques, Algèbre commutative, Masson, Paris 1983–85; English translation Addison-Wesley, Reading, Mass. 1972.
16. Bourbaki, N.: Élements de Mathématiques, Algèbre, Chap. 2 (Algèbre linéaire), Hermann, Paris 1962.
17. Bourbaki, N.: Élements de Mathématiques, Groupes et algèbre de Lie, Hermann, Paris 1960–1975 (Chap. I: 1960, Chap. IV VI: 1968, Chap. II–III: 1972, Chap. VII–VIII: 1975) and Masson, Paris (Chap. IX: 1982); English transla-

tion of Chap. 1–3: Lie groups and Lie algebras, Springer-Verlag, Berlin Heidelberg New York 1989.

18. Cartan, H.: Théorie élémentaire des fonctions analytiques d'une ou plusieurs variables complexes, Hermann, Paris 1961; English translation Elementary theory of analytic functions of one or several complex variables, Hermann, Paris 1963 and Addison Wesley, Reading, Mass., Palo Alto, London 1963; MR **26** #5138.

19. Cartier, P.: Équivalence linéaire des ideaux de polynomes, Séminaire Bourbaki 1964–65, Éxposé 283, Benjamin, New York Amsterdam 1966.

20. Chern, S. S.: Complex manifolds without potential theory, Van Nostrand, Princeton Toronto London 1967; MR **37** #940.

21. Clemens, C. H., Griffiths, P. A.: The intermediate Jacobian of the cubic threefold, Ann. of Math. (2) **95** (1972), 281–356; MR **46** #1796.

22. de la Harpe, P., Siegfried, P.: Singularités de Klein, Enseign. Math. (2) **25** (1979), 207–256; MR **82e**:32010.

23. de Rham, G.: Variétés différentiables. Formes, courants, formes harmoniques, Hermann, Paris 1965; English translation: Differentiable manifolds, Springer-Verlag, Berlin Heidelberg New York 1984. MR **16**–957.

24. Esnault, H.: Classification des variétés de dimension 3 et plus, Séminaire Bourbaki 1980–1981, Éxposé 586, Lecture Notes in Math. **901** 1981.

25. Fleming, W.: Functions of several variables, Springer 1965.

26. Forster, O.: Riemannsche Flächen, Springer-Verlag, Berlin Heidelberg New York 1977; English translation Lectures on Riemann surfaces, Springer-Verlag, Berlin Heidelberg New York 1981. MR **56** #5867.

27. Fulton, W.: Intersection Theory, Springer-Verlag, Berlin Heidelberg New York 1983.

28. Fulton, W.: Algebraic curves, Benjamin, New York Amsterdam 1969.

29. Gizatullin, M. Kh.: Defining relations for the Cremona group of the plane, Izv. Akad. Nauk SSSR, Ser. Mat. **46** (1982) 909–970; English translation,Math. USSR Iszvestiya **21** (1983), 211–268

30. Goursat, É.: Cours d'Analyse Mathématique, 3 vols., Gauthiers-Villar, Paris 1902; English translation: A course in mathematical analysis, 3 vols., Dover, New York 1959–1964; MR **21** #4889.

31. Griffiths, P. A., Harris, J.: Principles of Algebraic Geometry, Wiley, New York 1978.

32. Grothendieck, A.: Cohomologie locale des faisceaux cohérents et théorèmes de Lefschetz locaux et globaux (SGA 2), North-Holland, Amsterdam and Masson, Paris 1968.

33. Grothendieck, A.: Technique de descente et théorèmes d'existence en géométrie algébrique, IV, Séminaire Bourbaki t. 13 Éxposé 221, May 1961. V, Séminaire Bourbaki t. 14 Éxposé 232, Feb 1962. V, Séminaire Bourbaki t. 14 Éxposé 236, May 1962. Reprinted in Fondements de la géométrie algébrique (extraits du Séminaire Bourbaki 1957–1962), Secrétariat mathématique, Paris 1962. MR **26** #3566.

34. Gunning, G., Rossi, H.: Analytic functions of several complex variables, Prentice Hall, Englewood Cliffs, N. J. 1965; MR **31** #4927.

35. Hartshorne, R.: Algebraic Geometry, Springer-Verlag, Berlin Heidelberg New York 1977.

36. Hilbert, D.: Mathematical problems (Lecture delivered before the International Congress of Mathematicians at Paris in 1900), Göttinger Nachrichten, 1900, 253–297; English translation reprinted in Proc. of Symposia in Pure Math. **28**, AMS, Providence, R.I. 1976, 1–34.

37. Hironaka, H.: On the equivalence of singularities. I, in Arithmetic Algebraic Geometery (Proc. Conf., Purdue Univ., 1963), O. F. G. Schilling, Ed., Harper and Rowe, New York 1965, 153–200. MR **34** #1317.
38. Humphreys, J. E.: Linear algebraic groups, Springer-Verlag, Berlin Heidelberg New York 1975.
39. Husemoller, D.: Fibre bundles, McGraw-Hill, New York London 1966; 2nd Edition Springer-Verlag, Berlin Heidelberg New York 1975.
40. Iskovskikh, V. A.: A simple proof of a theorem of a theorem of Gizatullin, Trudy Mat. Inst. Steklov **183** (1990), 111–116; translated in Proc. Steklov Inst. Math. 1991, Issue 4, 127-133.
41. Iskovskikh, V. A., Manin, Yu. A.: Three-dimensional quartics and counterexamples to the Lüroth problem, Mat. Sbor. (N. S.) **86** (**128**) (1971), 140–166; English translation Math. USSR Sbornik **15** (1971) 141–166. MR **45** #266.
42. Kähler, E.: Über die Verzweigung einer algebraischen Funktion zweier Veränderlichen in der Umgebung einer singuläre Stelle, Math. Zeit. **30** (1929) 188–204.
43. Kawamata, Y.: Minimal models and the Kodaira dimension of algebraic fibre spaces, J. reine angew. Math. **363** (1985) 1–46; MR **87a**:14013.
44. Kawamata, Y., Matsuda, K., Matsuki, K.: Introduction to the minimal model problem, Proc. Sympos. Algebraic Geometry (Sendai 1985), Adv. Stud. Pure Math. vol. **10**, T. Oda ed., Kinokuniya and North Holland 1987, 283–360.
45. Kleiman, S., Laksov, D.: Schubert calculus, Amer. Math. Monthly **79** (1972) 1061–1082; MR **48** #2152.
46. Knutson, D.: Algebraic spaces, Lecture Notes in Math. **203** 1971; MR **46** #1791.
47. Koblitz, N.: p-adic numbers, p-adic analysis and zeta functions, Springer-Verlag, Berlin Heidelberg New York 1977. MR **57** #5964.
48. Kollár, J.: Shafarevich maps and the plurigenera of algebraic varieties, Invent. Math **113** (1993), 177–215.
49. Kollár, J.: Shafarevich maps and automorphic forms (lecture notes, Univ. of Utah preprint) to appear in Princeton Univ. Press.
50. Kostrikin, A. I., Manin, Yu. I.: Linear algebra and geometry, Moscow Univ. publications, Moscow 1980; English translation: Gordon and Breach, New York 1989.
51. Kurosh, A. G.: The theory of groups, Gos. Izdat. Teor.-Tekh. Lit., Moscow 1944; English translation: Vol. I, II, Chelsea, New York 1955, 1956. Zbl. 64, 251.
52. Lang, S.: Algebra, 2nd Edn., Addison-Wesley, Menlo Park, California 1984.
53. Lang, S.: Introduction to algebraic geometry, Wiley-Interscience, New York 1958.
54. Lang, S.: Introduction to the theory of differentiable manifolds, Wiley-Interscience, New York 1962; MR **27** #5192.
55. Matsumura, H.: Commutative ring theory, Cambridge Univ. Press, Cambridge New York 1986.
56. Milnor, J.: Morse theory, Princeton Univ. Press, Princeton, N. J. 1963; MR **29** #634.
57. Milnor, J.: Singular points of complex hypersurfaces, Princeton Univ. Press, Princeton, N. J. 1968. MR **39** #969.
58. Mumford, D.: Algebraic Geometry, I. Complex projective varieties, Springer-Verlag, Berlin Heidelberg New York 1976.
59. Mumford, D.: Introduction to Algebraic Geometry (Harvard notes, 1976). Reissued as The Red Book of Varieties and Schemes, Lecture Notes in Math. **1358** 1988.

60. Mumford, D.: Lectures on curves on a algebraic surface, Princeton Univ. Press, Princeton, N. J. 1966; MR **35** #187.
61. Mumford, D.: Picard groups of moduli problems, Arithmetical Algebraic Geometry, Harper and Rowe, New York 1965, 33–81. MR **34** #1327.
62. Mumford, D., Fogarty, J.: Geometric invariant theory, 2nd Edn., Springer-Verlag, Berlin Heidelberg New York 1982.
63. Pham, F.: Introduction à l'étude topologique des singularités de Landau, Mém. Sci. Math., Gauthiers-Villar, Paris 1967; MR **37** #4837.
64. Pontryagin, L. S.: Continuous groups, Gos. Izdat. Teor.-Tekh. Lit., Moscow 1954; English translation: Topological groups (Vol. 2 of selected works), Gordon and Breach, New York 1986.
65. Saltman, D. J.: Noether's problem over an algebraically closed field, Invent Math. **77** (1984), 71–84.
66. Seifert, G., Threlfall, V.: Lehrbuch der Topologie, Chelsea, New York 1934; English translation Acad. Press, New York 1980.
67. Shafarevich, I. R. and others, Algebraic surfaces, Proceedings of the Steklov Inst. **75**, Nauka, Moscow, 1965. English translation: Amer. Math. Soc., Providence, R.I. (1967); MR **32** #7557.
68. Shokurov, V. V.: Numerical geometry of algebraic varieties, Proc. Int. Congress Math. (Berkeley 1986), A. M. S., Providence, R.I. 1988, Vol. 1, 672–681.
69. Siegel, C. L.: Automorphic functions and Abelian integrals, Wiley-Interscience, New York, 1971.
70. Siegel, C. L.: Abelian functions and modular functions of several variables, Wiley-Interscience, New York 1973.
71. Springer, G.: Introduction to Riemann surfaces, 2nd Edition, Chelsea, New York 1981
72. Springer, T.: Invariant Theory, Springer-Verlag, Berlin Heidelberg New York 1977.
73. van der Waerden, B. L.: Moderne Algebra, Bd. 1, 2, Springer-Verlag, Berlin, 1930, 1931; I: Jrb, 56, 138. II: Zbl. 2, 8. English translation: Algebra, Vol. I, II, Ungar, New York 1970.
74. Walker, R.J.: Algebraic curves, Springer-Verlag, Berlin Heidelberg New York 1978
75. Wallace, A.: Differential topology: first steps, Benjamin, New York Amsterdam 1968.
76. Weil, A.: Introduction à l'étude des variétés kählériennes, Publ. Inst. Math. Univ. Nancago, Hermann, Paris 1958; MR **22** #1921.
77. Wilson, P. M. H.: Towards a birational classification of algebraic varieties, Bull. London Math. Soc. **19** (1987), 1–48.
78. Zariski, O., Samuel, P.: Commutative Algebra, 2 vols, Springer-Verlag, Berlin Heidelberg New York 1975.

Index

Printed in the United States
120616LV00002B/136-144/P

9 783540 548126